T0340208

# Science and Engineering of Hydrogen-Based Energy Technologies

# Science and Engineering of Hydrogen-Based Energy Technologies

Hydrogen Production and Practical Applications in Energy Generation

*Edited by*

## Prof. Paulo Emilio V. de Miranda

Metallurgy and Materials Engineering
Transportation Engineering
Coppe-Federal University of Rio de Janeiro
Rio de Janeiro
Brazil

**ACADEMIC PRESS**

An imprint of Elsevier

Academic Press is an imprint of Elsevier
125 London Wall, London EC2Y 5AS, United Kingdom
525 B Street, Suite 1650, San Diego, CA 92101, United States
50 Hampshire Street, 5th Floor, Cambridge, MA 02139, United States
The Boulevard, Langford Lane, Kidlington, Oxford OX5 1GB, United Kingdom

**Notices**
Knowledge and best practice in this field are constantly changing. As new research and experience broaden our understanding, changes in research methods, professional practices, or medical treatment may become necessary.

Practitioners and researchers must always rely on their own experience and knowledge in evaluating and using any information, methods, compounds, or experiments described herein. In using such information or methods they should be mindful of their own safety and the safety of others, including parties for whom they have a professional responsibility.

To the fullest extent of the law, neither the Publisher nor the authors, contributors, or editors, assume any liability for any injury and/or damage to persons or property as a matter of products liability, negligence or otherwise, or from any use or operation of any methods, products, instructions, or ideas contained in the material herein.

**Library of Congress Cataloging-in-Publication Data**
A catalog record for this book is available from the Library of Congress

**British Library Cataloguing-in-Publication Data**
A catalogue record for this book is available from the British Library

ISBN: 978-0-12-814251-6

For information on all Academic Press publications visit our website at
https://www.elsevier.com/books-and-journals

Working together
to grow libraries in
developing countries

www.elsevier.com • www.bookaid.org

*Publisher:* Joe Hayton
*Acquisition Editor:* Raquel Zanol
*Editorial Project Manager:* Jennifer Pierce
*Production Project Manager:* Surya Narayanan Jayachandran
*Cover Designer:* Victoria Pearson

Typeset by TNQ Technologies

# Contents

# List of Contributors

**Marcelo Carmo**, Forschungszentrum Jülich GmbH, Jülich, Germany

**Alberto Coralli**, Federal University of Rio de Janeiro, Rio de Janeiro, RJ, Brazil

**Debabrata Das**, Indian Institute of Technology, Kharagpur, West Bengal, India

**Paulo Emílio V. de Miranda**, Federal University of Rio de Janeiro, Rio de Janeiro, RJ, Brazil

**Sergio P. de Oliveira**, National Institute of Metrology, Quality and Technology, Duque de Caxias, Brazil

**Makoto R. Harada**, Research Adviser, National Institute of Advanced Industrial Science and Technology (AIST), Research Institute for Chemical Process Technology

**David Hart**, E4tech, London, United Kingdom and Lausanne, Switzerland; Imperial College London, London, United Kingdom

**Luigi Osmieri**, National Renewable Energy Laboratory, Golden, CO, United States

**Nguyen Q. Minh**, University of California, San Diego, La Jolla, CA, United States

**Newton P. Neves Jr.**, H2 Technical Analyses and Expertise in Gases, Capivari, Brazil

**Beatrice Sampson**, University of Birmingham, Birmingham, United Kingdom

**Bernardo J.M. Sarruf**, Federal University of Rio de Janeiro, Rio de Janeiro, RJ, Brazil

**Vaishali Singh**, Indian Institute of Technology, Kharagpur, West Bengal, India

**Stefania Specchia**, Politecnico di Torino, Torino, Italy

**Robert Steinberger-Wilckens**, University of Birmingham, Birmingham, United Kingdom

**Detlef Stolten**, Forschungszentrum Jülich GmbH, Jülich, Germany; RWTH Aachen University, Aachen, Germany

**Andrei V. Tchouvelev**, A.V. Tchouvelev & Associates, Mississauga, Canada

**Hirohisa Uchida**, Professor, School of Engineering, Tokai University/President & CEO, KSP Inc., Japan

# Foreword

I am delighted to write this foreword, not only because Paulo Emílio V. de Miranda has been a friend and colleague for the past several years, but also because I strongly believe in the way he has been serving in the areas of science and engineering of hydrogen-based energy technologies. Interestingly, the title of this edited book has become the same. Paulo has tried his best to bring the key people on board to contribute to this edited book with their distinct chapters, covering basic aspects of hydrogen and hydrogen energy, fuel cells, hydrogen production methods, energy storage via hydrogen storage, hydrogen transportation, hydrogen stations, hydrogen deployment, hydrogen energy engineering practices and demonstrations, hydrogen energy policies, hydrogen energy roadmaps, hydrogen safety, hydrogen marketing, etc.

This book has educative and training values of interpretive discussions in many subjects for all readers, including students, engineers, practitioners, researchers, scientists, etc. I am sure this book will inspire many young and senior people to advocate about hydrogen energy for sustainable economies. The specific details and discussions on every specific topic make this particular book even more appealing to all readers in every age group. In addition, there is a remarkable set of science-related material in every hydrogen-related subject to emphasize the importance of the applied nature of science. Furthermore, there is huge engineering-related material from A to Z type covering entire spectrum of hydrogen energy from the production to the deployment in various sectors, ranging from residential to industrial and from industrial to utility sectors.

In closing, I am quite satisfied with the authorship of each specific subject and the material presented as well as the science- and engineering-related examples and case studies tailored for the readers, and I am sure that this will be an excellent asset to the hydrogen energy literature.

Last, but not least I warmly congratulate Paulo and his contributors who have brought this unique edited book to fruition.

**Prof. Dr. Ibrahim Dincer**
*Vice President for Strategy*
*International Association for Hydrogen Energy*
*Vice President, World Society of Sustainable Energy Technologies*

# Preface

The world is experiencing its steepest ever-observed growth of energy consumption and population. These two factors coupled with progressively growing urbanization rates at very high levels have threatened the planet and life on it. This has motivated the vision of a sustainable society capable of implementing the creative and innovative concept of a circular restorative economy. Such a vision can be implemented only by transitioning to a new energy era in which hydrogen and renewable energies play a main role. Hydrogen energy is about utilizing hydrogen and hydrogen-containing compounds to generate and supply energy for all practical uses with high-energy efficiency, overwhelming environmental and social benefits, and economic competitiveness. The implementation of hydrogen energy for widespread utilization requires the use of currently available technologies that resulted from intense long-lasting scientific developments involving fuel cells, hydrogen production methods, selection of specific application options, safety and regulations, policies and planning for early adoption, as well as market introduction.

The dawn of hydrogen energy, which brought the practical deployment of sustainable devices and mobility, called for the present text on science and engineering of hydrogen-based energy technologies. Its content was structured in such a way that both knowledgeable professionals in the area, as well as newcomers possessing a strong basis on engineering, energy, or sustainability, will be attracted and interested.

The general approach to hydrogen energy establishes a broad vision of technological possibilities and future prospects. It explores emerging technologies to show that additional efficiency gains and environmental benefits will be progressively achievable as conventional technologies are surpassed. Upon unveiling the occurrence of natural hydrogen on earth, once believed nonexistent, hydrogen energy was presented as sustainable and perennial.

Fuel cells are deeply discussed, with an emphasis on polymeric membrane electrolyte fuel cells and on solid oxide fuel cells and their technological variants, either to generate electrical energy or to consume it for hydrogen production, independently or reversibly.

Biomass is one of the major renewable sources for energy generation and acts as a natural medium for sunlight energy storage. Its enormous availability as residues and wastes (agricultural, industrial, domestic household, municipal) makes its energetic use very beneficial to society. Focus on the production of hydrogen from biomass through biological routes using fermentation processes, in special dark fermentation, is provided.

The progressive increase in the use of renewable energies, which represents the main future prospects of countries and regions for environmental and energy-security reasons, motivates and facilitates the large-scale production of green hydrogen by water electrolysis. Electrolyzer technologies and hydrogen storage methods are thoroughly analyzed and discussed.

Hydrogen utilization applications from the dawning of the hydrogen energy era to the present include transitional technologies, such as the use of hydrogen as fuel in turbines and internal combustion engines. However, the use of hydrogen to feed fuel cells will ubiquitously dominate engineering, mobile, and stationary applications. These options are explored and exemplified with emphasis on technological and engineering procedures.

New regulations, codes, and standards (RC&S) and the adaptation of existing codes are necessary to introduce such technologies into use by the society. This also requires thoroughly understanding and systematizing the roles of the different active regulating institutions as well as establishing and guaranteeing safety protocols. RC&S, metrology, and safety are defined and understood within the ample variety of official world and regional active actors.

Hydrogen may be made available in different world regions depending on local specificities. Also, early deployment of hydrogen energy technologies may be based on niche applications for a specific society. It results that identifying suitable sectors, actions, timing and actors is mandatory for planning and to create adequate policies. Roadmapping techniques were defined, exemplified and discussed as an important tool to make the necessary transition to the hydrogen energy era.

Most of hydrogen, effectively produced in large scale, has a captive use as a chemical product. To be traded as a world energy commodity and also in order to guarantee the market entrance of hydrogen-based technologies, alternative approaches to determining total cost of ownership must be adopted. This approach is explored by taking into consideration externalities associated with the use of conventional technologies that include hidden costs of environmental and societal damage.

I apologize for not being able to include so many other topics of interest, and I hope the selected content will fill the gap of scientific and technological information to understand and foster engineering applications of hydrogen-based energy technologies.

**Rio de Janeiro**
*Brazil, June 2018*
**Paulo Emílio V. de Miranda**

Chapter 1

# Hydrogen Energy: Sustainable and Perennial

Paulo Emílio V. de Miranda
*Federal University of Rio de Janeiro, Rio de Janeiro, RJ, Brazil*

## OVERVIEW

Hydrogen energy unveils perennial and sustainable energy production and utilization methods to fulfill all needs required by the human society. It represents an opportunity to utilize a great possible variety of raw materials and an energy source, such as electricity, heat, or mechanical work, to obtain fuel to be used in energy-efficient devices and therefrom to generate the same energy elements, such as electricity, heat, or mechanical work, with very limited noise and no aggressive wastes. This circular path concerning energy production and use had not been achieved and made possible for large-scale utilization until the advent of hydrogen energy became a reality.

The adoption and implementation of hydrogen energy makes more clean and sustainable energy available and introduces the creative and innovative concept of a circular economy, restorative by nature. This requires the use of intermittent renewable energy and the adequate control of seasonal energy storage. It also calls for a transition from the present fossil fuel−based energy system that is hitherto characterized for being structured in such a way that it possesses fuel ownership spotted in selective geographical locations, that it presents growing consumption of known reserves to depletion, that its exploitation devastates the environment, and that its utilization is made using energy-inefficient and pollutant engineering procedures and technologies. Such transition is able to create a system that is based on renewable energies, such as hydroelectric, solar, wind, geothermal, and oceanic energies, and is also based on a host of raw materials as source of the energy carrier that includes water and virtually any type of biomass. It makes the selection of primary energy and raw material to be adopted under judicious local possibilities and the possession of fuel to be widely distributed throughout the world, potentially decreasing concentrated ownership for market control. It also introduces the utilization of the most efficient energy converter known,

Science and Engineering of Hydrogen-Based Energy Technologies. https://doi.org/10.1016/B978-0-12-814251-6.00001-0

the fuel cell. It abandons the sequential and inefficient conversion of energy forms used by heat engines, turbines, and motors, to make the direct, unique, and highly efficient electrochemical energy conversion of the chemical energy contained in the fuel into electric energy and heat, thereby producing water.

Other energy transitions have been experienced before. Since several thousand centuries ago, biomass, mainly wood, has been used as a source of energy, and watermills and windmills were known since several thousand years ago. Renewable energies and fuels dominated the scene for the period the human kind continuously developed to enter modern times. A transition to the fossil fuel era was made with the use of coal and the steam engine and characterized the Industrial Revolution from the 18th century. The peak supply of world energy with wood occurred around 1850 and that of coal to transition to petroleum happened by 1930. The internal combustion engine was the invention that accelerated the use of oil derivatives and the supply of world energy with oil peaked in 2000. Curiously, the world's first automobile powered with an internal combustion engine used hydrogen as fuel, which was designed and demonstrated by François Isaac de Rivaz in 1806. Due to political and economic crises related to the commercialization of oil, since the years 1970 the consumption of natural gas increased steadily and is expected to peak by 2050, when the hydrogen economy will be installed and will have paved the way to take the world lead for energy supply. The transition from one fuel to another has not eliminated the use of previous ones. Instead, their utilization has been superimposed with progressive higher amounts. Wood, coal, oil, and natural gas are all simultaneously supplied, as well as the electricity from hydropower, nuclear, thermal, and geothermal plants. In complement to that, much electricity is also generated using modern windmills and photovoltaic solar cells.

It is remarkable to observe what the fuel transitions are able to tell. Wood is chemically more complex and has smaller specific energy (20.6 MJ/kg) than coal (23.9 MJ/kg). Conversely, coal is also chemically more complex and has smaller specific energy than oil (45.5 MJ/kg), which, in turn, repeats this trend with natural gas (52.2 MJ/kg). In addition to that, one may observe that there is an ongoing progressive decarbonization of fuels. The carbon content decreases from wood, to coal, to oil, to natural gas. It is also amazing to realize that the content of hydrogen increases continuously from wood, to coal, to oil, to natural gas, to reach the ultimate, perennial, and carbon-free fuel, hydrogen, for which specific energy equals 142.2 MJ/kg for its high heating value. Fig. 1.1 shows the carbon to hydrogen ratio for selected fuels that demonstrates the spontaneous decarbonization that is taking place, as our society aggregates new fuels to massive utilization [1]. Wood, not shown in Figure 1, has an average carbon to hydrogen ratio of 0.68. Among these, ethanol is usually obtained as a biofuel, made in large scale from sugarcane in Brazil and from corn in the United States. In this case, the photosynthesis process that consumes $CO_2$ from the environment during crop growth compensates the

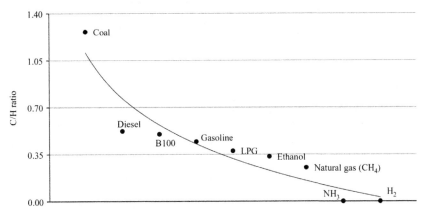

**FIGURE 1.1** Carbon to hydrogen ratio in selected fuels. Reproduced from [1].

waste carbon resultant from its utilization for energy production. It is noticeable that the use of clean hydrogen and, eventually, ammonia does not involve carbon emission.

The ability to harvest fuels of all types and the talented development of ingenious forms for utilizing them have resulted in very steep, ever-growing, world fuel consumption since the Industrial Revolution. This is depicted in Fig. 1.2 and shows that while the world needed an annual energy delivery of 43.5 EJ for all needs in the year 1900, about 575 EJ was required for the year

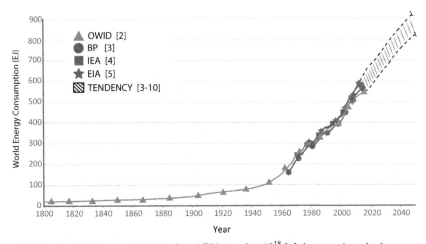

**FIGURE 1.2** World energy consumption. 1 EJ is equal to $10^{18}$ J. It is approximately the energy contained in 7 million tons of gaseous hydrogen or in 170 million barrels of oil. The historical data to the present was gathered from [2–5], including data started in 1800 [2] and the curves showed closed together depicting values for the second half of the twentieth century to the present [3-5]. The shaded area presents extrapolations gathered from [3–10].

2016 and the amazing amount of 760 EJ is expected to be necessary for the year 2040. The later result is deduced using data from several different origins, which were extrapolated based on their historical series of actual measurements.

This is pushed by a five- to sixfold increase in world population in the total period mentioned, but also because of the widespread use of energy for food production, storage, and distribution, for all sorts of vehicles, appliances, and devices, as well as for closed ambient heating or cooling, which has raised the level of practical, comfortable, and healthy living. However, life on the planet and the planet itself were threatened by such a huge kinetics of liberation to the environment of the carbon contained in fossil fuels, producing greenhouse effect gases and other contaminants.

The world has collectively awakened to worries regarding the environment during the United Nations' 1992 Conference on Environment and Development (Rio-92). An analysis made of the world environmental situation before that moment and from it onto the present shows alarming results, as depicted in Fig. 1.3 [11]. Although the use of products that are sources of allogeneic stratospheric gases under ultraviolet solar radiation, which destroy the ozone layer, has decreased 68%, allowing forecasting that a significant recovery of the ozone layer will occur by 2050, other results have markedly worsened. Per capita freshwater availability decreased 26% in the period especially because of the population increase of 35.5%. In addition, coastal dead zones that are mainly caused by fertilizer runoff and fossil fuel use increased 75.3%, killing large swaths of marine life. The latter, coupled with an annual increase in $CO_2$ emissions of 62.1% and a reduction of 2.8% in total area of forests, has markedly affected the biodiversity with a decrease of 28.9% on vertebrate species abundance. Moreover, the 10 warmest years from a 136-year record have occurred since 1998, and the most recent year of the data treated, 2016, ranks as the warmest on record. The ample recognition of these effects increased the relative importance of environment mitigation actions for controlling climate changes and preserving life in comparison with previous worries about fossil fuels shortage.

In fact, in addition to the pollutants already mentioned, the indiscriminate use of fossil fuels also produces small particulate material with size of up to 2.5 $\mu$m, PM2.5, which are especially present in large world metropolitan areas, affecting local population. Dispersed in the air, they are easily inhaled, going through the whole respiratory system until the alveolus, being mainly responsible for the occurrence of respiratory and cardiac illnesses and eventually contributing to injure human life. Made of micrometric solid carbon particles with condensed hydrocarbons on their surfaces, they still bear adhered particles of liquid hydrocarbons that are soluble in organic media, hydrated sulfates, and eventually, small particles of toxic heavy metals. In addition to that, they play the role of bacteria, viruses, and toxic chemical product carriers and pollute the water, soil, plants, food, and also the air. The

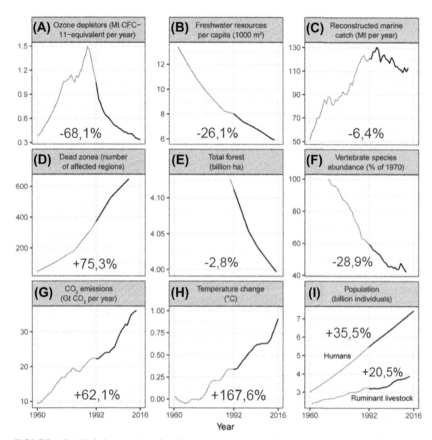

**FIGURE 1.3** Variation on the planet's environmental variables before (*faded line*) and after (*black line*) the United Nations' 1992 Conference on Environment and Development Convention, named Rio-92, which was held in Brazil. *Reproduced from [11].*

World Health Organization, WHO, establishes that a safe level of PM2.5 contamination in urban centers is below 10 $\mu g/cm^3$. Fig. 1.4 presents data gathered in previous publications [12,13] built with data taken from Pascal et al. [14]. It unveils a correlation between mortality peaks and greater levels of air pollution with particulate materials with sizes up to 2.5 $\mu m$.

The inexorable transition to the hydrogen energy era is taking place with a marked participation of renewable energies. Their inherent intermittent output is well complemented with the production and storage of hydrogen, projecting a perennial renewable circular sustainable cycle. Hydrogen, as an energy carrier, is versatile, clean, and safe. It can be used to generate electricity, heat, and power and still finds many applications as a raw material for the industry. It can be stored and transported with high energy density in the liquid or gaseous states and may be produced from raw materials where it is contained

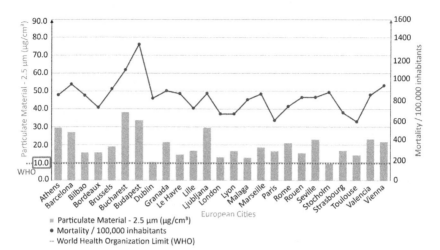

**FIGURE 1.4** Average level of environmental contamination with particulate material in air suspension with sizes up to 2.5 μm (*green bars* [light gray in print version]) and mortality per 100,000 inhabitants owing to respiratory and cardiac problems (*red curve* [dark gray in print version]) in the European cities indicated. *[12], using information from [13] and data from [14].*

and an energy source. Hydrogen is mostly produced from natural gas, by the steam reforming of methane, which constitutes its cheaper fabrication procedure. Direct consequences of that are the possibilities to include carbon capture, storage, and utilization technologies during the energy transition and/ or the use of biomethane to mitigate carbon emissions when producing hydrogen. This can also be achieved by water electrolysis technologies to produce hydrogen, using renewable energies, or yet by gasification of biomasses that opens an enormous opportunity for different regions in the world with their own local specificities in terms of raw materials and energy sources. Natural sources of hydrogen, once thought nonexistent, have been proved in several geographical spots with huge amounts, some with high purity and others combined with methane, nitrogen, helium, and other gases.

Fuel cells that use hydrogen as fuel and oxygen from the air as oxidant constitute the most energy-efficient devices known to date to generate electricity. Their market entry represents an era in the 21st century in which their utilization assumes importance comparable with the one the computers had for the 20th century. They are being applied in all sorts of electric energy—powered applications, with emphasis on the distributed stationary generation of electricity, heavy-duty vehicles, and automobiles. The descriptions about hydrogen energy that follow will inevitably involve the use of fuel cells.

This chapter will describe what hydrogen energy is about, what it is needed to implement its use, the types of applications possible, the challenges faced, and the benefits gained by using it. New possibilities for harvesting hydrogen for perennial energetic use will also be unveiled.

## WHAT HYDROGEN ENERGY IS ABOUT

The annihilation of matter by its collision with antimatter is the most energetic per unit mass energy conversion envisioned, though out of our reach. Interstellar stars, such as our sun, make incredibly effective energetic use of hydrogen; the fusion of four atoms of such light element to produce helium liberates an enormous amount of energy. A small fraction of the hydrogen atoms mass is not converted into mass of the helium atom, and it is enough to generate much energy following the well-known Einsteinian equation. Although nuclear fusion has been developed and experimented for long, it is not practical, economical, or easily feasible to date as an energy generation procedure for large-scale utilization by mankind. Although the energy produced in a fusion reaction is measured in millions of electron volts, the ionization energy to displace an electron from a hydrogen atom in a typical chemical reaction is only 13.6 eV. Even though nuclear fusion might eventually become viable, it is not expected to happen soon. Chemical and electrochemical reactions are, however, accessible as feasible procedures to generate energy. And that is not bad. In fact, it is very good compared with the inefficient 20th century's thermal machines that burned fossil fuels to generate energy using successive conversion steps. The energy-efficient direct, one-step, conversion of the chemical energy contained in the fuel into electric and heat energies using fuel cells is well established to dominate the energy scenario in this century.

Hydrogen energy is about utilizing hydrogen and hydrogen-containing compounds to generate energy to be supplied to all practical uses needed with high energy efficiency, overwhelming environmental and social benefits, as well as economic competitiveness.

The dawning of the hydrogen energy era revolutionizes several aspects of civilized life to allow a circular path concerning energy production and use, giving convenient and beneficial utilization to the huge amount of wastes resultant from developed life style, decarbonizing different sectors of intense energy consumption, making viable to implement large-scale production of renewable energy, better homogenizing the distribution of energy throughout different regions of the world, and facilitating the access to it. To transition from the fossil fuel—based economy to the hydrogen energy economy, provided that technology is available, new approaches have to be put in place, as depicted in Fig. 1.5. These include

1. circular, clean, and beneficial path for energy production and use;
2. widespread use of renewable energies, including
   2.1 production and storage of hydrogen to stabilize the delivery of electric energy, regulating the inherent intermittence associated with renewable energies;

Use of sewage and of urban and rural organic wastes

Use of hydrogen to decarbonize activities

Production and storage to stabilize delivery of electric energy to act as a buffer to increase resilience

Energy distribution across sectors, countries and regions
Hydrogen trading as an energy commodity

Circular, clean and beneficial path for energy production and use

Facilitating the access to energy due to local specific options of primary energy
Production of natural hydrogen

**FIGURE 1.5** The various characteristics and possibilities related to hydrogen energy technology application.

**2.2** production and storage of hydrogen to act as a buffer to increase resilience of a country or region energy system;

**3.** use of sewage and of urban and rural organic wastes to produce hydrogen and hydrogen-rich gases and compounds;

**4.** use of hydrogen to decarbonize activities in sectors such as
   **4.1** the industry
      **4.1.1** supplying electrical and thermal energies;
      **4.1.2** supplying renewable feedstock produced by conveniently reacting hydrogen with biomasses;

**4.2** energy supply, as combined heat/cooling and power, to buildings and households, thereby introducing the distributed generation of electrical and thermal powers;

**4.3** transportation, including light-duty and heavy-duty vehicles and automobiles for terrestrial, nautical, and aeronautical applications;

**5.** energy distribution across sectors, countries, and regions using hydrogen and hydrogen-rich gases and compounds as carriers and also hydrogen trading as an energy commodity;

**6.** facilitating the access to energy in different countries or regions because of local specific options of primary energy source and raw materials to produce hydrogen and also local production of natural hydrogen.

When hydrogen is produced from water electrolysis and used in fuel cells, water appears once again, as a by-product, closing an advantageous circular cycle. Such hydrogen production is considered environmentally friendly when renewable energies are used as the source of energy. This is facilitated in countries such as China, Brazil, Canada, the United States, and others, where much electrical energy is generated by hydroelectric power plants, in which in some periods of the year there is a surplus of turbinable water creating availability of turbinable discharge energy, because of difficulties associated with storing electrical energy once produced. That is, when demand to dispatch electricity decreases, either water is accumulated in a dam, which cannot be done with run-of-river hydroelectric plants, or water is spilled aside, not going through the turbines. Conversely, electricity generated may be used for hydrogen production as a way to store energy. Hydrogen then produced may be used for the various energetic or conventional chemical applications it finds. Similarly, hydrogen production from water electrolysis may also be done using electrical energy originated from other renewable energies, such as wind energy, solar energy, or ocean energy. In such cases, there are two additional benefits concerning the production of hydrogen by water electrolysis using renewable energies. The first one is that hydrogen may be used to complement and adjust electrical energy—delivering issues related to the inherent intermittence of renewable electric energy production. When there is shortage of water, wind, sun or ocean activity but the demand for electricity consumption exists, the hydrogen already produced and stored is available to generate electricity using fuel cells and/or turbines. The second benefit of storing hydrogen is that it can act as a buffer to increase resilience of the whole energy system of a country or region, considering all procedures used to generate electricity, either renewable or not, thereby stabilizing a regional electric energy distribution network.

The world population, once dispersed in rural areas, has modernly been concentrated in large, intensively built and structured environment and densely populated areas, where the exceeding production of wastes challenges the quality of life, threatens the health of living beings, and harms the local ecosystem. Table 1.1 presents data on measured and simulated population and urbanization rate in selected parts of the world. Although there is an ongoing tendency of decreasing the world population average annual growth rate, the world will have more than 9 billion inhabitants in 2040 with a very high rate of urbanization. It is remarkable that a very young society, such as the Brazilian one, is expected to reach 90% of urbanization rate in 2040. It is also important to observe that highly populated countries, such as India and China, will possibly move to urban centers until 2040 about 300 million or 230 million people, respectively, which is a significant increase of urban population, amounting more than the whole population of other countries.

All these urban concentrations generate a host of waste, and an important fraction of the waste collected is organic. In addition to this, there is also an important production of sewage, which in many places is largely untreated, or only primarily treated, before being discarded to the environment. Sewage usually contains and is a carrier of bacteria, viruses, protozoa, and parasites. All urban organic wastes and wastewater (sewage) may be treated to produce hydrogen or hydrogen-rich gases, giving a useful destination to a huge urban problem. Similarly, rural and agribusiness wastes may also potentially be used for the production of hydrogen, hydrogen-rich gases, and solid fertilizers.

A fundamental advantage of the widespread use of hydrogen energy is the possibility to help to decarbonize different sectors of activity, with inherent social and environmental benefits. Such decarbonization affects industry in two different aspects. One is related to the supply of clean electricity and high-grade thermal energy. Powering of all sorts of equipment and systems may be made using fuel cells, and the supply of low-grade and high-grade heat may be achieved, feeding hydrogen to burners and heat exchangers. The other concerns the supply of feedstock, mainly for chemical industries, once produced using fossil fuels and consisting of a host of hydrocarbons, which can be alternatively made out of hydrogen and biomasses, thus becoming of renewable origin, while allowing the use of the same industrial methodologies already in place to fabricate the end products.

Two additional sectors of activities are likely to be decarbonized by the use of hydrogen energy and gain very much importance for being, not only, but also located in urban environment, where there is dense concentration of human lives. Their decarbonization causes therefore direct and strong social benefit. One of them is the supply of combined heat/cooling and power to all sorts of residential, business, and public buildings using fuel cells. It presents the additional advantage of shifting from centralized production of electricity

**TABLE 1.1** World Population and Urbanization Rate [10]

| | Compound Average Annual Growth Rate (%) | | | Population (million) | | Urbanization Rate (%) | |
|---|---|---|---|---|---|---|---|
| | 2000–16 | 2016–25 | 2016–40 | 2016 | 2040 | 2016 | 2040 |
| **North America** | 1.0 | 0.8 | 0.7 | 487 | 570 | 81 | 86 |
| United States | 0.8 | 0.7 | 0.6 | 328 | 378 | 82 | 86 |
| **Central and South America** | 1.2 | 0.9 | 0.7 | 509 | 599 | 80 | 85 |
| Brazil | 1.1 | 0.7 | 0.5 | 210 | 236 | 86 | 90 |
| **Europe** | 0.3 | 0.1 | 0.1 | 687 | 697 | 74 | 80 |
| European Union | 0.3 | 0.1 | 0.0 | 510 | 511 | 75 | 81 |
| **Africa** | 2.6 | 2.4 | 2.2 | 1216 | 2063 | 41 | 51 |
| South Africa | 1.5 | 0.7 | 0.6 | 55 | 64 | 65 | 75 |
| **Middle East** | 2.3 | 1.7 | 1.4 | 231 | 321 | 69 | 76 |
| **Eurasia** | 0.4 | 0.3 | 0.1 | 230 | 236 | 63 | 67 |
| Russia | −0.1 | −0.2 | −0.3 | 144 | 133 | 74 | 79 |
| **Asia Pacific** | 1.1 | 0.8 | 0.6 | 4060 | 4658 | 47 | 59 |
| China | 0.5 | 0.3 | 0.0 | 1385 | 1398 | 57 | 73 |
| India | 1.5 | 1.1 | 0.9 | 1327 | 1634 | 33 | 45 |
| Japan | 0.0 | −0.3 | −0.4 | 127 | 114 | 94 | 97 |
| Southeast Asia | 1.2 | 1.0 | 0.7 | 368 | 763 | 48 | 60 |
| **World** | 1.2 | 1.0 | 0.9 | 7421 | 9144 | 54 | 63 |

UN Population Division databases; IEA databases and analysis.

and distribution across long distances, which involves energy losses causing high overall inefficiency, to the distributed generation of combined heat/cooling and power. This procedure benefits from the legal framework already put in place for using the well-established distributed generation of electricity with wind and solar energy systems and also the technological adaptation already made to avoid and control the dispersion of harmonics into the grid. Harmonics are unwanted voltage and/or current frequencies eventually generated in distributed energy devices that overload wiring and transformers, heating them up and possibly even causing fires in extreme cases. Such voltage and/or currents are harmful to equipment, decreasing usage reliability and life expectancy. The other sector mentioned for which decarbonization bears magnificent importance for postmodern society is transportation. The ever-growing number of automobiles in the cities and heavy-duty vehicles covering high distances introduces a heavy environmental and social hurdle to human kind, as depicted in Fig. 1.4. Although automobiles call so much attention because of the personal use made with them and the mobility freedom they represent, light-duty vehicles, such as forklift, and heavy-duty vehicles, such as buses, have gained market rapidly powered by hydrogen with fuel cells.

The hydrogen energy era brings an innovation with respect to fuel availability that was not imaginable during the fossil fuel era. The latter has clearly established a strong ownership relation between fuel, countries, and regions. Regional or world political and market-based power has been directly correlated to be proprietary of petroleum, natural gas, and once important coal reserves. Particular land and marine extensions possessing the privilege of such resources have been the motivation for harsh disputes and military actions all throughout the 20th century. In certain cases, politically unstable or unprepared civil organizations in specific countries have easily enriched and threatened other countries with their petro power, contributing to world instability. Conversely, powerful countries have found momentary excuses to occupy and explore certain such regions, once again creating world political instabilities. The hydrogen energy economy changes completely this situation because any country or any world region is able to find its own particular options to combine primary energy source, preferably renewable, and raw material for local hydrogen production to satisfy its own needs while harvesting natural hydrogen has not yet become a reality. Water electrolysis with renewable energy varieties represents an option that gains economic viability with the increase of the magnitude of the particular undertaking in terms of installed power. The enormous experience already accumulated on reforming natural gas to produce hydrogen may also be used with biogases. The immense availability of biomasses in several world regions facilitates to implement gasification or biodigestion processes for hydrogen production. The extra need of hydrogen fuel in a country or region has easy solution by transnational distribution and trading across the world of hydrogen, hydrogen-rich gases, or hydrogen-rich compounds, as new energy carriers, reestablishing a specific world fuel market. The existence of such energy commodity that could eventually recreate the previous unbalanced fuel proprietary condition of the fossil fuel era is, nevertheless, alleviated by local possibilities of hydrogen production for partial supply of the total amount needed. And, in addition to all that, amazingly, natural hydrogen wells begin to have their existence proved in specific world regions.

## FULL IMPLEMENTATION OF HYDROGEN ENERGY TECHNOLOGIES

### Green Hydrogen Production

An important aspect concerning the utilization of hydrogen energy technologies is that its main fuel, hydrogen, has been produced and used in very large scale by different industrial sectors for many years. Although hydrogen has been used as a chemical product and not as a fuel for energy production, the main methodologies for hydrogen production, storage, and transportation are well dominated. Fig. 1.6 presents prospects for hydrogen utilization until 2050, when hydrogen energy technologies will be well established. It shows energy values expressed in EJ ($10^{18}$ J) calculated from the energy contained in the amount of hydrogen that is expected to be necessary, using the higher heating value of hydrogen, equal to 142.18 MJ/kg [15]. All energy sectors are expected to increase the amount of hydrogen used, especially transportation, industry, and feedstock production.

The increase in the amount of hydrogen required in the next decades is so significant that a variety of technological options, raw materials, and energy sources for hydrogen production must be made viable, taking into

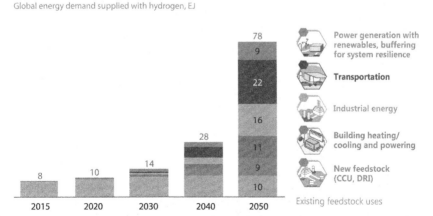

FIGURE 1.6    Global energy demand supplied with hydrogen. Energy quantities are shown in EJ. 1 EJ = $10^{18}$ J. *CCU*, carbon captured and utilized; *DRI*, direct reduced iron. *Reproduced from [15].*

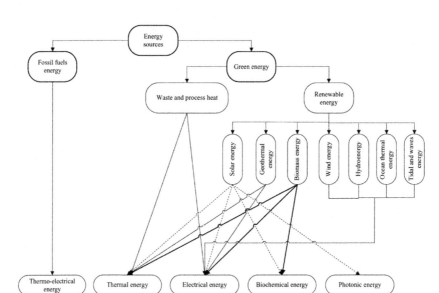

**FIGURE 1.7** Energy conversion paths for hydrogen production. *Reproduced from [1].*

consideration that low environmental impact technologies will be preferred. A great variety of energy conversions will be possible, as depicted in Fig. 1.7 [1]. The chemical energy contained in fossil fuels that undergo thermoelectrical energy conversions will follow on playing an important role, however, resulting in hydrogen that is not considered green because of the deleterious emissions associated with such technologies. These represent, however, well-established mass production approaches that are cheaper than others because of mass production and also for not taking into account externalities associated with their processing and use, such as those discussed in Fig. 1.3.

There is a host of green energy possibilities for hydrogen production, associated with wastes from several origins and process heat, as well as with renewable energies, all of them used to harvest four forms of energy that are thermal, electrical, biochemical, and photonic. Wastes and process heat may be used for thermal energy and electrical energy generation, which are then utilized for green hydrogen production. The renewable energies such as solar energy, geothermal energy, biomass energy, wind energy, hydroenergy, and ocean thermal, tidal, and wave energies, give origin to all four forms of energies aforementioned for green hydrogen production. Solar energy, of course, is dominant with the generation of all four forms of energy mentioned for green hydrogen production and is expected to undergo large-scale utilization throughout the world. This is foreseen by the data presented in Fig. 1.8. Fig. 1.8A shows how the global average levelized[1] cost of electricity from utility-scale solar photovoltaic (PV) has been decreasing and is expected to vary until 2040. It declined 70% from 2010 to 2016 and is projected to decline an additional 60% to 2040, period in which the solar PV electricity costs will become competitive to those of other fuels described in Fig. 1.8B, thereby progressively facilitating its adoption by consumers. According to Fig. 1.7, biomass energy also presents versatility for generation of thermal, electrical, and biochemical energies for green hydrogen production, gaining special importance because of its great availability all throughout the planet. The ensemble of renewable energies and the energy contained in wastes and process heat may all generate electrical energy that is required for electrolytic, electrochemical, electrophotochemical, and electrothermochemical processes, which, in addition to photobiochemical methods, give origin to an ample variety of technologies for green hydrogen production.

The total Gibbs free energy necessary to break the water molecule for hydrogen production by water electrolysis is composed by a smaller portion of electricity that is converted to work if there is heat energy concomitantly available to complement the work to be done. At temperatures above steam formation, the increase in heat energy offered to the system decreases progressively the requirement of work to be done by the electric current. That is the reason why high-temperature electrolysis is more advantageous. Fig. 1.9 [16] represents the basic thermodynamic variables associated with this process, showing that it is energetically more beneficial to process steam than liquid water. In addition to that, the higher the temperature, the smaller the amount of electric energy needed, $\Delta G$ (the molar Gibbs energy of the reaction), to convert directly steam into hydrogen and oxygen. This happens because the total energy needed, $\Delta H$ (the molar enthalpy of the reaction), remains approximately constant while the heat, $T\Delta S$ (absolute temperature multiplied by the variation in entropy), is progressively augmented. Consequently, Fig. 1.9 shows that less electricity is required for the electrolysis of steam per cubic meter of hydrogen produced compared with the electrolysis of liquid water. However, heat must be provided and it is costly to be produced. If electricity from renewable source is available, the solid oxide electrolysis cell

---

1. The levelized cost of an energy carrier, either electricity or hydrogen, is calculated in a consistent way that takes into account the initial capital, discount rates and costs of operation, feedstock, and maintenance. It represents an economic assessment of the average total cost to build and operate the power generation or hydrogen production infrastructure over its entire lifetime divided by the total electrical energy or hydrogen output using such asset over its lifetime.

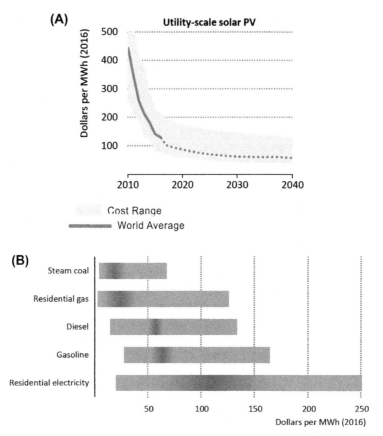

**FIGURE 1.8**  (A) Global average levelized cost of electricity from utility-scale solar photovoltaic (PV), *blue solid line* (dark gray in print version), world costs range, shaded, and cost projections to 2040, *dotted line*; (B) Fossil fuels and residential electricity range of world average prices paid by consumers, 2015. *Notes: MWh*, megawatt-hour. A utility-scale solar facility is one which generates solar power and feeds it into the grid, supplying a utility with energy. The areas shaded in blue (dark gray in print version) represent the range of reference prices used for the purposes of calculating energy consumption subsidies. Variations in quality may explain a part of the variations in price, especially for electricity where differences in reliability of services mean that it is not a homogenous product across countries. *Reproduced from [10].*

(SOEC) is a more energy-efficient technology for green hydrogen production. It is a high-temperature electrochemical device, which when coupled with an electrical energy source promotes the reduction of a gaseous reactant to generate products such as hydrogen and oxygen [17−22].

The SOEC operating at high temperatures, 700−900°C, has the advantage that the unavoidable Joule heat inherent to its operation is used in the

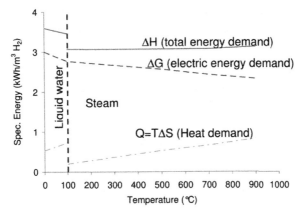

**FIGURE 1.9** Thermodynamic approach of the high-temperature water/steam electrolysis process. *Reproduced from [16].*

electrolysis process. It consists of an apparatus similar to a solid oxide fuel cell (SOFC), in which the oxygen ion—conducting electrolyte may be made of yttria-stabilized zirconia (YSZ); the air electrode, the anode, may be composed of perovskites such as lanthanum strontium manganese oxide (LSM), lanthanum strontium cobalt iron oxide (LSCF), and nickel-YSZ cermet for the hydrogen electrode, the cathode. The electrochemical reactions involved are as follows:

$$\text{SOEC fuel electrode, cathode: } H_2O + 2e^- \rightarrow H_2 + O^{2-} \qquad (1.1)$$

$$\text{SOEC air electrode, anode: } O^{2-} \rightarrow \tfrac{1}{2} O_2 + 2e^- \qquad (1.2)$$

The SOEC benefits from the intense development being made for SOFCs because it basically utilizes similar materials and systems. The special advantage is that SOECs and SOFCs may be used reversibly, for the production of hydrogen or the generation of electricity, respectively, in the same device, alternating the role of anode and cathode, the fuel utilized, as well as either the input or the output of electricity as it is used to produce hydrogen or to generate electricity.

Other technologies of water electrolysis are presently very well developed and have successfully reached market application [23]. Like fuel cells, they are subdivided according to the type of electrolyte used. In an alkaline electrolysis cell (AEC), the electrolyte is made of liquid solutions of NaOH or KOH, thereby possessing $OH^-$ as charge ion conductor; it uses carbon and transition or noble metals as catalysts, electrodes, and interconnectors and is operated at low temperatures, 40—90°C. A polymeric electrolyte membrane electrolysis cell (PEMEC) has electrolyte composed of a hydrated polymeric membrane that possesses $H^+$ as charge ion conductor, moving through the membrane by the aid of another charge ion conductor that is $H_3O^+$, using carbon and platinum as electrode and catalyst, carbon—metal as interconnector with an operation temperature ranging between 20 and 150°C. The main intrinsic and operational features for the AEC, PEMEC, and SOEC are summarized in Table 1.2 that also simulates a specific comparison between these devices, in which the AEC and PEMEC are set to be operated at 80°C and the SOEC at

**TABLE 1.2** Summary of Different Electrolyzer Types, Their Particular Features, and a Specific Comparison Among Them

| Electrolyzers | AEC | PEMEC | SOEC |
|---|---|---|---|
| Electrolyte | Solution of NaOH or KOH | Hydrated polymeric membranes | Ceramic |
| Charge ion conductor | $OH^-$ | $H^+$, $H_3O^+$ | $O^{2-}$ |
| Cathode reaction | $2H_2O + 2e^- \rightarrow$ $H_2 + 2OH^-$ | $2H^+ +$ $2e^- \rightarrow H_2$ | $H_2O + 2e^- \rightarrow$ $H_2 + O^{2-}$ |
| Anode reaction | $2OH^- \rightarrow H_2O +$ $\frac{1}{2}O_2 + 2e^-$ | $H_2O \rightarrow \frac{1}{2}O_2$ $+ 2H^+ + 2e^-$ | $O^{2-} \rightarrow \frac{1}{2}O_2$ $+ 2e^-$ |
| Electrodes | Ni, C | C | Ceramic/cermet |
| Catalyst | Ni, Fe, Pt | Pt | Ni, perovskites (LSM, LSCF) |
| Interconnector | Metal | Carbon—metal | Stainless steel, ceramic |
| Operating temperatures (°C) | 40—90 | 20—150 | 700—900 |
| **Specific Comparison** | | | |
| Operating temperature (°C) | 80 | 80 | 800 |
| Operating potential (V) | 1.9 | 1.7 | 1.15 |
| Internal resistance ($\Omega$ cm$^2$) | 2.5 | 0.5 | 0.15 |
| Hydrogen production rate (mol $H_2$/m$^2$h) | 50 | 175 | 211 |
| Hydrogen production per energy consumed (mol $H_2$/kWh) | 27 | 40 | 110 |

*AEC*, alkaline electrolysis cell; *PEMFC*, polymeric electrolyte membrane electrolysis cell; *SOEC*, solid oxide electrolysis cell.
Adapted from [24].

800°C [24]. It is remarkable that the SOEC presents outstanding results, including smaller operating potential and internal resistance, as well as higher hydrogen production rate and hydrogen production per energy consumed. In a situation in which the electrical energy needed to produce hydrogen using the AEC or PEMEC varies from about 1 to 2 kWh/Nm$^3$H$_2$, the SOEC may require less than 0.5 kWh/Nm$^3$H$_2$. This exemplifies the advantageous energy efficiency characteristics of the SOEC compared with the present conventional technologies for hydrogen production by electrolysis. In addition to that, the SOEC has also prospects to become a cheaper technology than the present conventional ones [25]. Fig. 1.10 presents levelized cost of hydrogen production using SOEC current and future advanced technologies in comparison with the use of the AEC and PEMEC as a function of the electricity cost. It shows that within all ranges of electricity costs the SOEC presents better future prospects to produce cheaper hydrogen.

There are novel, yet to become commercial, electrolysis methods for hydrogen production. Among them, the microbial electrolysis cell (MEC) utilizes, just like the AEC, the PEMEC, and the SOEC, similar fixture as the microbial fuel cell (MFC) does [26]. The MEC makes use of exoelectrogen bacteria that are microorganisms with ability to transfer electrons extracellularly. In an MEC, they generate electric current in a suitable medium where an extra external power is also supplied and combined to reduce protons, thereby producing hydrogen. Fig. 1.11 [27] depicts schematically the MEC setup and functioning as a two-chamber reactor. Electrochemically active bacteria (EAB) colonize in the MEC's anode, consume an energy source and a substrate, such as organic waste materials, and produce protons that diffuse through the electrolyte solution, trespassing the separator membrane toward the MEC's cathode. They also generate electrons that are driven to an external circuit and CO$_2$ as reaction waste that is discarded. At the MEC's cathode,

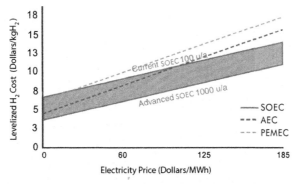

**FIGURE 1.10**  Levelized cost of hydrogen production as a function of electricity cost using an alkaline electrolysis cell (AEC), a polymeric electrolyte membrane electrolysis cell (PEMEC), or a solid oxide electrolysis cell (SOEC). *Adapted from [25].*

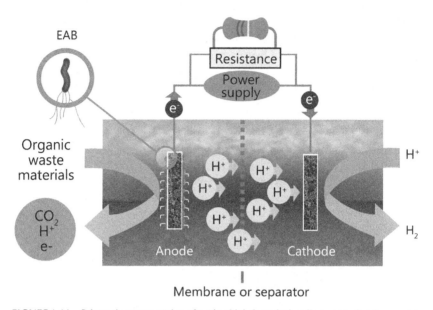

**FIGURE 1.11** Schematic representation of a microbial electrolysis cell as a two-chamber reactor, showing the role of electrochemically active bacteria (EAB) in the anode chamber on waste material substrate to produce protons that diffuse through the separator membrane and reach the cathode where combination with electrons promotes hydrogen production. *Reproduced from [27].*

protons are reduced to produce gaseous hydrogen, what is achieved by superimposing an external electrical potential to the system. In the SOEC case, heat was supplied by the cell's electrochemical exothermic reactions to vaporize water and to decrease the work to be done by the electric current imposed to the system, thereby allowing electrolysis to be performed with a much smaller external power than that of conventional water electrolysis. In the MEC case, the external power needed for electrolysis is even smaller. The electrochemical potential needed for hydrogen production by electrolysis using the different technologies previously discussed are 1.9 V for the AEC, 1.7 V for the PEMEC, and 1.15 V for the SOEC for the specific comparison example presented in Table 1.2. Conversely, the electrical power needed for hydrogen production in an MEC can be as small as 0.25 V [26] and may result in energy consumption as low as that of the SOEC, of the order of 0.6 kWh/$Nm^3H_2$ [28]. Virtually, any biodegradable organic electron donor may be used as a substrate in MECs, which include domestic wastewater, ocean and marine sediments, anaerobic sewage sludge, acetate, butyrate, glucose, ethanol, polymeric materials such as cellulose and proteins, complex mixtures such as dairy manure, swine wastewater, brewery wastewater, and a host of industrial and agribusiness wastes. Table 1.3 presents a list of possible EAB and their substrates used in MECs. MECs use microbes as biocatalysts to produce hydrogen with purity level that does not require expensive hydrogen purification

**TABLE 1.3** Electrochemically Active Bacteria and Substrates Used in MECs [27]

| Electrogenic Microorganisms | Substrates | References |
|---|---|---|
| *Rhodopseudomonas palustris* DX-1 | Volatile acids, yeast extract thiosulfate | [29] |
| *Ochrobactrum anthropi* YZ-1 | Acetate, lactate, propionate, butyrate, glucose, sucrose, cellobiose, glycerol, ethanol | [30] |
| *Acidiphilium* sp. strain 3.2 Sup 5 | Ferric iron, Ferrous iron | [31] |
| *Rhodoferax ferrireducens,* *Citrobacter* sp. SX-1 | Glucose, citrate, lactose, sucrose, acetate, glycerol | [32,33] |
| *Shewanella putrefaciens* MR-1, IR-1, SR-21 | Lactate, pyruvate, acetate, glucose | [34] |
| *Shewanella oneidensis* MR-1 | Lactate | [35] |
| *Klebsiella pneumoniae* strain L17, *Enterobacter cloacae* | Glucose, starch, cellulose | [36,37] |
| *Aeromonas hydrophila* KCTC 2358 | Acetate | [38] |
| *Aeromonas* sp. strain ISO2-3, *Geobacteraceae* | Glucose, acetate | [39,40] |
| *Geobacter metallireducens,* *Geobacter sulfurreducens* | Acetate | [41–44] |
| *Desulfobulbus propionicus* | Pyruvate, acetate | [45] |
| *Propionibacterium freudenreichii* ET-3 | Acetate, lactate | [46] |
| *Arcobacter butzleri* strain ED-1 | Sodium acetate | [47] |
| *Clostridium beijerinckii,* *Clostridium butyricum* EG3 | Starch, glucose, lactate, molasses | [48,49] |
| *Firmicutes* Thermincola sp. strain JR | Acetate | [50] |
| *Geothrix fermentans,* *Gluconobacter oxydans* | Acetate; glucose | [51,52] |

*MEC,* microbial electrolysis cell.

procedures, and specially, they integrate pollution treatment and hydrogen production with the advantages of cleanness, energy saving, and waste utilization [27]. Future development of MECs is very much dependent on the development prospects of MFCs, presenting the benefit of a wide variety of different applications that MFC-based technologies may reach. In addition to electricity generation and remote power supply, these include alternative technologies such as carbon capture, desalination, resource recovery, wastewater treatment, bioremediation that is a process used to treat contaminated media, electrochemical biosensors, medical diagnostics tool, and on-chip power sources and ecobots, that is, ecological robots that present a self-sustaining operation using, for example, waste as raw material for energy production [53].

## Natural Hydrogen

In spite of all the well-known or the new and sophisticated technologies for hydrogen production and the belief kept for so long that these are the possible ways to make hydrogen fuel viable for use on earth, exploration of natural hydrogen, not once considered possible, is beginning to become reality. There is no doubt that hydrogen is the most abundant element in the Universe. On earth it was thought to exist only bound to compounds, into any hydrocarbons and water, being one of the constituents of all flora and fauna. The gas hydrogen was not considered to be available on earth, either mixed with other gases or in high proportion, almost pure, because it is composed of such a reactive chemical element. However, recent evidence proves the contrary. Fig. 1.12 shows a circular geological structure on the earth's surface in Brazil where measurements are made to detect continuous outgassing of natural hydrogen [54]. These circular, sometimes elliptical, structures that may possess a few meters or kilometers of diameter are zones of deformation of the soil, resulting from basement faults, bounded by rounded depressions of a few meters, presenting inside a flat bottom.

They have also been found elsewhere, such as in North America, the Sultanate of Oman, Philippines, Mali, Turkey, New Caledonia, and Russia [54–56]. The following are considered as characteristics and mechanisms related to natural hydrogen existence on earth [55]:

**FIGURE 1.12** Geological structure composed of a circular depression on a craton zone formation in Brazil where hydrogen gas is detected flowing out. *Reproduced from [54].*

1. Natural hydrogen outgassing is now understood to appear in craton formation regions that are rock formations on the earth's continental crust that have remained stable for a period of time as extended as 500 million years.
2. Hydrogen is found at the earth's free surface and in fairly shallow depths of up to about 500 m [56].
3. Natural molecular hydrogen also occurs in ophiolitic[2] formations, eventually associated with nitrogen and abiotic methane, whose generation is not linked with organic matter thermal cracking but by reduction of any source of carbon. That is, there is no organic matter accumulation associated with the sites of occurrence, situation in which hydrocarbons would rather be produced.
4. The effect of serpentinization[3] in peridotite, a very dense, coarse-grained, olivine-rich $[(Mg^{2+}, Fe^{2+})_2SiO_4]$ ultramafic rock, which is a silicate mineral rich in magnesium (forsterite end-member[4]) and iron (fayalite end-member), is twofold [55]:

   a. The hydration of the forsterite end-member of olivine ($Mg_2SiO_4$) produces much hydroxide ion, such as in Eq. (1.3), to make it an ultrabasic rock:

   $$2Mg_2SiO_4 + 3H_2O \rightarrow Mg_3Si_2O_5(OH)_4 + Mg^{2+} + 2OH^- \qquad (1.3)$$

   Fosterite + water → serpentine + magnesium + hydroxide ion

   b. Since $Fe^{2+}$ is by far the most important electron donor in ultrabasic rocks, hydration of the iron end-member (fayalite) of olivine minerals induces the formation of $Fe^{3+}$ minerals such as magnetite, leading to the formation of hydrogen as depicted in Eq. (1.4):

   $$3Fe_2SiO_4 + 2H_2O \rightarrow 2Fe_3O_4 + 3SiO_2 + 2H_2 \qquad (1.4)$$

   Fayalite + water → magnetite + silica + hydrogen

   The existence of aquifers under the earth's surface in geologically stable craton formation regions may represent a mechanism of continuously promoting the formation of natural hydrogen, as far as ferrous iron is present in their surrounding (as olivine, or as any other mineral containing ferrous iron being able to decompose and liberate soluble $Fe^{2+}$ in water). This may eventually be consistent with the amazing possibility of replenishing natural hydrogen wells with new-formed gas.

---

2. Ophiolite is a stratified igneous rock complex in the earth's oceanic crust and the underlying upper mantle that has been uplifted and exposed above the sea level, emplaced onto continental crustal rocks. Meaning snake rock, from Greek, it contains *serpentinized* rocks that present a scaly, greenish brown—patterned surface resembling snakeskin.
3. Serpentine rocks are formed as a result of hydration processes, such as serpentinization, when the spreading tectonic plates in the earth's crust lift them up from the ocean and they are chemically altered by water or, alternatively, when a similar process is induced by the presence of the underlying aquifers that promote water movement.
4. *End-member* is a mineral that is at the extreme end of a mineral series in terms of purity. Fayalite $Fe_2SiO_4$ and forsterite $Mg_2SiO_4$ are *end-members* of the olivine series $(Mg,Fe)_2SiO_4$.

**FIGURE 1.13** Localization of Bourakebougou, in Mali, Africa, where the first world hydrogen wells have been producing natural hydrogen, used locally for electricity generation.

The first hydrogen wells actually producing natural hydrogen in the world are being explored in Bourakebougou, in Mali, Africa, in the region depicted in Fig. 1.13, by the company Petroma Inc. It was the search for water in that region that unveiled the presence of gaseous occurrence composed of 98% of pure natural hydrogen 1% of nitrogen and 1% of methane that is explored and used locally for electricity generation [54]. The natural hydrogen wells are a little over 100 meters below the earth's surface in that region, confirming that this energy resource may be available in shallow wells for which the technological setup for exploration becomes simpler and cheaper. This may result on the extraordinary possibility of harvesting natural hydrogen at a cost smaller than that of hydrogen produced by any of the methods known to date, from the conventional natural gas steam reforming and the well-developed water electrolysis to the innovative technologies hitherto discussed.

Because it has been herein showed that the first world's natural hydrogen wells are under production in Mali, on the northeast of Africa (Fig. 1.13) and because natural hydrogen occurrence has also been proved to exist in the northeast of South America, in Brazil (Fig. 1.12), considering that it is known that the South American and African continents once joined together, in old geological era, forming the Pangea supercontinent as depicted in Fig. 1.14, and that they may share similar geological structures, an analysis was made of the potentiality of discovering simultaneous occurrence of craton formation regions on the top of the underlying aquifers in these continents, which would bring the possibility to identify other possible regions where natural hydrogen wells would likely be found. The resulting analysis is shown in Fig. 1.14. It is absolutely amazing to verify that there are several coincident occurrences of craton rock formations onto the underlying aquifers both in South America and in Africa, Fig. 1.14. The formation verified in the Amazon region calls attention for its enormous extension.

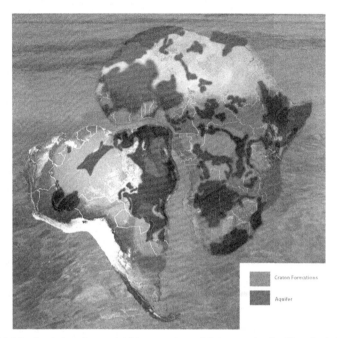

**FIGURE 1.14**   South American and African continents joining together in old geological era that once formed the Pangea supercontinent. Regions where craton rock formations are found on the top of the underlying aquifers in South America and Africa are indicated with different color contrasts.

In addition, taking into account that natural hydrogen-rich gaseous formations were found on the paleo ocean floor in New Caledonia (ophiolite) such as depicted in Fig. 1.15 [55], it is conceivable to admit that in the vast intercontinental oceanic extensions, there might exist several sites where natural hydrogen is likely to be found and explored, the feasibility of which will be very much dependent on the depth and local conditions but will benefit from the experience already accumulated with the exploration of hydrocarbons in the ocean. It is also important to remark that an eventual future exploration of subsea natural hydrogen will never submit such sites to the danger of extraordinary environmental disasters such as the ones already occurred with the exploration of hydrocarbons, oil and natural gas, simply because hydrogen would be partially absorbed by water and partially vented to be consumed in open air, thereby producing water.

## HYDROGEN ENERGY APPLICATION

Once hydrogen is available, either industrially produced or naturally harvested, it becomes the 21st-century ultimate fuel commodity for clean,

**FIGURE 1.15** Natural bubbling of a gaseous mixture of nitrogen, hydrogen, and methane on the ocean floor in the bay of Carénage, New Caledonia. *Reproduced from [55].*

environmentally friendly, energy commercial application. Fig. 1.16 presents future prospects for hydrogen energy technologies start of commercialization to mass market acceptability in the sectors of transportation, industry energy, building heating/cooling and power, industry feedstock, and power generation.

Transportation is the sector that calls so much attention because it includes automobiles, the 20th century's star of the vehicles because of the freedom associated with personal mobility. However, dense-populated urban regions and their inhabitants suffer the effects of deleterious externalities never taken into account when new technologies for mobility are considered more expensive than the conventional ones. In the transition period to a new and cleaner energy era, different options are considered in parallel to the intro-duction of the ultimate solution represented by fuel cell automobiles. In particular, both battery and fuel cell vehicles possess electric drivetrains, presenting the advantage of coupling well with growing renewable energy sources utilization and sharing the burden of striving for a new refueling infrastructure to be installed, requiring heavy investments [57]. Intermediate transitional efforts also include a variety of hybrid vehicles and the use of renewable fuels such as ethanol [58] in conventional internal combustion engines. The advancement concerned with the direct utilization of methane or ethanol in fuel cells without previous reforming [59] may open new strategies for clean vehicles with an electric power train.

It is interesting to see that fuel cell—powered forklifts that are used in confined ambient of companies have already reached plenty application and represent brand new hydrogen energy technology that needs no subsidies for acceptation. Moreover, fuel cell hydrogen buses are a class of vehicles comprising a niche market for reasons such as:

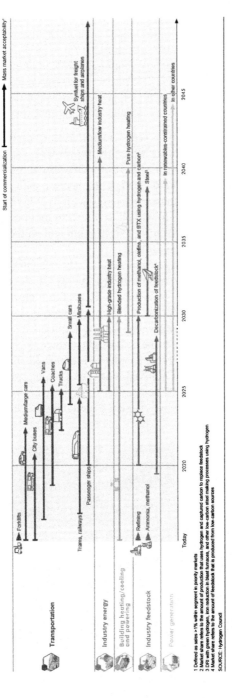

**FIGURE 1.16** Future prospects for hydrogen energy technologies start of commercialization to mass market acceptability in the sectors of transportation, industry energy, building heating/cooling and power, industry feedstock, and power generation. *From [15].*

1. they are very much used in huge urban areas throughout the world, where local pollutant emissions represent a problem requiring urgent solution;
2. they represent mass transportation mode and their use contributes to decrease the need for automobile utilization;
3. they are much more silent than conventional buses, making their use to decrease local noise pollution;
4. they are refueled in their central garage, which simplifies and makes infrastructure needs much cheaper than for automobiles, dispersedly used.

Fig. 1.17 presents a version of a series hybrid battery-dominant electric hydrogen fuel cell plug-in city bus for which the power train and the auxiliary system were developed and demonstrated [12]. In this case, emphasis has been given to the design of the hybridization energy engineering with predominance of power in batteries and predominance of energy with hydrogen. It succeeded to take 46.6% of the total energy embarked to effective motion at the vehicle axle, resulting in a fuel economy of 6.7 kg $H_2$/100 km and on the achievement of working ranges normally above 300 km.

Hydrogen energy is ready to positively and directly influence the industrial sector in a number of ways. In addition to the supply of electric energy, it opens the possibility to offer high-grade heat for a host of industrial processes. However, a certain time span will be necessary to modify industrial infrastructure, which means that several years will be required for such a transformation. Creative adaptations are being designed such as the case of the chemical industry that has industrial procedures established for the use of a variety of hydrocarbons normally derived from fossil fuels, oil, and natural gas. A huge facility is designed to be created for hydrogen production from water electrolysis using renewable wind energy, including massive amounts of

**FIGURE 1.17** Example of a series hybrid battery-dominant electric hydrogen fuel cell plug-in city bus.

hydrogen storage in underground wells and transportation to an industrial site, where an infrastructure is being built to produce the same feedstock hydrocarbons presently obtained from fossil fuels by locally reacting hydrogen with selected biomasses as sources of carbon [60]. The advantage of such an endeavor is to transform the chemical industry into renewable, while keeping in use the same present industrial methodologies because the same hydrocarbon feedstock will continue to be used.

With respect to the steel industry, there is also an extraordinary transformation in course. In this case, renewable energy is also the energy source chosen to produce hydrogen by water electrolysis to displace the use of an ancient industrial apparatus that is the blast furnace for metallic iron production from iron ore. Fig. 1.18 compares the procedures presently used for

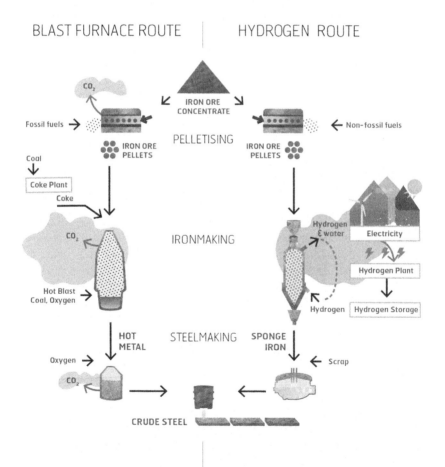

**FIGURE 1.18** Conventional environmentally unfriendly, on the left-hand side, and hydrogen-based clean, on the right-hand side, procedures to fabricate metallic iron from iron ore. *From [61].*

metallic iron production with the new hydrogen-based procedure under development by a particular industry [61], which is concomitantly under development with similar engineering approaches by other companies in a few different European and Asian sites. The left-hand side of Fig. 1.18 shows that iron ore concentrate is pelletized using fossil fuels, so that the mineral pellets and coke are fed into a blast furnace, which produces hot metal that is, subsequently, used to be transformed into crude steel. Coke and fossil fuels are used, and greenhouse effect gases and much particulates and ashes are emitted with this conventional procedure. Alternatively, the right-hand side of Fig. 1.18 presents a "hydrogen route" in which iron ore concentrate is also used to produce pellets, but no fossil fuel is used for that. Hydrogen is produced from water electrolysis on site or nearby using renewable electricity and is stored in large amounts to be used for two purposes: one is for the production of high-grade industrial heat and the other one is for the procedure of direct reduction of iron ore into metallic iron, which gives origin to sponge iron without using coke as a reducing agent, without using fossil fuels for heating, without deleterious environmental emissions, and also without having to convert the iron ore, the raw material, into liquid form as it has to be done in the blast furnace. The sponge iron is then used for crude steel production. Such a hydrogen route for the production of direct reduced iron from pelletized iron ore is very innovative and encompasses a future vision of using hydrogen energy to clean the ancient and pollutant steel industry into an environmentally friendly one.

Among the different possibilities available to speed up the commercialization of hydrogen energy technologies, the small-power distributed generation of electricity and heating/cooling devices are effectively ready for use. Although this has called the attention of and has motivated organizations throughout the world, it was in Japan that it succeeded to reach impressive numbers, as a result of an enduring public—private partnership, with the introduction and programmed withdrawn of subsidies. It gave a solid demonstration of how a well-programmed and well-implemented long-term road mapping for the introduction of a new and disruptive technology may be accomplished. Two types of fuel cells were chosen for utilization in urban homes connected to the city gas distribution system, the polymeric electrolyte membrane fuel cell, PEMFC, and the solid oxide fuel cell, SOFC. Fig. 1.19 presents the steep increase of the number of systems installed throughout the years to reach more than 220,000 in 2017. However, this will remain small compared with the prospects of installing 1.4 million units by 2020 and 5.3 million units up to 2030 [62].

The increase on the number of units to be installed in the near future will certainly contribute to force down the decreasing trend already observed for the devices' retail prices. Although the distributed generation of electricity is a known tendency for the near future and hydrogen energy technologies may effectively contribute to it, the centralized generation of electricity and its distribution to consuming centers coupled with hydrogen energy technologies

**FIGURE 1.19** Evolution as a function of the years of the number of low-power fuel cell—based devices effectively installed for distributed generation of electricity and heat, as well as their selling prices. *Modified from [62]. Currency was converted from Yens to Dollars.*

is likely to be developed in conjunction to power generation with renewable energies. This will benefit from the significant capacity presented by hydrogen for long-term carbon-free seasonal energy storage [15] and the importance energy storage gains when much renewable energy is produced.

## CONCLUDING REMARKS

The world has gone through a few energy transitions and presently experiments the dawning of a new one with much better prospects concerning the energy availability, energy efficiency, and environmental impact on the planet resulting from intensive use. With a smaller population and fewer methods to make use of energy, the total amount of energy used in the world during the 19th century was equal to 22 EJ. However, the Industrial Revolution was so intense that the two first decades of the 20th century were enough to use the same amount of energy consumed during the whole 19th century. From thereon, with a steady and important increase in human and ruminant livestock population, the century of marked evolution in the fields of science and technological innovation has eagerly demanded more and more energy, however, using it inefficiently, with regrettable losses, unequally shared by the enormous variety of the world's ethnicity, and causing an environmental disaster on the planet.

In the 21st century a harsher trend of energy requirement and use is imposed. Once again, a little over three decades will have used energy in an amount equivalent to the whole previous century. By 2050 the world might need 900 EJ of energy supply. This calls for more than a simple energy transition to make more energy available. Instead, a perennial, renewable, environmentally friendly, and prone to be extensively shared form of energy enters the scene all throughout the world. In addition to that, it has been conceived for use with very efficient energy conversion devices, the fuel cells. Hydrogen energy is the solution for the huge present and future energy requirements. It consists of utilizing hydrogen and hydrogen-containing compounds to generate energy to be supplied to all practical uses needed with high energy efficiency, overwhelming environmental and social benefits, as well as economic competitiveness.

In spite of the inventiveness that has given origin to a host of procedures for hydrogen production from water, biomasses, and hydrocarbons, making use of well-known as well as of innovative technologies with the necessary action of primary energy forms, it is herein unveiled that hydrogen is recognized to be no longer only an important energy carrier, such as electricity. It is also a primary fuel. It is available on earth to be harvested and used. It may be found with very high purity or mixed to other gases. It evolves from the earth's surface on several, probably all, continents from fairly shallow reservoirs, and it also degases in varied locations yet to be explored of the huge oceans' bottom extensions. It contains no carbon. It is not the result of biomass transformation, having rather an inorganic origin. It does not harm the planet. It may be renewable and perennial, through the permanent serpentinization, hydration, of ophiolitic rocks, existent all through the Planet, by water from aquifers that are replenished by rainwater percolation. The first hydrogen wells are already being explored. Others will also be as the 21st century goes by.

The hydrogen energy era is born.

The consumption of natural hydrogen by human beings and its transformation into water somehow mimics nature's superficial biological activity and atmospheric oxidation on Earth in terms of circular water cycles. While the use of natural hydrogen by humans is a modern subject of the present times, natural hydrogen was a precursor of life on Earth, before the apparition of chlorophyll and photosynthesis, through a succession of conditions during which the Earth's atmosphere changed from a generally reducing to a strongly oxidizing one. An era is being inaugurated in which, just like the fossil fuels coal, oil, and natural gas, hydrogen is a primary fuel and, just like electricity, it is also an energy carrier. Hydrogen carries with it only energy, not the environmentally harmful carbon. Moreover, in terms of energy it is the fuel that carries the largest specific density among all others.

Because of the facts of being able to be produced by a host of different raw materials and renewable energy sources, projecting its availability to become ubiquitous in the world; of being able to be harvested in natural form on earth without the requirement of any further treatments or conversions to other compounds for practical use; of, eventually, being perennial, being continuously processed in the earth guts; of possessing the highest energy density among all known fuels; of being chemically the simpler fuel among all others; of being environmentally friendly and not possessing carbon; of acting as a buffer for short- or long-term energy storage in conjunction with and solving issues associated with the intermittent renewable energies; of, if naturally or if artificially produced from water, thus being originated by reactions occurring with water and, whenever used, either conventionally burned or efficiently converted into electricity and heat by feeding a fuel cell, giving origin once again to water, hydrogen is the suitable fuel for a circular economy, that is, a perennial renewable circular sustainable fuel utilization cycle that will characterize the highly efficient engineering and the energy technological choices of the 21st century.

The interaction of humanity with natural existent hydrogen is not new, but it was not known to be natural hydrogen.

To understand it, it is first important to recall that only in the 20th century it was accepted that the movement of tectonic plaques made the earth's oceanic crust and the underlying upper mantle to be uplifted and exposed above sea level, emplaced onto continental crustal rocks, giving rise to many present mountains. That is the reason why ophiolite is frequently found on top of mountains, sometimes associated with limestone rocks. Subsequently, it is needed to consider that the natural formation of hydrocarbons such as methane, ethane, ethylene, and propane may occur without participation of biomasses, with inorganic origin, by reaction of hydrogen with $CO_2$ or bicarbonate ions dissolved in water. Methane formation follows the well-known Sabatier equation:

$$CO_2 + 4H_2 \rightarrow CH_4 + 2H_2O \qquad (1.5)$$

It could even be possible that this would be the source of hydrocarbon compounds giving inorganic origin to natural shale gas formations without the intervention of biomass. It could also be possible to develop new hydrogen and hydrocarbons production methods and $CO_2$ sequestration procedures by mimicry of the natural processes.

Once this explanation is taken into consideration, one can refer to Yanartaş, flaming rock in Turkish. It is situated close to Olympos, in Lycia, in the province of Antalya, Turkey, where the Chimaera Mountain presents ever-lasting flames, known for more than 2500 years. Fig. 1.20 presents the site localization and a photograph of the flames. The composition of the gas that gives origin to these flames was found to be 12% of hydrogen, mainly methane, and traces of nitrogen and helium [63].

However, there is an alternative explanation for the ever-lasting flames of Yanartaş on the Chimaera Mountain that was given by Homer, in the Iliad, as follows:

*Chimaera, daughter of Echidna, breathed raging fire, a creature fearful, great, swift-footed and strong, who had three heads, one of a grim-eyed lion; in her hinderpart, a dragon; and in her middle, a goat, breathing forth a fearful blast of blazing fire. Her did Pegasus and noble Bellerophon slay.*

**FIGURE 1.20** Map of Turkey with indication of the localization of the Chimaera Mountain on the Southeast and Yanartaş, ever-lasting flames from sustainable perennial natural gases.

Defeated by Pegasus and Bellerophon, Chimaera (Fig. 1.21) was sent to the guts of the earth, from where she breathes fire through the rocks for ever...

**FIGURE 1.21** Bellerophon fighting the Chimaera. Side A of an attic black-figured "overlap" Siana cup, ca. 575–550 BC. Found in Camiros (Rhodes). Courtesy Louvre Museum.

## ACKNOWLEDGMENTS

The author acknowledges the financial support to his research work by Furnas/Aneel and Tracel Ltda. as well as by BNDES and the enterprises Oxiteno S.A. and EnergiaH Ltda. The effort made by Aline Lys and by Alberto Coralli on the preparation of figures is also acknowledged.

## REFERENCES

[1]  I. Dincer, C. Zamfirescu, Sustainable Hydrogen Production, Elsevier, 2016.
[2]  Our World in Data, OWID, https://ourworldindata.org/energy-production-and-changing-energy-sources/.
[3]  BP — Statistical Review of World Energy, https://www.bp.com/content/dam/bp/en/corporate/excel/energy-economics/statistical-review-2017/bp-statistical-review-of-world-energy-2017-underpinning-data.xlsx.
[4]  International Energy Agency, IEA — Headline Energy Data, 2017. http://www.iea.org/media/statistics/IEA_HeadlineEnergyData_2017.xlsx.
[5]  United States Energy Information Administration, EIA, https://www.eia.gov/beta/international/data/browser/#/?c=4100000002000006000000000000000g000200000000000000001&vs=INTL.44-1-AFRC-QBTU.A&vo=0&v=H&end=2015.
[6]  Exxon Mobil Corporation 2017, Outlook for Energy: A View to 2040, 2017.
[7]  H. Chen, Q. Ejaz, X. Gao, et al., "Food.Water.Energy. Climate Otlook — Perspectives from 2016", MIT Joint Program on the Science and Policy of Global Change, 2016. https://globalchange.mit.edu/sites/default/files/newsletters/files/2015_Outlook_projection_tables.xlsx. https://globalchange.mit.edu/sites/default/files/newsletters/files/2016-JP-Outlook.pdf.
[8]  GWEC Solar Power Europe Greenpeace, Energy [r]evolution - a Sustainable World Energy Outlook 2015, 2015. https://www.greenpeace.de/sites/www.greenpeace.de/files/publications/studie_energy_revolution_2015_engl.pdf.
[9]  Asia/world Energy Outlook 2016, The institute of Energy Economics, Japan, 2016. http://eneken.ieej.or.jp/data/7199.pdf.
[10]  World Energy Outlook 2017, International Energy Agency, https://www.iea.org/weo2017/.
[11]  W.J. Ripple, C. Wolf, M. Galetti, T.M. Newsome, M. Alamgir, E. Crist, M.I. Mahmoud, W.F. Laurance, World scientists' warning to humanity: a second notice, BioScience 67 (12) (01/Dec./2017) 1026—1028. https://doi.org/10.1093/biosci/bix125.
[12]  P.E.V. de Miranda, E.S. Carreira, U.A. Icardi, G.S. Nunes, Brazilian hybrid electric-hydrogen fuel cell bus: improved on-board energy management system, Int. J. Hydrogen Energy 42 (2017) 13949—13959. https://doi.org/10.1016/j.ijhydene.2016.12.155.
[13]  P.E.V. de Miranda, Particulate materials: threatening products resulting from burned fuels, Rev. Materia 18 (4) (2013) IV—VI.
[14]  M. Pascal, M. Corso, O. Chanel, C. Declercq, C. Badaloni, G. Cesaroni, et al., Assessing the public health impacts of urban air pollution in 25 European cities: results of the Aphekom Project, Sci. Total Environ. 449 (2013) 390—400.
[15]  Hydrogen Council, Hydrogen Scaling Up — A Sustainable Pathway for the Global Energy Transition, http://hydrogencouncil.com/wp-content/uploads/2017/11/Hydrogen-Scaling-up_Hydrogen-Council_2017.compressed.pdf.
[16]  W. Doenitz, R. Schmidberger, E. Steinheil, Hydrogen production by high temperature electrolysis of water vapour, Int. J. Hydrogen Energy 5 (1980) 55—63.

[17] M. Boaro, A.S. Aricò (Eds.), Advances in Medium and High Temperature Solid Oxide Fuel Cell Technology, Springer, CISM International Centre for Mechanical Sciences, 2017.

[18] N.Q. Minh, in: D. Stolten, B. Emonts (Eds.), Hydrogen Science and Engineering: Materials, Processes, Systems and Technology, Wiley-VCH, 2016 (Chapter 16).

[19] N.Q. Minh, M.B. Mogensen, ECS Interface 22 (4) (2013) 55.

[20] N.Q. Minh, P.E.V. de Miranda, High-temperature electrosynthesis of hydrogen and chemicals, ECS Trans. 75 (43) (2017) 49−58. https://doi.org/10.1149/07543.0049ecst.

[21]] D. Ferrero, A. Lanzini, P. Leone, M. Santarelli, Reversible operation of solid oxide cells under electrolysis and fuel cell modes: experimental study and model validation, Chem. Eng. J. 274 (2015) 143−155.

[22] S.C. Singhal, K. Kendall, High Temperature Solid Oxide Fuel Cells: Fundamentals, Design and Applications, Elsevier, 2003.

[23] Fuel Cell Today, Water Electrolysis & Renewable Energy Systems, 2013.

[24] A. Tarancón, C. Fábrega, A. Morata, M. Torrell, T. Andreu, Power-to-Fuel and artificial photosynthesis for chemical energy storage, in: X. Moya, D. Muñoz-Rojas (Eds.), Materials for Sustainable Energy Applications, CRC Press, 2016.

[25] A. Godula-Jopek (Ed.), Hydrogen Production by Electrolysis, Wiley-VCH Verlag GmbH & Co., 2015, p. 259.

[26] J.M. Regan, H. Yan, Bioelectrochemical systems for indirect biohydrogen production, in: D. Zannoni, R. De Philippis (Eds.), Microbial BioEnergy: Hydrogen Production, Springer, 2014, pp. 225−233.

[27] A. Kadier, M.S. Kalil, A. Mohamed, et al., Microbial electrolysis cells (MECs) as innovative technology for sustainable hydrogen production: fundamentals and perspective applications, in: M. Sankir, N.D. Sankir (Eds.), Hydrogen Production Technologies, Wiley, 2017, pp. 407−470.

[28] K.J.J. Steinbusch, E. Arvaniti, H.V.M. Hamelers, C.J.N. Buisman, Selective inhibition of methanogenesis to enhance ethanol and n-butyrate production through acetate reduction in mixed culture fermentation, Bioresour. Technol. 100 (2009) 3261−3267.

[29] D. Xing, Y. Zuo, S. Cheng, J.M. Regan, B.E. Logan, Electricity generation by *Rhodopseudomonas palustris* DX-1, Environ. Sci. Technol. 42 (2008) 4146−4151.

[30] Y. Zuo, D.F. Xing, J.M. Regan, B.E. Logan, Isolation of the exoelectrogenic bacterium *Ochrobactrum anthropi* YZ-1 by using a U-tube microbial fuel cell, Appl. Environ. Microbiol. 74 (2008) 3130−3137.

[31] M. Malki, A.L. De Lacey, N. Rodriguez, R. Amils, V.M. Fernandez, Preferential use of an anode as an electron acceptor by an acidophilic bacterium in the presence of oxygen, Appl. Environ. Microbiol. 74 (2008) 4472−4476.

[32] S.K. Chaudhuri, D.R. Lovley, Electricity generation by direct oxidation of glucose in mediatorless microbial fuel cells, Nat. Biotechnol. 21 (2003) 1229−1232.

[33] S. Xu, H. Liu, New exoelectrogen *Citrobacter* sp. SX-1 isolated from a microbial fuel cell, J. Appl. Microbiol. 111 (2011) 1108−1115.

[34] H.J. Kim, H.S. Park, M.S. Hyun, I.S. Chang, M. Kim, B.H. Kim, A mediator-less microbial fuel cell using a metal reducing bacterium, *Shewanella putrefaciens*, Enzym. Microb. Technol 30 (2002) 145−152.

[35] O. Bretschger, A. Obraztsova, C.A. Sturm, I.S. Chang, Y.A. Gorby, S.B. Reed, D.E. Culley, C.L. Reardon, Current production and metal oxide reduction by *Shewanella oneidensis* MR-1 wild type and mutants, Appl. Environ. Microbiol. 73 (2007) 7003−7012.

[36] L. Zhang, S. Zhou, L. Zhuang, W. Li, J. Zhang, N. Lu, L. Deng, Microbial fuel cell based on *Klebsiella pneumoniae* biofilm, Electrochem. Commun. 10 (2008) 1641−1643.

[37]   F. Rezaei, D. Xing, R. Wagner, J.M. Regan, T.M. Richard, B.E. Logan, Simultaneous cellulose degradation and electricity production by *Enterobacter cloacae* in a microbial fuel cell, Appl. Microbiol. Biotechnol. 75 (2009) 3673–3678.

[38]   C.A. Pham, S.J. Jung, N.T. Phung, J. Lee, I.S. Chang, B.H. Kim, H. Yi, J. Chun, A novel electrochemically active and Fe(III)-reducing bacterium phylogenetically related to *Aeromonas hydrophila*, isolated from a microbial fuel cell, FEMS Microbiol. Lett. 223 (2003) 29–134.

[39]   K. Chung, S. Okabe, Characterization of electrochemical activity of a strain ISO2–3 phylogenetically related to Aeromonas sp. isolated from a glucose-fed microbial fuel cell, Biotechnol. Bioeng. 104 (2009) 901–910.

[40]   D.E. Holmes, J.S. Nicoll, D.R. Bond, D.R. Lovley, Potential role of a novel psychrotolerant member of the family Geobacteraceae, *Geopsychrobacter electrodiphilus* gen. nov., sp. nov., in electricity production by a marine sediment fuel cell, Appl. Environ. Microbiol. 70 (2004) 6023–6030.

[41]   B. Min, S. Cheng, B.E. Logan, Electricity generation using membrane and salt bridge microbial fuel cells, Water Res. 39 (2005) 1675–1686.

[42]   F. Caccavo, D.J. Lonergan, D.R. Lovley, M. Davis, J.F. Stolz, M.J. McInerney, Geobacter sulfurreducens sp. nov., a hydrogen- and acetate-oxidizing dissimilatory metal-reducing microorganism, Appl. Environ. Microbiol. 60 (1994) 3752–3759.

[43]   D.R. Bond, D.E. Holmes, L.M. Tender, D.R. Lovley, Electrode-reducing microorganisms that harvest energy from marine sediments, Science 95 (2002) 483–485.

[44]   D.R. Bond, D.R. Lovley, Electricity production by Geobacter sulfurreducens attached to electrodes, Appl. Environ. Microbiol. 69 (2003) 1548–1555.

[45]   D.E. Holmes, D.R. Bond, D.R. Lovley, Electron transfer by *Desulfobulbus propionicus* ton Fe(III) and graphite electrodes, Appl. Environ. Microbiol. 70 (2004) 1234–1237.

[46]   Y.F. Wang, M. Masuda, S. Tsulimura, K. Kano, Electrochemical regulation of the end-product profile in *Propionibacterium freudenreichii* ET-3 with an endogenous mediator, Biotechnol. Bioeng. 101 (2008) 579–586.

[47]   V. Fedorovich, M.C. Knighton, E. Pagaling, F.B. Ward, A. Free, I. Goryanin, A novel electrochemically active bacterium phylogenetically related to *Arcobacter butzleri*, isolated from a microbial fuel cell, Appl. Environ. Microbiol. 75 (2009) 7326–7334.

[48]   H.S. Park, B.H. Kim, H.S. Kim, H.J. Kim, G.T. Kim, M. Kim, I.S. Chang, Y.K. Park, A novel electrochemically active and Fe(III)-reducing bacterium phylogenetically related to *Clostridium butyricum* isolated from a microbial fuel cell, Anaerobe 7 (2001) 297–306.

[49]   J. Niessen, U. Schroder, F. Scholz, Exploiting complex carbohydrates for microbial electricity generation-a bacterial fuel cell operating on starch, Electrochem. Commun. 6 (2004) 955–958.

[50]   K.C. Wrighton, P. Agbo, F. Warnecke, K.A. Weber, E.L. Brodie, T.Z. De Santis, P. Hugenholtz, G.L. Andersen, A novel ecological role of the Firmicutes identified in thermophilic microbial fuel cells, ISME J. 2 (2008) 1146–1156.

[51]   D.R. Bond, D.R. Lovley, Evidence for involvement of an electron shuttle in electricity generation by *Geothrix fermentans*, Appl. Environ. Microbiol. 71 (2005) 2186–2189.

[52]   S.A. Lee, Y. Choi, S. Jung, S. Kim, Effect of initial carbon sources on the electrochemical detection of glucose by *Gluconobacter oxydans*, Bioelectrochemistry 57 (2002) 173–178.

[53]   D. Das (Ed.), Microbial Fuel Cell, Springer, 2018.

[54]   I. Moretti, A. Dagostino, J. Werly, C. Ghost, D. Defrenne, L. Gorintin, L'Hydrogène Naturel, un Nouveau Pétrole ? Pour la Sci. (March 2018) 24–26.

[55] E. Deville, A. Prinzhofer, The origin of $N_2$-$H_2$-$CH_4$-rich natural gas seepages in ophiolitic context: a major and noble gases study of fluid seepages in New Caledonia, Chem. Geol. 440 (2016) 139−147.

[56] N. Larin, V. Zgonnik, S. Rodina, E. Deville, A. Prinzhofer, V.N. Larin, Natural molecular hydrogen seepage associated with surficial, rounded depressions on the european craton in Russia, Nat. Resour. Res. 24 (3) (2015) 369−383. https://doi.org/10.1007/s11053-014-9257-5.

[57] M. Robinius, J. Linssen, T. Grube, et al., Comparative Analysis of Infrastructures: Hydrogen Fueling and Electric Charging of Vehicles, Publisher Jülich Research Center Central Library, 2018.

[58] A.S. Santosa, L. Gilioa, V. Halmenschlagerb, T.B. Diniza, A. Nç Almeida, Flexible-fuel automobiles and $CO_2$ emissions in Brazil: parametric and semiparametric analysis using panel data, Habitat Int. 71 (2018) 147−155.

[59] S.A. Venâncio, P.E.V. de Miranda, Direct utilization of carbonaceous fuels in multifunctional SOFC anodes for the electrosynthesis of chemicals or the generation of electricity, Int. J. Hydrogen Energy 42 (2017) 13927−13938.

[60] A. v. Wijk, The Green Hydrogen Economy in the Northern Netherlands, The Northern Netherlands Innovation Board, 2017. www.noordelijkeinnovationboard.nl.

[61] Hybrit, http://www.hybritdevelopment.com/steel-making-today-and-tomorrow.

[62] S. Kawamura, Japan Country Up-date at the IPHE Steering Committee Meeting, The Hague, , Netherlands, November 21, 2017.

[63] A. Prinzhofer, E. Deville, Hydrogène Naturel la Prochaine Révolution Énergétique?, Belin, Paris, 2015.

# Chapter 2

# Fuel Cells

Alberto Coralli[1,*], Bernardo J.M. Sarruf[1,*], Paulo Emílio V. de Miranda[1,*], Luigi Osmieri[2,†], Stefania Specchia[3,†], Nguyen Q. Minh[4,‡]

[1]*Federal University of Rio de Janeiro, Rio de Janeiro, RJ, Brazil;* [2]*National Renewable Energy Laboratory, Golden, CO, United States;* [3]*Politecnico di Torino, Torino, Italy;* [4]*University of California, San Diego, La Jolla, CA, United States*

## INTRODUCTION TO FUEL CELLS

Fuel cells (FCs) are electrochemical energy converters that transform chemical energy into electricity and heat and are characterized by high efficiency and low pollutant emissions. Two electrodes, anode and cathode, compose every cell in an FC, where fuel oxidation (anode) and oxidant reduction (cathode) take place. The electrodes are separated by an electrolyte able to conduct ions but not electrons. The electrons released in the anode semireaction migrate through an external circuit to the cathode, where they participate in the cathode semireaction, thus generating electric current. Fig. 2.1 depicts the general operation of FCs both anionic (Fig. 2.1A) and cationic (Fig. 2.1B) electrolyte conducting. The general electrochemical equations in Fig. 2.1 denote fuel and oxidant intakes ($A\chi_f$ and $B\chi_o$, respectively), in which $B$ is the molar amount of oxidant required to react with $A$ moles of fuel, thus generating the molar amount $C$ of products $\chi_p$. The reactions of these compounds are related to the generation of the molar amount $k$ of ions $\Phi$ with charge $\eta$. The production of an amount $k \cdot \eta$ of electrons is associated with this generation of ions. The electrons circulate through an external circuit from the anode toward the cathode. In this way, the electrochemical reactions are responsible to generate the electric current and the products in either the anode or cathode chambers for the cases of anionic or cationic electrolyte conductors, respectively.

In Fig. 2.1A the anode general reaction depicts the fuel being electrochemically oxidized by anions migrating through the electrolyte. These negative ions are a result of the cathodic reaction of oxidant reduction.

---

* Authors contributed Introduction.
† Authors contributed Polymer Electrolyte Membrane Fuel Cells.
‡ Author contributed Solid Oxide Fuel Cells.

Science and Engineering of Hydrogen-Based Energy Technologies. https://doi.org/10.1016/B978-0-12-814251-6.00002-2

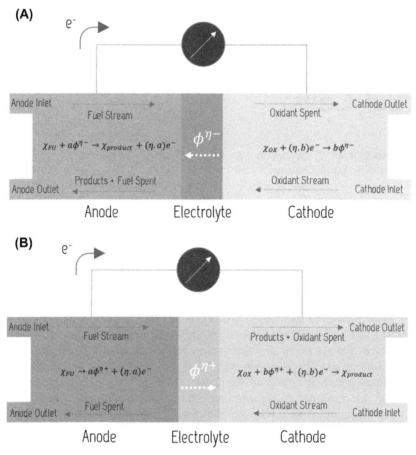

**FIGURE 2.1** General operation of a fuel cell (A) with an anion-conducting and (B) cation-conducting electrolyte. *Elaboration of the authors.*

Fig. 2.1B presents a similar set of events, alternating the direction of ionic conduction because of the use of positively charged ions that displace the products generation to the cathode chamber. In this case, the reduction of the oxidant, promoted in the cathode, occurs due to the supply of positive ionic charges originated from fuel oxidation in the anode side. The overall reaction for both cases is represented in Eq. (2.1).

$$A\chi_f + B\chi_o \rightarrow C\chi_p \tag{2.1}$$

The main advantages of FCs as energy converters are the low environmental impact, high-energy efficiency, low sound emissions, few maintenance requirements, and modularity, being FC's efficiency almost constant with FC system's rated power. Despite the high technology used in modern FCs, these devices have been known since more than 150 years. The electrochemical principle of FCs was first enunciated by Christian Friedrich Schoenbein; however, William Groove developed in 1839 the first documented FC, using a sulfuric acid electrolyte. In 1889, Ludwing Mond and Charles Langer named this kind of device "fuel cell." In 1893, Wilhelm Ostwald experimentally tested several different FC components (electrodes, electrolytes, fuels, and oxidant compounds), and in 1896, William Jacques created the first FC with practical applications [1]. The first large-scale industrial development of FC technology happened in the 1960s, in connection with the Apollo space program, in which alkaline FCs were used as auxiliary power units (APUs) for spacecrafts. Since then, FC R&D activities have been intense, leading to the production of several prototypes and commercial products with a wide range of applications, such as portable and stationary power generation, uninterruptible power supplies, APUs, and automotive power trains. In practical applications, FCs are normally arranged in stacks, with several cells electrically connected in series. The connection between adjacent cells is usually done, interposing between them a metallic plate as bipolar plate, which homogeneously distributes the reagents on the electrode surfaces and transport electrons from the anode of a cell to the cathode of the other one. Many variants of this basic structure exist, normally classified according to the kind of the electrolyte used. It is important to emphasize that all FCs need several auxiliary devices to function. These devices are required to feed the reagents, remove the reaction products, maintain stack's proper working conditions (pressure, temperature, etc.), and transform the produced electric power and other indispensable functions. The whole of these auxiliary devices is called balance of plant (BoP) and is different for every type of FC, varying in cost and complexity and being an important element in FC systems project.

The majority of FCs are fed with hydrogen-rich gases or hydrocarbons and therefore need a fuel reformer in the BoP, which represents a considerable increase in complexity and energy consumption of the system. Therefore, research works that aim to eliminate the need for fuel reforming, by the direct utilization of carbonaceous fuels such as ethanol or methane, represent innovative FC development [2,3].

Table 2.1 lists the main characteristics of some of the most common types of fuel cells.

In the following paragraphs, a short summary of the main features of these fuel cell types is reported. Special sections are devoted to polymer electrolyte membrane (PEMFCs) and solid oxide fuel cells (SOFCs), because of their prominence in research activities and market applications in the past years, and to the positive outlook for these technologies in the near future. In fact, data regarding FC shipments by FC type in 2017 clearly show that PEMFCs largely dominated the market with around 62% of the shipments, followed by SOFCs (33% of the shipments with a 48% increase from 2016 to 2017) [4].

**TABLE 2.1** Types of Fuel Cells

| Types | Electrolytes | Ion $\Phi^\eta$ | Power (kW) | Temperature (°C) | Stack Efficiency (%) | System Efficiency (%) | Status |
|---|---|---|---|---|---|---|---|
| Polymer electrolyte | Polymeric membrane | $H^+$ | 0.001–500 | 50–100 Nafion 125–200 PBI | 50–70 | 30–50 | Commercial / Research |
| Solid oxide | Ceramic ionic conductor | $H^+$ or $O^{2-}$ | <100,000 | 500–1100 | 60–65 | 55–60 | Commercial / Research |
| Alkaline | Aqueous alkaline solution | $OH^-$ | 10–200 | <80 | 60–75 | 62 | Commercial / Research |
| Phosphoric acid | Molten phosphoric acid | $H^+$ | <10,000 | 150–200 | 55 | 40 | Commercial / Research |
| Molten carbonate | Molten carbonate | $CO_3^{2-}$ | <100,000 | 600–650 | 55 | 47 | Commercial / Research |
| Solid acid | Proton-conducting oxyanion salt | $H^+$ | 0.001–1 | 200–300 | 55–60 | 40–45 | Research |
| Microbial | Polymeric membrane or humic acid | $H^+$ | | <40 | 80 | | Research |
| Enzymatic | Any that will not denature the enzyme | $H^+$ | | <40 | | | Research |

*PBI, polybenzimidazole.*

## Alkaline Fuel Cell

The electrolyte of this kind of FC is an alkaline aqueous solution, usually potassium or sodium hydroxide, contained in a porous matrix or continuously recirculated. The electrodes are usually made with porous carbon or nickel [5]. These FCs operate in a wide pressure range and at low temperatures (60−250°C) and present a high efficiency (up to 75%) [6]. An important advantage of this FC is represented by the possibility of using as catalysts low-cost elements such as nickel (in the anode) and silver (in the cathode), lowering considerably the total cost of the system [7]. On the other hand, this kind of FC does not tolerate the presence of $CO_2$, neither in the fuel nor in the oxidant; therefore, it needs to operate with high-purity gases [8]. In addition to that, it is difficult to impede the leakage of the alkaline electrolyte from its containing material. As a result, currently alkaline fuel cells are not one of the most successful types of FC. The carbon dioxide sensitivity greatly reduces the range of practical applications of the cell, impeding its use with gases produced by hydrocarbon reforming and atmospheric air. Therefore, the overall activity on this technology is quite small, especially after the creation of Nafion, a material that could be easily used at lower temperatures [9]. Some recent research studies concerning solid-state alkaline polymer electrolytes showed promising results [10]. Alkaline electrolytes are also used in some FCs still in the R&D phase, such as the metal hydride FC [11], the direct borohydride FC [12], and the zinc−air FC [13,14].

## Phosphoric Acid Fuel Cell

Phosphoric acid fuel cells (PAFCs) use an electrolyte composed by concentrated phosphoric acid ($H_3PO_4$) dispersed in a silicon carbide matrix. The electrodes are composed by dispersed platinum particles on a porous graphitic substrate [15]. The presence of platinum increases substantially the cost of this kind of FC and makes it CO-sensitive, impeding the use of fuel mixture with more than 1.5% of CO. PAFCs operate at temperatures around 200°C and show system electrical efficiencies around 40% [5].

PAFCs were the first FC technology to be developed for use in high-power generators (up to some megawatts of electric power), thanks to some positive characteristics as the easiness of maintaining the internal water balance and the possibility to produce cells of high area (up to 1 $m^2$). This kind of systems enjoyed a moderate commercial success, with the installation of hundreds of units worldwide. However, the interest in this technology is gradually disappearing because of the high production and operation costs and the lack of long-term operation reliability (related to cathode corrosion problems).

## Molten Carbonate Fuel Cell

In this kind of cells, the electrolyte is a molten carbonate, usually composed by alkaline metals such as lithium and potassium ($Li_2CO_3/K_2CO_3$) or lithium and sodium. The corrosive solution is contained in a solid porous matrix, generally MgO or $LiAlO_2$ [16]. The most common electrodes are porous NiO structure doped with lithium (cathode) and an alloy of nickel and chromium or aluminum. Because the melting temperature of the electrolyte is around 500°C, the operating temperature is usually between 600 and 650°C. The efficiency of a molten carbonate fuel cell (MCFC) system is typically around 50% [5]. The working of an MCFC system is peculiar in one aspect: Carbon dioxide takes part in the electrode reactions, being consumed at the cathode and produced at the anode.

MCFCs present some interesting characteristics, such as the opportunity to use carbon monoxide as fuel and to make possible the internal reforming of hydrocarbons. This is mainly due to the high operating temperature and to the anode composition (the main element is nickel, a good catalyst for steam reforming reaction). MCFC systems are therefore able to use syngas or even untreated hydrocarbons as fuel, not requiring the use of complex and costly external devices for pure hydrogen production and allowing for a greater fuel flexibility of the power generator. The high operating temperature permits to avoid costly noble metals as catalysts, greatly reducing the cell cost, and allows for the use of the high-quality residual heat for other purposes, as heating of buildings and additional electricity production by cogeneration with turbines. On the down side, the high temperature also impedes the fast startup and shutdown of MCFC devices. Another operational difficulty of MCFCs is related to the presence of a $CO_2$ management system, necessary to supply the cathode with the carbon dioxide used in cell reactions. This system increases the complexity and cost of MCFC generators. Developed MCFC and PAFC systems present similar characteristics: high power level (200−1000 kW) and high cell area (up to 1 $m^2$). Short system lifetime is a problem also for MCFC devices and constitutes the main limit to their commercial success. Existing equipment demonstrated only a few thousand hours durability, while an acceptable value for stationary generators is around 40,000 h.

## Solid Acid Fuel Cell

The electrolyte of these FCs is an intermediate between a salt and an acid, a "solid acid." Compounds of interest for application as electrolytes are based on oxyanion groups ($SO_4^{2-}$, $PO_4^{3-}$, $SeO_4^{2-}$, $AsO_4^{3-}$) linked by hydrogen bonds one to the other and with large cations ($Cs^+$, $Rb^+$, $NH_4^+$, $K^+$), to balance the charge [17]. The low-temperature structure of solid acids is crystalline, like most salts, and their ionic conductivity is low. At higher temperatures (between 50 and 150°C, depending on the compound [5]), some of these substances undergo a phase transition that leads to a disordered structure, increasing the proton conductivity of some orders of magnitude. Solid acids can be used as FC electrolytes in the temperature window between the aforementioned phase transition and the decomposition temperature (usually higher than 250°C [5]). The first solid acid fuel cells (SAFCs) were developed using $CsHSO_4$, but the formation of by-products that damage the anode led to discard the use of acid sulfates. SAFC systems were also operated with $CsH_2PO_4$ and $KH(PO_3H)$ electrolytes [18], using humid gas flows to prevent dehydration of the solid acids.

SAFCs combine some positive characteristics of the PEMFCs (the use of a thin solid electrolyte with a quite good mechanical resistance) and of the PAFCs (the operation at intermediate temperatures, which implies a greater tolerance to CO and other poisoning substances). SAFCs running on hydrocarbons such as propane [19] and diesel have been demonstrated. The most important R&D challenges regarding SAFC technology are the prevention of electrolyte degradation in long-term operation and the reduction of the amount of noble metal catalysts used in the electrodes.

## Microbial Fuel Cells

Microbial FCs are characterized by the use of living microbes as catalyst for fuel oxidation inside the anode chamber of the cell. The microorganisms, dispersed in planktonic state or forming a biofilm, are able to oxidize the supplied substrate (lower alcohols, simple sugars, or even wastewaters), producing protons and electrons as metabolites. The microbial FC makes use of exoelectrogen bacteria that are microorganisms with ability to transfer electrons extracellularly. The electrons are transferred to the metallic electrode in contact with the microbes and from there to an external electric circuit that connects the anode to cathode. This transfer can be direct (mediator-free, through conductive structure on the cell membrane or electron-conducting substances produced by the microbes) or mediated by inorganic redox-active dye molecules, which can exist in both oxidized and reduced states (thionine, methyl blue, humic acid, or neutral red) [5]. The produced protons diffuse to the cathode, where they combine with oxygen and the electrons conducted through the external circuit, in a way similar to what happens in other kinds of FCs. Anodic and cathodic chambers may be separated by ion-exchange membrane (usually a polymeric membrane), but single-chamber fixtures without a separator also exist. The most common electrode materials are carbon-based, providing key qualities such as high conductivity, biocompatibility, chemical stability, and good mechanical resistance. At the cathode, catalysts such as Pt, $MnO_2$ and polyaniline are often applied.

A great advantage of microbial FCs over other biological wastewater treatments aimed to produce electricity is the high efficiency. In fact, the direct conversion of waste to electric power in microbial FCs can be as high as 80% [20], while the conversion through production of biofuels does not exceed 40%.

## Enzymatic Fuel Cells

Enzymatic FCs take advantage of biologic processes to assist the electro-chemical reactions of the cell, in a similar way to what happens in the microbial FCs. However, instead of living whole cells, the enzymatic FCs use as catalysts only the redox enzymes separated from the living organisms and purified. The advantage of this approach is that it avoids the need for continuous maintenance of the microbial population, which imposes an accurate control of the working conditions of the cell. However, also the enzymes need to be protected from degradation when exposed to the environment, applying special methods of stabilization and utilization. In contact with the anode, the enzymes are able to oxidize various substrates (alcohols, sugars, organic acids [21]), transferring electrons to the electrode. "Full" enzymatic cells also exist, which use enzymes as catalysts not only at the anode but also at the cathode, to favor the reduction process of the oxidant. It is possible to fix on the electrodes highly selective enzymes, which catalyze only a specific reaction (oxidation or reduction) with a specific substrate. This implies the possibility to construct open-chamber FCs without separator, which can be miniaturized down to the micrometer scale [22].

Prospective applications for enzymatic FCs range from powering portable electronic devices to produce electricity for implanted devices, e.g., pacemakers and glucose monitors. However, power density and operational stability of these cells have to be improved to make commercial application viable.

## POLYMER ELECTROLYTE MEMBRANE FUEL CELLS

FCs have received more and more attention because of their high efficiency and potential use in many different applications where energy is required, from few microwatts to megawatts. In fact, FCs are electrochemical devices able to transform the chemical energy of a fuel directly into heat and electricity. Compared with batteries, FCs can continuously produce electricity, as long as fuel and oxidant (either air or oxygen) are available [23]. If the fuel is hydrogen, this fuel can be electrochemically oxidized to water, without production of pollutant emissions, according to the following overall reaction:

$$H_2 + \frac{1}{2} O_2 \rightarrow H_2O \qquad (2.2)$$

Historically, FCs have been discovered first by Sir William R. Gove in 1839 [24]. He developed a wet cell gaseous voltaic battery, consisting of glass tubes containing hydrogen and oxygen, with platinum electrodes immersed in acidulated water with sulfuric acid, as described for the first time in 1942 on *Philosophical Magazine* and represented in Fig. 2.2A [25]. As Sir Grove wrote, "this battery is peculiar in having the current generated by gases, and by synthesis of an equal but opposite kind at both anode and cathode; it is therefore, theoretically, more perfect than any other form" [25].

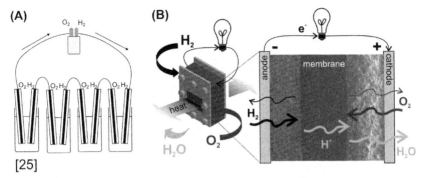

[25]

**FIGURE 2.2**    (A) Grove's scheme of the first voltaic fuel cell, published in 1842 on *Philosophical Magazine*. (B) Operative principle of a PEMFC. *PEMFC*, polymer electrolyte membrane fuel cell.

FCs are subdivided into several types, depending on the operative conditions and, consequently, the electrolyte used. PEMFCs consist of a proton-conducting polymer electrolyte membrane (PEM) sandwiched between two electrodes, the anode and the cathode, which provide hydrogen and oxygen, respectively. Fig. 2.2B shows a basic operative sketch of a PEMFC. The overall reaction (Eq. 2.2) is split into two semireactions, anodic (negative pole) (Eq. 2.3) and cathodic (positive pole) (Eq. 2.4):

$$2H_2 \rightarrow 4H^+ + 4e^- \tag{2.3}$$

$$O_2 + 4H^+ + 4e^- \rightarrow 2H_2O \tag{2.4}$$

Hydrogen, fed at the anode, diffuses into the anodic catalytic layer and is oxidized into protons and electrons. Electrons move into the external electrical circuit from the anode to the cathode, providing electricity, while protons cross the electrolyte, the PEM, to reach the cathode, diffuse into the cathodic catalytic layer, and, with the electrons, reduce oxygen to water, the only product of reaction.

PEMFCs have very high efficiency values, ranging from 50% to 70%, depending on the operative temperature [24,26]. Such efficiency values are almost unaffected by the size of FCs, thus making PEMFCs very interesting systems for various applications, depending on the electrical output delivered [27,28]: portable systems (from microwatts to few watts [29−32]), APU and backup units (from few watts up to few kilowatts [33−36]), transportation (all sectors, from few kilowatts up to several hundred kilowatts [37−42]), and stationary systems (from few kilowatts up to several megawatts [28,43−46]).

Considering the zero emissions released by FCs when the fuel used is hydrogen, and the huge range of applications they can cover, FCs can be considered as the preferred energy conversion device of the 21st century.

Increasing demand for clean energy coupled with government initiatives in the form of funding can drive the market globally. Up to now, only few niche markets are already available, for example, forklifts (a niche market developed by Nuvera [47]), cars (both Toyota and Hyundai launched on the market their FC electric vehicles fed with hydrogen very recently, Mirai and Tucson [48−50]), and cogeneration systems (based on SOFCs, launched on the market by Solid Power [51]). Development of wider market applications and full commercialization is expected to increase more and more in the near future, facing these challenges: the search for renewable hydrogen, the expansion of hydrogen infrastructures, the durability of FC components, and the reduction of FC cost production.

## Polymer Electrolyte Membrane Fuel Cell Operation Mode

PEMFCs can operate in different temperature ranges, typically 60−90°C (low-temperature PEMFCs [52,53]), 100−120°C (intermediate temperature PEMFCs [54,55]), or higher, up to 200°C (high-temperature PEMFCs [56,57]), depending on the proton-conducting membrane used. Going deeper in the basic scheme of a PEMFC that had its operative principle unveiled in Fig. 2.2B, Fig. 2.3A shows an exploded view of a PEMFC single-cell. Two ending plates, together with anodic and cathodic flow fields, hosting a series of channels to deliver both hydrogen and oxygen, sandwich the membrane electrode assembly (MEA), the key component of a PEMFC. The MEA is constituted by a proton-conductive membrane sandwiched between two gas diffusion layer (GDL) electrodes, which host the catalytic layers. Anode and cathode present different kinds of catalytic layers. In the MEA, the electrochemical reactions (Eqs. 2.3 and 2.4) take place, with delivery of electricity and release of heat. The membrane has different roles, such as conducting protons from the anode to the

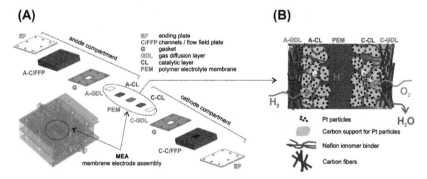

**FIGURE 2.3** (A) Exploded view of a PEMFC single-cell. (B) Scheme of an MEA. *MEA*, membrane electrode assembly; *PEMFC*, polymer electrolyte membrane fuel cell.

cathode but insulating electrons (to force them moving in the external circuit), separating hydrogen from oxygen, and physically supporting the anodic/ cathodic catalysts. The most widespread membranes are based on per-fluorosulfonic acid ionomers, typical for low-temperature PEMFCs, which are usually used as benchmark materials [52,53,55,58−60]. Different polymer−nanocomposite electrolytes, mainly based on acid−base polybenzimidazole (PBI) [57−59,61,62] or sulfonated polyetheretherketone [62−64], are used in case of intermediate- and high-temperature PEMFCs. Channel and flow-field plates distribute hydrogen and oxygen to the MEA, while releasing water as reaction product. They must be impermeable to gases, electron-conductive to allow for electrons flowing, thermally conductive to allow for heat management, and mechanically and chemically resistant. Mostly, they are made with carbon-based materials, eventually reinforced with tunable fibers, easily machinable to fabricate microchannels for gas distribution [65−68]. Sealing through gaskets is necessary to prevent leakage of both hydrogen and oxygen, while ensuring electrical contact between the MEA and channel and flow-field plates [69−71].

In the key component MEA, schematically represented in Fig. 2.3B, the occurring electrochemical reactions have a heterogeneous character. In fact, these reactions occur at the surface of the so-called three-phase boundary, constituted by platinum supported on carbon bonded to ionomer acting as a binder, where electrons, ions, and uncharged species are involved [72,73]. The kinetic mechanism includes adsorption/desorption steps, chemical steps, and one or more charge transfer steps. The typical catalyst is based on platinum over a high specific surface area carbon (Pt/C) for both sides, the hydrogen oxidation reaction (HOR) at the anode (Eq. 2.3) and the oxygen reduction reaction (ORR) at the cathode (Eq. 2.4), but with different catalyst loadings because of the different nature of the two reactions. In fact, the kinetic of HOR is simple and fast, with very high exchange current density [74−77], while the ORR has a very complex kinetic mechanism, mainly because of the high strength of the O−O bond, which makes this reaction sluggish [78−82].

The electrochemical behavior of a PEMFC can be represented by the polarization curve, together with the power density curve, as shown in Fig. 2.4A. The polarization curve represents the cell potential versus the current density, and it is affected by a series of potential losses:

$$V_{cell} = V_{OCV} - \Delta V \tag{2.5}$$

$$\Delta V = \Delta V_{act} + \Delta V_{ohm} + \Delta V_{conc} \tag{2.6}$$

$$V_{OCV} < E_{cell} \tag{2.7}$$

The open circuit voltage ($V_{OCV}$) of the cell is always lower compared with its thermodynamic value, or reversible voltage, $E_{cell}$, which depends on the Nernst equation. The theoretical amount of energy produced by a PEMFC fed

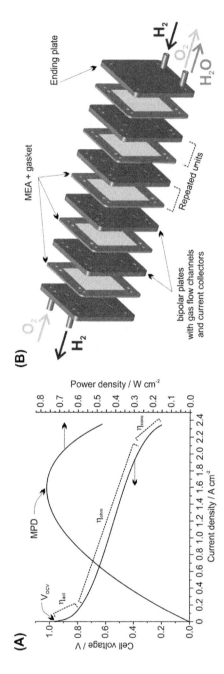

**FIGURE 2.4**   (A) Typical polarization and power density curves of a PEMFC. (B) Exploded view of a PEMFC stack. *PEMFC*, polymer electrolyte membrane fuel cell; *MPD*, maximum power density; *MEA*, membrane-electrodes assembly.

with hydrogen can be calculated from the changes of the Gibbs free energy of products and reactants, according to Eq. (2.8):

$$\Delta G = G_{products} - G_{reactants} = G_{H_2O} - G_{H_2} - G_{O_2} = \Delta G^0 - RTln\left[\frac{p_{H_2} \cdot p_{O_2}^{1/2}}{p_{H_2O}}\right]$$

$$(2.8)$$

with $\Delta G^0$ the variation of the Gibbs free energy at standard working pressure (1 bar, which varies with the operative temperature $T$), $R$ the universal constant for ideal gases, and $p$ the partial pressure of the gases involved in the reaction. Thus, the reversible voltage of a hydrogen PEMFC, $E_{cell}$, can be expressed as

$$E_{cell} = \frac{\Delta G}{nF} = \frac{\Delta G^0}{2F} - \frac{RT}{2F}ln\left[\frac{p_{H_2} \cdot p_{O_2}^{1/2}}{p_{H_2O}}\right] = E^0 - \frac{RT}{2F}ln\left[\frac{p_{H_2} \cdot p_{O_2}^{1/2}}{p_{H_2O}}\right] \quad (2.9)$$

with $n$ the number of electrons involved in the reaction (2 electrons, for Eq. 2.1) and $F$ the constant of Faraday, and depends on the operative temperature. $E^0$ represents the theoretical voltage at standard conditions (1 atm, 298 K), which is equal to 1.229 V [83].

The lower value of $V_{OCV}$ compared with $E_{cell}$ depends on the onset of the mixed potential of the two semireactions, but mainly of the cathodic reaction (Eq. 2.4), because of a series of parasitic processes that reduce the cell voltage [83]. When the electric circuit is closed and both the fuel and oxidant are fed to the cell, the cell voltage $V_{cell}$ further decreases, because of a series of irreversible phenomena known as activation overpotentials ($\Delta V_{act}$, linked with electrochemical limitations on the surface of the electrodes), ohmic overpotentials ($\Delta V_{ohm}$, associated with ohmic losses in both the ionic and electronic conductors), and concentration overpotentials ($\Delta V_{conc}$, related to the change of the concentration of the reactants at the surface of the electrodes as soon as the reactions proceed), highlighted in Fig. 2.4A [83]. The product of the cell voltage and current density provides the power density curve, which allows determining the maximum power density (MPD) of the cell itself, also highlighted in Fig. 2.4A.

The power output of a PEMFC depends on its size. Up to now, the best achieved catalysts, based on platinum, can reach 1200 mW cm$^{-2}$ as maximum power density [84]. Thus, to reach the required power, individual cells can be stacked together to achieve a higher voltage and power tailored for specific applications. The assembly obtained is a stack, as shown in Fig. 2.4B. Each MEA is sandwiched by two bipolar plates, carrying on gas flow channels on both sides (for hydrogen and oxygen distribution). The stack is finished with two end plates and connections. Increasing the number of MEAs in a stack means increasing the voltage, whereas increasing the surface area of the MEAs means increasing the current [83].

The efficiency of a PEMFC device is defined as the ratio between the useful energy output, which is the electrical energy produced, $\Delta G^0$, and the variation and energy input, which is the enthalpy of hydrogen, $\Delta H^0$. Thus, the maximum theoretical efficiency of a PEMFC fed with hydrogen, working at 298 K and 1 bar, can be calculated as follows:

$$\zeta = \frac{\Delta G^0}{\Delta H^0} = \frac{237.2 \ \text{kJmol}^{-1}}{285.8 \ \text{kJmol}^{-1}} \cong 0.83 = \frac{\dfrac{\Delta G^0}{nF}}{\dfrac{\Delta H^0}{nF}} = \frac{E^0}{E_H^0} = \frac{1.229 \ \text{V}}{1.482 \ \text{V}} \tag{2.10}$$

with $E^0$ the theoretical voltage at standard conditions and $E_H^0$ the potential corresponding to the higher heating value of hydrogen, even known as the thermoneutral potential (the resulting voltage if all the enthalpy of hydrogen is converted into electric energy). Such a theoretical efficiency, also known as thermodynamic efficiency, decreases with temperature and represents the maximum efficiency limit of an FC. For temperatures lower than 1000 K, this maximum efficiency limit is always higher than the limit Carnot efficiency of a heat engine (which, instead, increases with the temperature) [83].

## Polymer Electrolyte Membrane Electrolysis Cell Operation Mode

PEMFCs can operate in reverse mode, as polymer electrolyte membrane electrolysis cells (PEMECs). In this way, the system is able to produce both oxygen and hydrogen by splitting electrochemically water, according to the reverse of Eq. (2.2). In PEMECs, anodic and cathodic reactions are reversed compared with PEMFCs. It means that the water fed to the anode is oxidized, producing oxygen, protons and electrons, as represented by the inverse pathway of the reaction in Eq. 2.4. In fact, water in liquid or gaseous state, fed at the anode, is oxidized producing oxygen, protons, and electrons (according to the reverse of Eq. 2.4). Protons cross the PEM, and electrons move to the cathode, to reduce the crossed-over protons to hydrogen (according to the reverse of Eq. 2.3). Because the overall process is thermodynamically unfavorable, the reactions occur only when a direct current voltage across the membrane is applied [85]. Fig. 2.5A shows a schematic representation of a PEMEC working principle. Even if electrolysis is an endothermic process, the overall temperature of the cell increases because of the electrical resistances of the system while in operation [86,87].

Historically, two Dutch gentlemen, the merchant Adriaan Paets van Troostwijk and the medical doctor Johan Rudolph Deiman, demonstrated first the principle of electrolysis using a glass tube closed at its bottom, filled with water, with a thin gold wire inside, and placed under water [88], publishing the first results in 1789−90. Connecting the gold wire with an electrostatic generator, they were able to evolve gas from water [88]. At the very beginning of 1800, the German physicist-chemist Johann Wilhelm Ritter realized an apparatus to evolve oxygen and hydrogen, whose original scheme is represented in

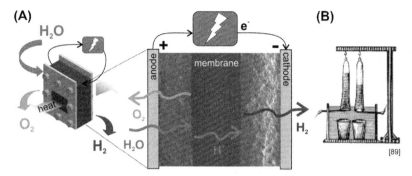

**FIGURE 2.5**  (A) Operative principle of a PEMEC. (B) Ritter's scheme of the first electrolytic cell, published in 1800 on *Voigts Magazin für den neuesten Zustand der Naturkunde* (vol. 2, 1800, pp. 356−400) [89]. *PEMEC*, polymer electrolyte membrane electrolysis cell.

Fig. 2.5B [89], capable to measure exactly the amount of gases evolved (hydrogen/oxygen = 2/1).

PEMECs are realized exactly as PEMFCs. Compared with PEMFCs, commercial electrolyzers, based on either alkaline or PEM electrolytes, are available on the market, with a relatively good energy efficiency, ranging from 60% to 75% [90]. However, the actual production costs of hydrogen by electrolysis are still not competitive with the costs of the most used industrial processes, such as natural gas (NG) steam, dry or partial reforming [91−94]. The reason mainly resides in the high anodic overvoltages (oxygen evolution side), which require high cell voltages [95]. Thus, the practical energy required to produce hydrogen is much higher than the theoretical one (more than $50 \text{ kWh kg}^{-1}$ compared with $33 \text{ kWh kg}^{-1}$ [85]).

The best way to overcome the economic problem, and then boost the true hydrogen economy locally [96,97], consists in making competitive the production and use of hydrogen through the use of PEMFC reversible systems [98], schematically depicted in Fig. 2.6. These systems, usually called unitized

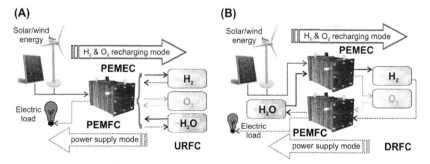

**FIGURE 2.6**  Basic schemes of a URFC system (A) and a DRFC system (B). *DRFC*, discrete regenerative fuel cell; *PEMFC*, polymer electrolyte membrane fuel cell; *URFC*, unitized reversible fuel cell.

reversible fuel cells (URFCs, Fig. 2.6A), are based on electrochemical devices able to operate in direct and reverse mode, as FCs or electrolyzers, depending on the energy input and output (according to customers' demand). In fact, in PEMFC mode, URFCs use oxygen (either from air or stored) and stored hydrogen to produce electricity and water, which is stored. While in PEMEC mode, URFCs split stored water into hydrogen and oxygen by using renewable electricity (either solar or wind energy, or a combination of the two, or other renewable energy sources). If air is used in PEMFC mode, the oxygen produced in the electrolysis mode is purged to the atmosphere. Technically, when operation in FC mode and that in electrolysis mode are not nearly balanced in terms of power density and operating times, the discrete regenerative fuel cells (DRFCs, Fig. 2.6B) represent a better solution, even if of bigger size compared with URFCs. In fact, DRFC design consists of two separate PEM electrochemical devices, an electrolysis stack and a FC stack, with a shared hydrogen/oxygen/water storage system. Keeping the electrolyzer and FC separate, DRFCs can work simultaneously in both ways, with optimized individual performance and better overall efficiency [99]. Hydrogen can be stored in either pressurized vessels or metal hydride cylinders. Up to now, URFCs/DRFCs can provide an integrated solution for energy production/storage, with many different potential applications, even if full commercial units are still missing from the market [99].

## Polymer Electrolyte Membrane Fuel Cell—Technical Targets

Considering that PEMFCs can be used in a wide range of applications, each market requires specific needs to meet customer desiderata. Such a technology provides yet enough performance to compete against already available technologies, but the high cost and a not sufficient durability still delay a full commercialization of this technology. Moreover, hydrogen availability, including storage and distribution [100,101], contributes to hinder commercialization, even if the rapid development and market entry of hydrogen-powered cars [48,50], together with the realization of specific hydrogen roadmaps, is pushing its penetration [97,102,103].

Since many years, the US Department of Energy (US DOE) Hydrogen and Fuel Cells Program [104] has been considered one of the most representative entity dictating targets for FCs, overcoming technical barriers still existing, and favoring full commercialization. In fact, cost, performance, and durability are the key challenges for FCs industry.

Cost reduction is a priority in FC research and development. According to the DOE, the cost of a hydrogen-fed 80-kW$_{net}$ stack for automotive application was estimated as 26 US\$ kW$^{-1}$ in 2015, for a production volume of 500,000 units per year (stack efficiency of 53%, cell voltage of 0.661 V, MEA area power density of 746 mW cm$^{-2}$, total platinum group metal (PGM) loading of 0.142 mg$_{PGM}$ cm$^{-2}$, and price of Pt considered equal to 1500 US\$ per troy ounce) [105]. The target for 2020 was set at 20 US\$ kW$^{-1}$, with an ultimate value equal to 15 US\$ kW$^{-1}$. Such a price doubles considering the entire PEMFC system, not only the stack. Almost 45% of the estimated price belongs to catalyst usage and its application [105], as shown in Fig. 2.7A, whereas the other components account for much less. The reason resides mainly in the high cost of platinum, whose price has a huge and unpredictable variability at the stock exchange, as visible in Fig. 2.7B [106]. Thus, platinum price is one of the most stringent requirements, which is driving most of the research and development efforts to increase both activity and platinum utilization in MEAs, as well as to instigate alternative catalysts based on PGM alloys and PGM-free materials [107].

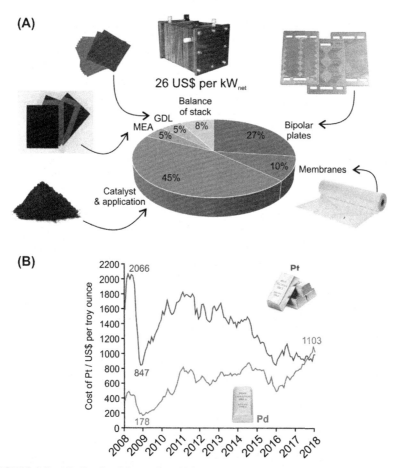

**FIGURE 2.7** (A) Cost breakdown of an 80-kW$_{net}$ stack, estimated by the US Department of Energy for a production of 500,000 pieces per year [105]. (B) Stock exchange price of Pt and Pd, at the first useful day of each month, from January 2008 to January 2018 [106]. *MEA*, membrane electrode assembly.

**TABLE 2.2** 2020 Technical Targets for an 80-kW$_{net}$ Polymer Electrolyte Membrane Fuel Cell Stack Operating With Direct Hydrogen for Transportation Applications [104]

| Characteristics | Targets |
| --- | --- |
| Stack power density | 2250 W L$^{-1}$ |
| Stack specific power | 2000 W kg$^{-1}$ |
| Performance at 0.8 V | 300 mA cm$^{-2}$ |
| Durability in automotive driving cycle | 5000 cycles |
| Cold startup time at −20 and +20°C (50% of rated power), for the integrated system | 30 and 5 s |
| Startup and shutdown energy from −20 and +20°C, for the integrated system | 5 and 1 MJ |
| Stack cost (500,000 units per year) | 20 US$ per kW$_{net}$ |
| MEA cost (500,000 units per year) | 14 US$ per kW$_{net}$ |
| Bipolar plate cost (500,000 units per year) | 3 US$ per kW$_{net}$ |
| Cost of the integrated system (500,000 units per year) | 40 US$ per kW$_{net}$ |

MEA, membrane electrode assembly.

To improve PEMFCs' performance, most of the research concentrates on the development of ion-exchange membrane electrolytes, to produce MEAs with enhanced efficiency and reduced costs, and fabrication of low-cost bipolar plates and subsystems tailored for each specific application. Durability is the third key challenge to assure a lifetime under realistic operation conditions to make PEMFC competitive on the market.

Considering that the most stringent application for PEMFCs is the transportation sector, the DOE suggests a series of targets to be reached for commercialization, which are summarized in Table 2.2 for an 80-kW$_{net}$ PEMFC stack operating with direct hydrogen [104]. These targets translate in specific requirements, and costs, of each component of the stack, specifically, electrocatalysts, membranes, MEAs, and bipolar plates.

## Materials—Electrocatalysts for Polymer Electrolyte Membrane Fuel Cells in Transportation Applications

Electrocatalysts represent the key elements enabling the electrochemical reactions occurring into PEMFCs. Several scientists drive their research efforts to reach the suggested DOE targets for electrocatalysts, summarized in Table 2.3 [104]. Concerning catalysts, the catalyst loading refers to both electrodes, even if the ORR occurring at the cathode is the more demanding one.

**TABLE 2.3** 2020 Technical Targets for Electrocatalysts for a Polymer Electrolyte Membrane Fuel Cell Stack Operating With Direct Hydrogen for Transportation Applications [104]

| Characteristics | Targets |
|---|---|
| PGM total content (both electrodes) | 0.125 g kW$^{-1}$ at 150 kPa, absolute |
| PGM total loading (both electrodes, electrode area) | 0.125 mg$_{PGM}$ cm$^{-2}$ |
| Mass activity | 0.44 A mg$_{PMG}^{-1}$ at 0.9 V$_{IR-free}$ |
| Loss in initial activity | <40% |
| Loss in performance at 0.8 A cm$^{-2}$ | <30 mV |
| PGM-free catalyst activity | <0.044 A cm$^{-2}$ at 0.9 V$_{IR-free}$ |

*PGM, platinum group metal.*

Platinum-based nanomaterials are the most commonly adopted electrocatalysts for both anode and cathode reactions. Platinum is highly dispersed over a carbon support. The anodic reaction of hydrogen oxidation (Eq. 2.3) is extremely fast, requiring usually a very low platinum loading, usually ranging between 0.1 and 0.05 mg cm$^{-2}$, while the sluggish cathodic reaction of oxygen reduction (Eq. 2.4) requires much higher platinum loading. Fig. 2.8A shows some of the best results reported in the open literature as maximum power density reached in testing MEAs in PEMFCs fed with hydrogen and oxygen (testing conditions and type of membrane slightly variable). The actual "best" performance still belongs to MEAs prepared with commercial Pt/C catalysts. In fact, Shao et al. [107] and Proietti et al. [113] reached 1360 and 1300 mW cm$^2$, respectively, with an overall Pt loading of 0.35 mg cm$^{-2}$ (0.30/0.05 mg$_{Pt}$ cm$^{-2}$ cathode/anode, Pt/C from Johnson Matthey, UK [107], or Tanaka Kikinzoku Kogyo, Japan [113]). Similarly, Jung and Popov [110] reached 1250 mW cm$^{-2}$ with an overall Pt loading of 0.26 mg cm$^{-2}$ (0.16/0.1 mg$_{Pt}$ cm$^{-2}$ cathode/anode, Pt/C from Tanaka Kikinzoku Kogyo, Japan). Depending on the nature of the catalyst, and on the preparation of the catalyst itself, the PEMFC performance is greatly affected. For example, very recently, Jung and Popov [110] developed a hybrid cathode catalyst with synergistic effect between a carbon composite, synthesized by pyrolysis of Fe-Co chelate compounds, and Pt, with ultralow Pt loading. With a Pt loading of only 0.14 mg cm$^{-2}$ (0.1/0.04 mg$_{Pt}$ cm$^{-2}$ cathode/anode), they reached an overall power density of 1025 mW cm$^{-2}$, which is the highest values reached for a not commercial Pt-based MEA. Such a catalyst is very close to the 2020 DOE target of 0.125 g$_{Pt}$ kW$^{-1}$, with a value of 0.136 g$_{Pt}$ kW$^{-1}$. Fig. 2.8B shows a high-resolution transmission electron microscopy (HR-TEM) picture of this hybrid cathode Pt/C catalyst, highlighting the fine

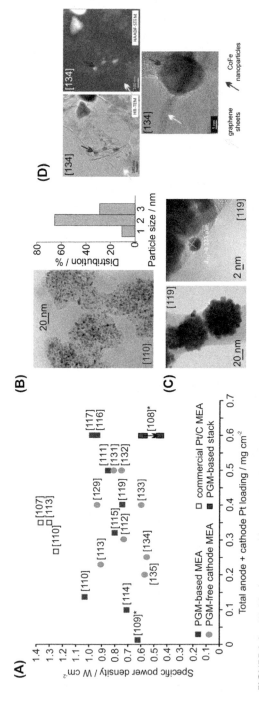

**FIGURE 2.8**   (A) Maximum specific power density values as a function of the overall Pt loading on the MEA for PEMFC single cells and stack [108] fed with hydrogen and oxygen ([108,109] with air). Testing conditions varied from 70 to 80°C, membranes used Nafion 211 or 212. (B) HR-TEM image of the Pt/C hybrid cathode catalyst tested in Ref. [110] and related Pt particle size average distribution. (C) HR-TEM images of the Pd@Pt3Co/C cathode catalyst tested in Ref. [111]. (D) HR-TEM and high-angle annular dark-field scanning TEM images of the CoFe-N-C cathode catalyst derived from polyaniline tested in Ref. [112]. *HR-TEM*, high-resolution transmission electron microscopy; *MEA*, membrane electrode assembly; *PEMFC*, polymer electrolyte membrane fuel cell; *PGM*, platinum group metal.

dispersion of Pt nanoparticles, having an average particle size of 2.5 nm, on the hybrid carbon-based support [110]. In 2005, Xiong and Manthiram [114] fabricated an MEA with a commercial Pt/C from Alfa Aesar with an ultralow overall Pt loading of 0.1 mg cm$^{-2}$, reaching a specific power density of 715 mW cm$^{-2}$. Using a commercial Pt/C from Johnson Matthey, Chen et al. [115] fabricated various MEAs by modulating the Pt/C loading along the thickness of the cathode side. Their best MEA, with an overall 0.32 mg cm$^{-2}$ (0.12/0.2 mg$_{Pt}$ cm$^{-2}$ anode/cathode, with a Pt/C ratio gradient 70% inner side and 40% outer side), provided a maximum specific power density of 800 mW cm$^{-2}$. Alloying Pt with other metals such as W and Cr and dispersing them on carbon, Shao et al. [116] and Rao et al. [117] reached very similar performance, 975 and 985 mW cm$^{-2}$, respectively, but with a higher overall Pt loading of 0.6 mg$_{Pt}$ cm$^{-2}$ (0.3/0.3 mg$_{Pt}$ cm$^{-2}$ cathode/anode for Shao et al. [116] and 0.4/0.2 mg$_{Pt}$ cm$^{-2}$ cathode/anode for Rao et al. [117]). Other research groups developed relatively well-performed cathode catalysts alloying platinum and palladium. In fact, those two elements belong to the same group and have similar electronic properties. Moreover, palladium has relatively good ORR properties [118] and an average lower price compared with Pt, as highlighted in Fig. 2.7B. As an example, Cho et al. [119] reached 750 mW cm$^{-2}$ with a PdPt/C catalyst, overall Pt loading of 0.4 mg$_{Pt}$ cm$^{-2}$ (0.2/0.2 mg$_{Pt}$ cm$^{-2}$ cathode/anode). Adding Co to Pd and Pt, Wang et al. [111] synthesized a Pd@Pt$_3$Co/C core−shell catalysts, reaching 854 mW cm$^{-2}$ with an overall Pt loading of 0.5 mg$_{Pt}$ cm$^{-2}$ (0.25/0.25 mg$_{Pt}$ cm$^{-2}$ cathode/anode). Such an MEA resulted very stable, losing only 3% of the performance after 100 h testing at a constant potential of 0.4 V. Interestingly, the Pd@Pt$_3$Co/C core−shell catalyst has a typical flowerlike shape, because of the strongly magnetic character of Pt$_3$Co nanoparticles, which are on the surface of Pd nanoparticles [111], as evident by HR-TEM images in Fig. 2.8C. Very recently, Hong et al. [109] prepared highly performing Pd/C@Pt$_{skin}$ catalysts, able to reach 620 mW cm$^{-2}$ (in a PEMFC fed with hydrogen and air) with an ultralow overall Pt loading of 0.019 mg$_{Pt}$ cm$^{-2}$. In addition, this MEA resulted stable, presenting its highest specific power density loss of only 4.8% after 30,000 cycles of potential variation from 0.6 to 1 V. Clearly, research efforts allowed reducing Pt content, maintaining a relatively high performance.

Few data are available for PEMFC stack performance. Li et al. [108] reported the performance of a 1.5-kW stack composed of 15 MEAs with 250 cm$^2$ as active area. Each MEA was fabricated with a carbon-supported platinum nanowire (PtNW/C) catalyst, used either as an anode or cathode catalyst, with an overall Pt loading of 0.6 mg$_{Pt}$ cm$^{-2}$ (0.4/0.2 mg$_{Pt}$ cm$^{-2}$ cathode/anode). The stack reached a specific peak power density of 575 mW cm$^{-2}$ feeding air, which decreased to 450 mW cm$^{-2}$ (∼ 22% performance loss) after 420 h of dynamic load cycling simulating a driving cycle (stack run at 70°C, 60 kPa, 100% RH, anode/cathode stoichiometry 1.2/4). A second stack realized with a commercial Pt/C catalyst from Johnson Matthey (same loading and testing conditions) reached a very similar performance, 600 mW cm$^{-2}$, which decreased to 460 mW cm$^{-2}$ after the dynamic load cycling.

Considering the suggestions provided by the DOE [104], many authors focused their research efforts on the development of PGM-free cathode catalysts for the ORR, trying to limit the use of platinum only at the anode side. In fact, starting from the discovery of Jasinski in the 1960s that Co-phthalocyanine has relatively good activity toward ORR [120], many scientists started to work on PGM-free cathode catalysts to replace the highly expensive platinum, as demonstrated by the incredibly high number of publications available in the literature [83,84,107,121−125]. Many different molecules have been investigated so far, from metal phthalocyanines to porphyrins, chalcogenides, and macrocycles. They require a specific heat treatment at high temperature, and mostly in controlled atmosphere, to favor the formation of the active sites for ORR. The ORR mechanism has been deeply studied, and most scientists converged that the catalytic process of oxygen reduction is related to the redox state of the metal ions involved, with the reaction favored in the order Cu < Ni < Co < Fe [123,126]. Moreover, those PGM-free catalysts have very high methanol and ethanol tolerance, which contributed to their success for direct alcohol FCs [127,128]. The actual "best" performance of MEAs prepared with PGM-free catalysts at the cathode side and tested in single-cell PEMFCs is reported in Fig. 2.8A. The highest activity has been reached very recently by Wang et al. [129]. Their MEAs reached 940 mW cm$^{-2}$ with a PGM-free loading at the cathode of 4 mg cm$^{-2}$ and at the anode of 0.4 mg$_{Pt}$ cm$^{-2}$ (higher specific peak power density reached increasing the backpressure). The developed Fe/N/C catalyst contains a certain amount of sulfur, which is able to change the electronic properties of carbon and increase the number of active sites, thus improving the ORR activity [130]. With a same platinum loading at the anode of 0.5 mg$_{Pt}$ cm$^{-2}$, but a different Fe/N/C catalyst (4 mg cm$^{-2}$, produced by ball-milling ZIF-8, tris-pyridyl-triazine, and iron acetate), Tian et al. [131] reached 750 mW cm$^{-2}$. Barkholtz et al. [132] reached 603 mW cm$^{-2}$ with a Fe/N/C catalyst synthesized with ZIF and iron phenanthroline, overall 4 mg cm$^{-2}$ at the cathode and 0.4 mg$_{Pt}$ cm$^{-2}$ at the anode. Reducing the overall platinum loading at the anode, with 0.3 mg$_{Pt}$ cm$^{-2}$, Yuan et al. [133] reached 730 mW cm$^{-2}$ with 4.1 mg cm$^{-2}$ Fe/N/C catalyst from polymerized iron-metalated tetrakis-di-thiophen-phenyl-porphyrin. With 0.25 mg$_{Pt}$ cm$^{-2}$ at the anode, Wu et al. [112] reached 550 mW cm$^{-2}$ with 4 mg cm$^{-2}$ Fe-Co/N/C catalyst derived from polyaniline, iron, and cobalt precursors. Fig. 2.8d shows HR-TEM and high-angle annular dark-field scanning TEM images of this Fe-Co/N/C, which highlight the typical shape of PGM-free catalysts involving carbon nanofibers and metal aggregates encapsulated in graphitic nanoshells with graphene sheet−like structure [112]. Such a catalyst revealed very promising stability as a cathodic catalyst after a 700 h testing in PEMFC at a constant cell voltage of 0.4 V (less than 10% performance loss), thanks to the presence of cobalt in the catalyst structure [112]. With 0.20 mg$_{Pt}$ cm$^{-2}$ at the anode, Serov et al. [134] reached 560 mW cm$^{-2}$ with 4 mg cm$^{-2}$ Fe/N/C catalyst synthesized with carbendazim and iron via a sacrificial support method with silica particles. Among all, the best performance in terms of g$_{Pt}$ kW$^{-1}$ belongs to the PGM-free cathodic Fe/N/C catalyst developed by Proietti et al. [113], using as precursors iron acetate, phenanthroline, and ZIF-8. In fact, such a catalyst reached 910 mW cm$^{-2}$ with 3.9 mg cm$^{-2}$ at the cathode and 0.23 mg$_{Pt}$ cm$^{-2}$ at the anode, which means 0.252 g$_{Pt}$ kW$^{-1}$. The total PGM content per kilowatt delivered is still far from the 2020 DOE target of 0.125 g$_{Pt}$ kW$^{-1}$, but the overall performance obtained with this PGM-free catalyst is not that far from the performance of the best PGM-based catalysts developed so far.

### Materials—Membranes for Polymer Electrolyte Membrane Fuel Cells in Transportation Applications

Systematic studies on membranes for PEMFCs started more than 60 years ago. Membranes in PEMFCs replaced liquid electrolyte used in the first FCs, avoiding problems of corrosion due to the strong acidic solutions used as electrolytes and simplifying the design of the stack through a more compact system [59]. The main roles of the electrolyte membrane reside in transferring protons from the anode to the cathode and being a physical barrier for electrons and fuel. The most famous and commonly used membranes have been developed first by E.I. du Pont de Nemours and Company (USA), with the trade name of Nafion. This membrane derives from the copolymerization of unsaturated perfluoroalkyl sulfonyl fluoride with tetrafluoroethylene with ether-linked side chains ending with sulfonated cation-exchange sites [53]. The resulting membrane has a semicrystalline structure resembling Teflon, which guarantees a long-term stability in both oxidative and reducing conditions, up to 60,000 h in stationary condition [59,135], and high ionic conductivity, ~0.1 S cm$^{-1}$ in fully hydrated conditions, provided by the strongly hydrophilic terminal side chains $-SO_3H$ functional groups [136]. The most performing working conditions of Nafion require wet conditions and a temperature lower than 90°C because its glass transition temperature is at around 120°C [136]. Thus, Nafion membranes suffer from various problems, among all water loss and viscous-elastic relaxation at high temperature, and high alcohol permeability, which is a big drawback for direct alcohol FCs. Moreover, they are still very expensive.

At present, several types of membranes are commercially available, the most important are listed in Table 2.4, together with their main properties, conductivity, and water uptake [34,59,137−139]. These membranes are characterized by fluorocarbon-based ion-exchange polymers and, as Nafion, have low conductivity at low water content, low mechanical strength at high temperature, and modest glass transition temperature [140]. The water uptake plays a crucial role because water favors proton conductivity. However, absorbed water also affects the mechanical properties of the membrane by

**TABLE 2.4** Main Properties of the Most Important Polymer Electrolyte Membrane Commercially Available [34,59,137–139]

| Membrane Types | Manufacturers | Equivalent Weight (g mol$^{-1}$) | Thickness in Dry Conditions (μm) | Conductivity (mS cm$^{-1}$) | Water Uptake (wt.%) |
|---|---|---|---|---|---|
| Nafion | DuPont (USA) | 1100 | 25–175 | 80–165 | 20–50 |
| FumaPEM | FuMA-Tech (Germany) | 930–1800 | 30–50 | 90–100 | 35 |
| Flemion | Asahi Glass (Japan) | 900 | 50–120 | 20–270 | 26–64 |
| Gore Select | Gore & Associates (USA) | 900–1100 | 5–20 | 28–96 | 32–43 |
| Aquivion PFSA | Solvay (Belgium) | 850–890 | 50–150 | >160–228 | 25–35 |
| Dow | Dow Chemicals (USA) | 800 | 125 | 114 | 54 |
| Aciplex | Asahi Chemicals (Japan) | 1000 | 90–245 | 110 | 18–44 |

acting as a plasticizer. Thus, a careful control of water uptake is critical for reducing both swelling and degradation of the membrane [140].

As an alternative, nonfluorinated hydrocarbon polymeric materials can be used as well, coupled with aromatic structures and functional groups. Typical hydrocarbon-based polymers are polyether ketones, polyarylene ethers, polyether sulfones, polyimides, and polyesters [140,141]. Compared with perfluorinated membranes, hydrocarbon-based membranes are cheaper. However, specific sulfonation treatments are needed to attain high proton conductivity without compromising mechanical properties [140]. Alternatively, membranes can be reinforced by adding to the base polymers organic or inorganic fillers, to improve their characteristics at high temperature [55,63,142] and reduce the alcohol permeability in case of direct alcohol FCs [143−145].

Membranes for FCs must guarantee a reliable lifetime for commercialization because their lifetime generally determines the lifetime of PEMFCs. For transportation application, the DOE suggests a series of targets, listed in Table 2.5, to be reached in 2020 for membranes to be integrated in a PEMFC stack operating with direct hydrogen [104]. Less stringent targets but longer cycles are required for stationary application [135]. Membrane degradation can occur for several reasons, including mechanical, thermal, chemical, and electrochemical degradation. Mechanical degradation, which affects particularly thin membranes, can happen in the form of cracks, pinholes, and tears. These failures can be caused by an incorrect or nonuniform MEA assembling within the bipolar plates, or inadequate humidification during operation, which render the membrane fragile and brittle [146−148]. Fig. 2.9A shows a cross-

**TABLE 2.5** 2020 Technical Targets for Membranes for a Polymer Electrolyte Membrane Fuel Cell Stack Operating With Direct Hydrogen for Transportation Applications [104]

| Characteristics | Targets |
|---|---|
| Maximum $O_2$ (or $H_2$) crossover | 2 mA cm$^{-2}$ (tested at 80°C, 1 atm total pressure, fully humidified gases) |
| Area specific proton resistance at a maximum operating temperature and water partial pressure from 40 to 80 kPa (30°C and up to 4 kPa; −20°C) | 0.02 Ω cm$^2$ (0.03 Ω cm$^2$; 0.2 Ω cm$^2$) |
| Maximum operating temperature | 120°C |
| Minimum electric resistance | 1000 Ω cm$^2$ |
| Combined chemical/mechanical durability | 20,000 cycles until >15 mA cm$^{-2}$ $H_2$ crossover or >20% loss in OCV |

*OCV*, open circuit voltage.

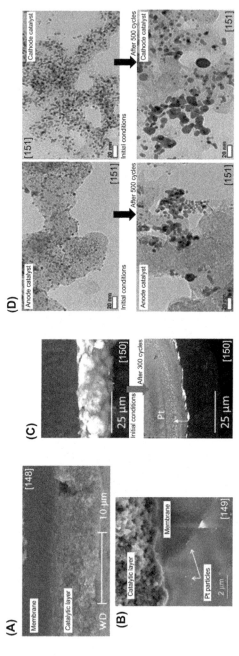

**FIGURE 2.9** (A) A cross-sectional SEM image of an MEA highlighting the degradation of the membrane and the cathode catalytic layer occurring during stack operation (PEMFC test performed on a 10 cm$^2$ single cell) [148]. (B) A cross-sectional field emission SEM image of an MEA highlighting the migration of Pt in the membrane after stack operation (PEMFC test performed on a 5 cm$^2$ single cell with 0.5 mg$_{Pt}$ cm$^{-2}$ on each electrode) [149]. (C) SEM images of a Pt/C cathode catalyst–membrane interface before and after 300 startup/shutdown cycles (PEMFC test performed on a 50 cm$^2$ single cell with 0.25 mg$_{Pt}$ cm$^{-2}$ on each electrode) [150]. (D) TEM images of anode/cathode Pt/C catalysts before and after 500 potential reverse cycling (PEMFC test performed on a 25 cm$^2$ single cell with 0.4 mg$_{Pt}$ cm$^{-2}$ on each electrode) [151]. *MEA*, membrane electrode assembly; *PEMFC*, polymer electrolyte membrane fuel cell; *SEM*, scanning electron microscopy; *TEM*, transmission electron microscopy.

sectional scanning electron microscopy (SEM) image of an MEA, which highlights the degradation occurring during stack operation at the interface membrane/cathode catalytic layer [148]. Moreover, potential cycling during operation and frequent startup/shutdown cycles can cause the migration of platinum catalyst particles into the membrane, as shown in the field emission SEM images of Fig. 2.9B and C, causing the further degradation of the membrane and loss of activity [135,149,150]. Fuel starvation causes cell reversal, with consequent formation of oxygen at the anode via electrolysis of water instead of hydrogen oxidation [151,152], and degradation of both catalytic layers [153]. In fact, a series of accelerated reversal potential tests demonstrated that Pt particles increased in size and coalesced with the number of reversal potential cycles, as shown by the TEM images of Fig. 2.9C [151]. This phenomenon is mainly due to the oxidation of the carbon-based support of Pt. Moreover, oxygen formation at the anode causes a localized increase of the temperature, which in turns causes pinhole formation in the membrane, up to the perforation of the membrane itself [71,154]. Membrane perforation causes further gases crossover and drop voltage [155,156].

On the chemical point of view, the harsh oxidizing/reducing environment of PEMFCs causes chemical degradation. In fact, highly reactive hydrogen (H*), oxygen-centered hydroxy (HO*), and hydroperoxy (HOO*) radicals generated locally can strongly react with the tertiary carbons of the membrane, causing an accelerated dissolution and failure of the membrane [147,157]. Moreover, also different degradation mechanisms leading to sulfonic acid or ether group cleavage can happen [158,159]. These effects can be mitigated by adding radical scavengers into the membrane by incorporating small amounts of single/double metal nanoparticles or Ce, Mn, Fe oxides. These nanoparticles are particularly effective in reducing the degradation of the membrane, acting as sacrificial elements reacting with radicals with a faster reaction rate, and avoiding thus the radical attack toward the carbon of polymeric structure of the membrane [160,161]. Fluorinated membranes can undergo defluorination, which makes the membrane more sensitive to peroxide radical attack. Usually, the measurement of fluoride released by PEMFCs can be considered as an indicator of the degradation of these membranes [162]. Fluoride release can be minimized by pretreating the polymer-based membrane with fluorine and substituting H-containing ending groups with F-containing ones. In fact, C$-$F bonds are stronger than C$-$H bonds; thus they are less susceptible to radical attack [163].

### Materials—Bipolar Plates for Polymer Electrolyte Membrane Fuel Cells in Transportation Applications

Bipolar plates are another major component of PEMFCs, which connect the MEAs physically, thermally, and electrically. Bipolar plates represent the majority of the stack weight, and they must be low cost and easily workable by casting, stamping, rolling, or molding. Although graphite [164] is the most

**TABLE 2.6** 2020 Technical Targets for Bipolar Plates for a Polymer Electrolyte Membrane Fuel Cell Stack Operating With Direct Hydrogen for Transportation Applications [104]

| Characteristic | Target |
|---|---|
| Plate weight | $0.4 \text{ kg kW}_{net}^{-1}$ |
| Plate $H_2$ permeation coefficient | $<1.3 \cdot 10^{-14} \text{ std cm}^3 \text{ s}^{-1} \text{ cm}^{-2} \text{ Pa}^{-1}$ at 80°C, 3 atm, 100% RH |
| Anode (or cathode) corrosion | $<1 \text{ μA cm}^{-2}$ |
| Electrical conductivity | $<100 \text{ S cm}^{-1}$ |
| Area specific resistance (included interfacial contact resistance) | $<0.01 \text{ Ω cm}^2$ |
| Flexural strength | $>25$ MPa |
| Forming elongation | 40% |

used material, carbon-composite materials [165,166], metallic materials (stainless steel (SS), TiN/Ti, coated aluminum [167–170]), and polymer composite materials [171], are also used to prevent cracks and limit the brittle nature of graphite. Table 2.6 lists the 2020 targets suggested by the DOE for bipolar plates to be integrated in a PEMFC stack operating with direct hydrogen for transportation applications [104].

The shape of the bipolar plates, and in particular the design of the gas flow-field channels, significantly affects the performance of a stack, both in the short- and long-term operation. In the literature, many different configurations have been studied, and each of them has specific advantages or disadvantages, depending on the operating conditions, in terms of pressure drop, reagents distribution, removal of the produced water. Fig. 2.10A shows the most common flow-field design, such as serpentine, parallel, interdigitated, grid, or spiral flow field [65–68], and the most innovative ones, inspired by nature through a heuristic approach, such as the lounge- or leaflike flow-field design [164,172–174]. The more complex geometry mainly serves to avoid misdistribution of reactants and increase the contact surface between the reactants and gas diffusion layer. The interdigitated configuration, based on a dead-end channel design forcing the reactants to pass through the gas diffusion layer, enhances the effective utilization of the catalyst, thanks to a more developed convective flow able to better remove the water forming at the cathode [173,177]. Instead, the grid flow-field design favored local flooding because of inhomogeneous distribution of reactants, notwithstanding a lower pressure drop [172]. The bioinspired flow field, such as the loungelike shapes, enhanced the performance of the entire stack, with a gain up to 30% on the specific peak

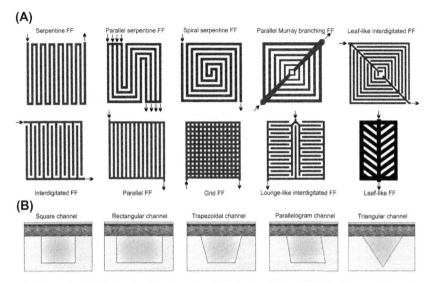

**FIGURE 2.10**   (A) Schemes of various flow-field designs [172–174]. (B) Geometrical configuration of various channel designs [172,175,176].

power density, compared with a classic serpentine flow field [173]. With the leaflike shapes, the gain on the specific peak power density reached similar values (+26% compared with serpentine flow field, +56% compared with parallel flow field), thanks to a more uniform gas velocity distribution along the channels [174]. Moreover, the fabrication of these leaflike flow fields requires less materials compared with the classical parallel or serpentine flow field, allowing a desired cost saving [174].

In addition, the geometry of the channels plays a crucial role on the overall performance of a stack [172]. Different cross-sectional geometries have been investigated so far, from the classic square channel to the trapezoidal or triangular channel, as visible in Fig. 2.10B [175,176]. Channel geometry can affect the volume of water formed at the cathode side, favoring its removal and avoiding flooding. Experimental studies demonstrated that triangular channels retain less water, minimizing flooding and favoring water droplet dispersion and not aggregation, as in the rectangular channels [175]. In fact, channels with a reduced depth promote water elimination [172]. Usually, long channels induce higher pressure drop; thus shorter channels and a larger number of parallel channels are usually preferable because they provide a better distribution of reactants. These aspects are of utmost importance, especially for automotive applications, where water formation, and especially removal, must be carefully controlled for operating the stack in subfreezing conditions.

The most interesting example of applied research used to produce commercial items refers to Toyota: A detailed design of both bipolar plates and MEA configuration allowed improving the stack design of the Mirai FC

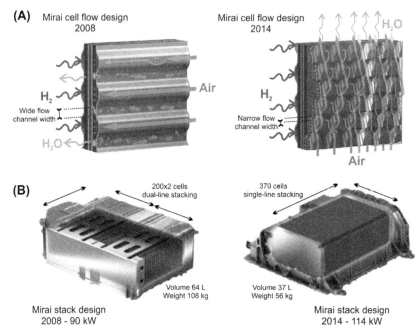

**FIGURE 2.11** (A) Design of the old and new flow fields of the Toyota Mirai FC vehicle [178]. (B) Design of the old and new stack of the Toyota Mirai FC vehicle [178]. *FC*, fuel cells.

vehicle and thus the performance of the car, which is able to operate from −30°C [178]. Fig. 2.11A shows the evolution of the flow-field design, from the model of 2008 to the model released in 2014 [178]. The larger distance between the ribs and channels of the bipolar plates resulted in lower cell voltage/current performance. Moreover, flooding of the MEA was more frequent because of the not uniform compression of the gas diffusion layer by the ribs themselves, with consequent water entrapped into the cells, oxygen starvation, and not uniform power generation. The new cell design, with a specific 3D fine mesh configuration, narrow flow channel width, and more hydrophilic materials, allows for an enhanced gas transport and a fast removal of the water formed, preventing accumulation [178]. Thus, power generation in such a cell results more uniform. Moreover, a specific targeted research allowed an optimization of the catalyst formulation, with a consequent enhancement of the ORR activity of 1.8-folds. Overall, in 2014, Toyota released a new stack, more compact, more powerful, and more efficient compared with the version released in 2008, as shown in Fig. 2.11B [178]. With the new improvements, the geometry of the stack became more compact and light, with a smaller number of cells and a better performance that increased from 1.4 to 3.1 kW L$^{-1}$ as volumetric power density and from 0.83 to 2.0 kW kg$^{-1}$ as mass power density [178]. Moreover, the adoption of titanium coated with an amorphous carbon instead of SS as a separator for the cells reduced the phenomena of corrosion, with an increased durability of the stack and a considerable cost reduction for production [178].

## Polymer Electrolyte Membrane Fuel Cell—Today and Tomorrow

PEMFC technology greatly improved in the past years, thanks to a massive incremental work of research and development from the academia and industry. Thus, PEMFCs are one of the most promising technologies for a sustainable green energy future. Today, the transportation sector is the most advanced on the point of view of commercialization, as shown in Fig. 2.12, with a series of hydrogen-powered zero emission vehicles (H-ZEVs) already available to the public, especially in North America, Japan, and Korea [47−50,179−183]. These vehicles are ranging from cars, minivans, buses, tricycles, bicycles, forklifts, and drones. Concerning passenger cars, also BMW and Daimler, will offer soon, expected in 2020, hydrogen-powered cars, together with Toyota, Honda, and Hyundai [184]. Prices are still higher than those of the counterpart internal combustion engine vehicles (ICEVs) powered by gasoline or diesel; thus a continuous research and development is imperative to reach a further cost reduction for a full penetration in the global market. Durability also must be increased to assure a longer life to H-ZEVs, comparable with the commercial ICEVs.

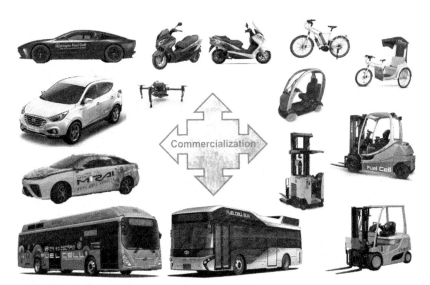

**FIGURE 2.12** First commercialization of PEMFC H-ZEVs [47−50,179−184]. *H-ZEV*, hydrogen-powered zero emission vehicle; *PEMFC*, polymer electrolyte membrane fuel cell.

Unfortunately, hydrogen is not easily available for the consumers. Hydrogen infrastructures, in particular hydrogen fueling stations, must penetrate the local territory to favor a capillary spread of hydrogen mobility at any level [101,103,185−188]. On this point of view, not only vehicle manufacturers but also governments and energy providers should sustain and provide financial support for the establishment of a hydrogen road map, together with the scientific efforts of the academia [178,189]. Otherwise, the spread of PEMFC H-ZEVs cannot be effective, quick, and economically feasible [40,190].

Research and development are going in parallel with many demonstration projects, with the aims to involve local population and make hydrogen a user-friendly fuel for consumers [189]. Public transport is the easiest field where it is possible to conjugate development with dissemination and outreach actions to increase consensus on hydrogen fuel. Fig. 2.13 shows the map of a European funded project, NewBusFuel [191], which is demonstrating that hydrogen refueling at very large scale is technically and economically viable for serving large hydrogen PEMFC bus fleet in different EU cities. Many other examples exist such as the use of a hybrid electric hydrogen FC bus during the Olympic Games in Rio de Janeiro 2016 [192] or the use of PEMFC H-ZEVs in

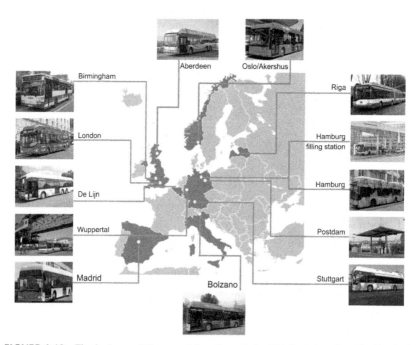

**FIGURE 2.13** The hydrogen FC-powered bus fleet of the EU funded project NewBusFuel, running in different EU cities [191]. *FC*, fuel cell.

cultural heritage areas in the south of Italy [193–195], as well in the North America [196,197] and in far-east Asia, China, Korea, and Japan [198–200].

The results of these demonstrative projects are helpful in policy decisions for favoring the commercialization of H-ZEVs and investment in hydrogen infrastructure. Very recently, Ballard announced that PEMFC electric hydrogen-powered buses in the city of London (UK) reached a new durability record, with more than 25,000 h of service (equivalent to operating a bus 14 h a day, 5 days a week, for 7 years) without significant maintenance for the FC stack [201]. The DOE recently announced that the use of FC hydrogen-powered buses is favoring not only the environment, with no harmful emissions compared with diesel-powered bus fleets, but also the fuel economy. In fact, these FC buses, which run for more than 23,000 h, run more miles per fuel consumed, resulting in 1.4 times higher fuel economy compared with diesel-powered buses [197]. Table 2.7 shows a brief comparison between ICEVs and H-ZEVs. The two main strengths, zero emissions and a higher efficiency, are still not enough to offsetting weaknesses [190]. Various technoeconomic scenarios and the biggest automotive manufacturer predict a 50% market penetration of hydrogen FC vehicles by 2050–70 [40,102,190,202]. Most of the studies conclude that the adoption of H-ZEVs can only be pushed by implementing local environmental, climate, and public policies and by struggling key stakeholders to invest higher capital costs [203,204]. Only strong and strategic policies can address the coevolution of the vehicles and

**TABLE 2.7** Strengths and Weaknesses of Internal Combustion Engine and Zero Emission Vehicles

|  | Strengths | Weaknesses |
| --- | --- | --- |
| ICEV | • Available in the market since many years<br>• High durability<br>• Long rangeability<br>• Fuel infrastructure available everywhere<br>• Continuous improvements in fuel efficiency | • Harmful emissions<br>• $CO_2$ emissions |
| ZEV | • Zero emissions<br>• Higher efficiency compared with ICEV | • Hydrogen infrastructure almost not available<br>• First and few commercial vehicles available only in North America and Japan<br>• Rangeability still lower compared with ICEV<br>• Expensive<br>• Small consensus among consumers |

*ICEV*, internal combustion engine vehicles.

fuels markets to reach a positive cost versus benefit analysis for the adoption of H-ZEVs and the related hydrogen fuel economy [205].

Thus, in conclusion, PEMFCs are one of the most promising and sustainable zero emission technologies available. PEMFC technology made enormous progress to reach commercialization, with the first FC hydrogen-powered vehicles launched in the market in 2015. In fact, the performance and durability reached acceptable levels, even if improvements are still necessary to become competitive with the actual markets and reduce the costs for consumers.

## SOLID OXIDE FUEL CELLS

An SOFC is an energy conversion device that generates electricity by electrochemical combination of a fuel with an oxidant across an oxide electrolyte. Present SOFCs, like other types of FCs, use mainly hydrogen as fuel and oxygen as oxidant: Hydrogen can be derived from common practical fuels such as NG, hydrocarbons, biogas, alcohols, and coal gas and oxygen is readily available from air. As an FC electrolyte must ionically conduct one of the elements present in the fuel and/or the oxidant, the electrolyte for the SOFC must be an oxygen ion—conducting oxide or a proton-conducting oxide [206,207]. An SOFC based on an oxygen ion—conducting electrolyte can be considered as an oxygen concentration cell, and an SOFC based on a proton-conducting electrolyte, as a hydrogen concentration cell. The main difference between these two SOFCs is the side in the FC in which water is produced (the fuel side in oxygen ion—conducting cells and the oxidant side in proton-conducting cells). In addition, certain gases, such as CO, can be used as fuel in oxygen ion—conducting SOFCs but not in proton-conducting SOFCs. Because of the conductivity requirement for the oxide electrolyte, present SOFCs operate in the temperature range of $500-1000°C$.

The distinctive features of the SOFC as compared with other types of FCs are its all-solid construction (ceramic and metal) and high operating temperature ($500-1000°C$). The combination of these two features leads to a number of attractive attributes for the SOFC but also presents several technical challenges [208]. The attributes of the SOFC include the following:

- *Design flexibility*: Because all the components are solid, the SOFC can be configured into thin films and/or unique shapes unachievable in other FCs (i.e., other than the planar configuration).
- *Multiple cell fabrication option*: The SOFC can be fabricated from a variety of processes ranging from thin-film deposition to conventional ceramic processing techniques.
- *Multifuel capability*: The SOFC can operate directly on reformates (without required cleanup of CO and $CO_2$) and on fuels other than hydrogen via internal reforming (on fuel feeds with significant amount of water) or direct fuel utilization (on fuel feeds with no water).

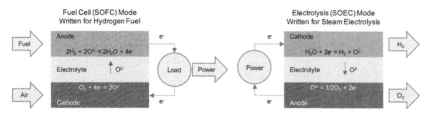

**FIGURE 2.14**  Power generation and electrolysis operation modes for SOFCs (written for oxygen ion–conducting electrolytes). *SOFC*, solid oxide fuel cell.

- *Range of operating temperature choice*: Depending on selected electrolyte materials and cell designs, the SOFC can operate at temperatures ranging from 500 to 1000°C.

The key technical challenges of the SOFC include the following:

- *Limited material selections*: The conductivity, stability, compatibility, and other requirements limit the number of suitable materials that can be selected for incorporation in the SOFC.
- *Undesirable chemical interactions*: The high operating temperature of the SOFC tends to promote and increase deleterious chemical interactions and elemental migration/interdiffusion between different components.
- *Operation and design restrictions*: The SOFC must be designed and operated under limited conditions that do not cause cracking and/or delamination of the electrolyte and other components, e.g., minimal reduction–oxidation (redox) cycling of Ni-based anodes.

The SOFC is an energy conversion device for power generation, but it can be operated efficiently in reverse mode (electrolysis mode) for chemical production; in this case, the cell is referred to as a solid oxide electrolysis cell (SOEC) [209]. Fig. 2.14 schematically shows the reactions occurring in cells operated in FC/power generation mode and in electrolysis/chemical production mode. The SOEC can be used to produce hydrogen (and oxygen) from $H_2O$, syngas (mixtures of $H_2$ and CO) from mixtures of $H_2O$ and $CO_2$, and oxygen from $CO_2$ [209]. It should be noted that syngas and oxygen (from $CO_2$) production is only applicable to SOECs with oxygen ion–conducting electrolytes.

## Solid Oxide Fuel Cell Technology

An SOFC cell or single cell is a fully functional unit composed of three components, an oxide electrolyte sandwiched between an anode and a cathode. The electrolyte, anode, and cathode can be a single layer or a multilayer, each made of one or more than one material. Like any FC technology, SOFC single cells are stacked and connected in electrical series (via a component referred to as interconnect) to build voltage. A multicell stack is the fundamental building block of any practical FC-based power systems.

## Component

*Requirement:* The principal components of an SOFC stack are the electrolyte, the anode, the cathode, and the interconnect. Each component serves several functions in the FC and must meet certain requirements for efficient and reliable operation. The basic requirements for the various components are summarized in Table 2.8. In addition to the requirements listed in Table 2.8, other desirable properties for the components from practical viewpoint are high strength and toughness, fabricability, and low cost. In addition, for certain cell/stack designs or specified fabrication processes, the components for the SOFC must be amenable to limited fabrication conditions because the process parameters cannot be selected independently for each component. Table 2.9 lists the materials commonly used for the SOFC at present. The components made from these materials have been shown to meet the basic requirements given in Table 2.8.

*Electrolyte*: Yttria-stabilized zirconia (YSZ) has been used almost exclusively as the electrolyte in SOFCs because this material possesses an adequate level of oxygen ion conductivity and exhibits desirable stability in both oxidizing and reducing atmospheres. In general, the conductivity requirement for the electrolyte determines the operating temperature of the SOFC. The operating temperature of the SOFC thus can be varied/reduced by modifying/ changing the electrolyte material and/or electrolyte thickness. Fig. 2.15 illustrates the two approaches commonly used to improve ionic conductivity to lower the operating temperature of YSZ-based cells. Examples include operating temperatures of 900–1000°C for thick ($>50\,\mu$m) YSZ electrolytes, 700–800°C for thin ($<15\,\mu$m) YSZ electrolytes [206–208] or doped lanthanum gallate electrolytes [214], 500–600°C for thin doped ceria electrolytes [215], and 400–500°C for thin-film YSZ [216] and thin doped ceria/bismuth oxide bilayer electrolytes [217]. It should be noted that, although having high ionic conductivities at reduced temperatures, $Bi_2O_3$ is unstable in fuel reducing environment, so the electrolyte in this case is a doped $CeO_2/Bi_2O_3$ bilayer. The doped $CeO_2$ of the bilayer electrolyte exposed to the fuel side protects the doped $Bi_2O_3$ from decomposing, while the $Bi_2O_3$ layer on the oxidant side blocks the electronic leakage current in the doped $CeO_2$ layer.

The SOFC uniquely has the characteristic of the possibility of mixed (ionic and electronic) conduction in the oxide electrolyte. Because the SOFC electrolyte is an ionic solid, its electronic conductivity cannot be absolutely zero. In general, the SOFC electrolyte is selected such as the electronic transference number is as small as possible under normal operating conditions to minimize electronic conduction losses. However, in some cases, depending on the partial

**TABLE 2.8** Basic Requirements for Solid Oxide Fuel Cell Components

| | | | Requirements | | |
|---|---|---|---|---|---|
| Components | Conductivity | Stability | Compatibility (Chemical and Thermal) | Electroactivity | Porosity |
| Electrolyte | High ionic conductivity Negligible electronic conductivity | Chemical, phase, electrical, morphological, and dimensional stability in fuel and oxidant environments (reducing and oxidizing atmospheres) | No damaging chemical interactions, elemental migration or interdiffusion with adjoining components Close thermal expansion match with adjoining components | (Not required) | Fully dense |
| Anode | High electronic conductivity | Chemical, phase, electrical, morphological, and dimensional stability in fuel environment (reducing atmosphere) | No damaging chemical interactions, elemental migration or interdiffusion with adjoining components Close thermal expansion match with adjoining components | Appropriate electroactivity for hydrogen oxidation reactions | Porous |

*Continued*

**TABLE 2.8** Basic Requirements for Solid Oxide Fuel Cell Components—cont'd

| Components | | Requirements | | | |
|---|---|---|---|---|---|
| | Conductivity | Stability | Compatibility (Chemical and Thermal) | Electroactivity | Porosity |
| Cathode | High electronic conductivity | Chemical, phase, electrical, morphological, and dimensional stability in oxidant environment (oxidizing atmosphere) | No damaging chemical interactions, elemental migration or interdiffusion with adjoining components Close thermal expansion match with adjoining components | Appropriate electroactivity for oxygen reduction reactions | Porous |
| Interconnect | High electronic conductivity Negligible ionic conductivity | Chemical, phase, electrical, morphological, and dimensional stability in fuel and oxidant environments (reducing and oxidizing atmospheres) | No damaging chemical interactions, elemental migration or interdiffusion with adjoining components Close thermal expansion match with adjoining components | (Not required) | Fully dense |

**TABLE 2.9** Common Solid Oxide Fuel Cell Materials

| Components | Most Common Materials | Other Materials |
| --- | --- | --- |
| Electrolyte | Yttria-stabilized zirconia (YSZ) | Sr,Mg-doped $LaGaO_3$ (LSGM), Sr,Mg-doped ceria (e.g., Gd-doped $CeO_2$ or GDC), proton-conducting oxides (e.g., doped $SrCeO_3$) [210] |
| Anode | Ni/YSZ | Cu/doped $CeO_2$, conducting oxides (e.g., doped $SrTiO_3$) [211] |
| Cathode | Sr-doped $LaMnO_3$ (LSM), Sr,Co-doped $LaFeO_3$ (LSCF) | Other perovskites (e.g., Sr-doped $LaCoO_3$ or LSC) and other oxides [212] |
| Interconnect | Stainless steel (SS), Sr-doped $LaCrO_3$ (LSCr) | Other high-temperature alloys and doped oxides [213] |

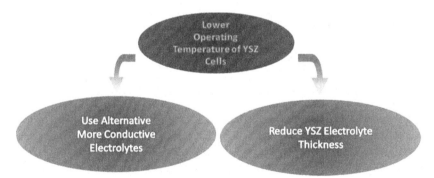

**FIGURE 2.15**   Approaches to lower operating temperature of YSZ-based SOFCs. *SOFC*, solid oxide fuel cell; *YSZ*, yttria-stabilized zirconia.

pressure of oxygen at the electrodes, the electronic conduction in the electrolyte becomes significant. For example, $CeO_2$ doped with divalent or trivalent oxide shows relatively high oxygen ion conductivity. Compared with YSZ, doped $CeO_2$ has a higher conductivity and lower conduction activation energy, thus suitable for use as the electrolyte material for the SOFC. Doped $CeO_2$, however, tends to undergo reduction at low oxygen partial pressures with the consequent introduction of electronic defects (electronic conductivity). Thus, bilayer electrolytes consisting of YSZ/doped $CeO_2$ have been considered and developed for the SOFC [218]. The YSZ layer of the YSZ/doped $CeO_2$ bilayer electrolyte is exposed to the fuel side to protect ceria from reduction (with attendant electronic conduction) by the fuel reducing environment.

Even if an electrolyte having some electronic conductivity is used, a high efficiency output can possibly be obtained for the mixed ionic electronic conducting electrolyte. It has been shown that SOFCs with presence of some electronic conductivity in the electrolyte can be designed to operate as efficiently as those made of purely ionic conducting electrolytes and even exhibit more stable operation [219]. Suitably designed mixed conducting cells consume fuel at the same rate, deliver the same power, and release the same amount of Joule heat as purely ionic conducting electrolyte cells [219].

*Anode*: Current SOFC cells based on YSZ electrolytes use almost exclusively Ni/YSZ compositions for the anode. Nickel is preferred as the SOFC anode material because of its stability in reducing environment at high temperatures (500−1000°C), high electrical conductivity, and excellent catalytic activity for the HOR. The functions of the YSZ are (1) to support the nickel metal to minimize coarsening of the metallic particles at the FC operating temperature, (2) provide a thermal expansion coefficient (CTE) for the anode acceptably close to those of the other cell components, (3) broaden contact areas between nickel and the YSZ, thus increasing anode active areas, and (4) improve adherence of the anode to the YSZ electrolyte. Ni/YSZ anodes have been extensively studied and have been used for cells in SOFC power systems of various sizes operated for tens of thousands of hours.

For fuels other than pure hydrogen, Ni/YSZ anodes can operate directly on reformates via external reformation or on fuels such as hydrocarbons, alcohols via internal reforming (on fuel feeds with significant amounts of water), or direct utilization (on fuel feeds with no water). Internal reforming on Ni/YSZ anodes is well known and has been demonstrated for the SOFC [220]. Instead of complete (100%) internal reforming, it is possible to have a portion of the fuel reformed in an external reformer (referred to as a prereformer), and the resulting reformate plus the remaining fuel are fed to the SOFC where the fuel is internally reformed (via steam reforming) within the FC. SOFC anodes have also been shown to have the capability for direct utilization of different types of fuel [221]. For direct fuel utilization operation, the anode material has been modified to address the carbon deposition issue associated with nickel commonly used in the anode composition (e.g., Cu/ceria instead of Ni/YSZ [222]). With modified anodes, high electrochemical performance can be achieved for direct SOFCs. For example, a peak power density of about 400 mW cm$^{-2}$ at 800°C was obtained with 7.3% ethanol (balance He) fuel and air oxidant for an SOFC with a dual layer anode consisting of a Cu-CeO$_2$-impregnated Ni/YSZ support outer layer and a Ni/YSZ electroactive inner layer [223] (Fig. 2.16). Long-term performance stability of direct SOFCs without significant carbon deposition, however, remains to be demonstrated.

There are several technological limitations of the Ni/YSZ anode: reduction−oxidation (redox) instability (chemomechanical instability of the anode under oxygen partial pressure changes during redox cycles at high temperatures), carbon deposition (deposition of carbon in the anode by

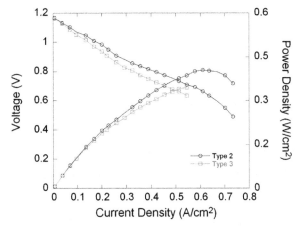

**FIGURE 2.16** Performance at 800°C with 7.3% ethanol (balance He) fuel and air oxidant for an SOFC (with a dual layer anode consisting of a Cu-CeO$_2$-impregnated Ni/YSZ support outer layer and a Ni/YSZ electroactive inner layer) (type 2 and 3 indicates different thermal treatments of infiltrated anodes). *SOFC*, solid oxide fuel cell; *YSZ*, yttria-stabilized zirconia.

catalytic cracking of fuels such as hydrocarbons or alcohols during internal reforming or direct utilization), and impurity poisoning (anode performance degradation by certain impurities such as sulfur present in the fuel). Among those limitations, redox instability is considered as the main disadvantage for the Ni/YSZ anode.

The Ni/YSZ anode can become unstable because of the oxidation of the Ni (i.e., redox instability) if fuel supply is interrupted, the cell is operated under extremely high fuel utilization/high current conditions, or seal leakage occurs. The oxidation of the Ni increases the volume of the anode and the volume expansion creates stresses in the different layers (compression in the anode and tension in the electrolyte). This can cause cracks in the electrolyte, especially the thin electrolyte in the case of anode-supported cells. The mechanism for the redox instability of the Ni/YSZ anode has been proposed [224,225] and is schematically shown in Fig. 2.17. Fig. 2.17 illustrates the different stages in the evolution of the Ni/YSZ anode microstructure during redox cycles [224,225].

Several approaches have been proposed to address the redox instability of the Ni/YSZ anode. These approaches can be categorized into three main groups depending on the focused area: (1) system approach (aims to keep an oxygen partial pressure low enough to protect the anode oxidation based on the system BOP, e.g., [226]), (2) Ni/YSZ anode modification approach (aims to modify Ni/YSZ anode microstructure or composition to minimize redox instability, e.g., [227]), and (3) alternative material approach (aims to use redox-resistant alternative materials, e.g., [228]).

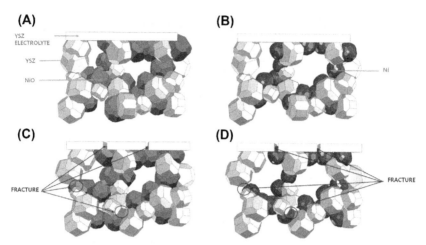

**(A)**

YSZ
ELECTROLYTE

YSZ

NiO

**(B)**

Ni

**(C)**

FRACTURE

**(D)**

FRACTURE

**FIGURE 2.17** Different stages in evolution of Ni/YSZ anode structure during redox cycling: (A) oxidation, (B) reduction, (C) reoxidation, (D) rereduction. *YSZ*, yttria-stabilized zirconia.

*Cathode*: Lanthanum strontium-doped manganite (LSM) (mainly for operating temperatures of $\geq 800°C$) and recently lanthanum strontium cobalt-doped ferrite (LSCF) (mainly for reduced operating temperatures of $\leq 800°C$) have been commonly used as cathode materials for the SOFC. The selection of these materials has been based primarily on two factors, namely high electrical conductivity in oxidizing atmospheres and adequate electrocatalytic activity for the ORR:

- LSM (CTE about $11 \times 10^{-6}\,K^{-1}$) has an acceptable thermal expansion that matches with YSZ one (CTE $10-11 \times 10^{-6}\,K^{-1}$). However, when in contact with YSZ, LSM tends to react to form an insulating lanthanum zirconate ($La_2Zr_2O_7$) phase at high temperatures [229]. Therefore, exposure of LSM/YSZ interfaces is often limited to temperatures below 1150°C, and La-deficient LSM is frequently used to minimize zirconate formation.
- LSCF (CTE about $15 \times 10^{-6}\,K^{-1}$) has a higher coefficient of thermal expansion than YSZ but exhibits excellent catalytic activity for the oxygen reduction at reduced temperatures because of its mixed conducting properties. LSCF tends to react with YSZ at SOFC operating temperatures. One common approach is the use of ceria barrier layers to prevent interactions between LSCF cathodes and YSZ electrolytes [230,231].

For SOFC single cells having minimal ohmic resistance contributions from the components, cathode polarization is generally the major contribution to cell performance losses (Fig. 2.18) [232]. Thus, many cathode studies have been conducted and approaches have been developed to improve cathode performance [233]. One example is the use of infiltration as a potent means to

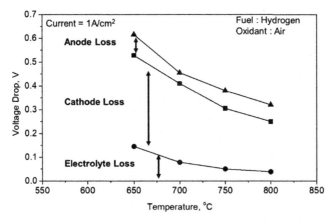

**FIGURE 2.18**  Contributions to SOFC cell performance loss. *SOFC*, solid oxide fuel cell.

form cathode nanostructures for electrode performance enhancement [234,235]. For example, infiltration of yttria-doped ceria into LSM/YSZ cathode increased peak power density from 208 to 519 mW cm$^{-2}$ at 700°C and power density at 0.7 V from 135 to 370 mW cm$^{-2}$ [234]. Fig. 2.19 shows an example of the increase in cell performance with infiltrated cathodes [236].

Infiltration of active components as dispersed particles or connected nanoparticulate networks to form nanostructures enhances cathode performance by modifying catalytic activities and/or conduction pathways of the electrode. The main issue is the stability of the nanostructure over extended periods of time at high operating temperatures. Operating the SOFC at reduced temperatures (e.g., <600°C) or stabilizing the nanostructure are potential approaches to maintain sufficient long-term stability [234,237].

*Interconnect*: The common interconnect materials for the SOFC are Sr-doped LaCrO$_3$ (LSCr) (mainly for operating temperatures of 900−1000°C) and SS (mainly for operating temperatures of ≤800°C). The particularly suitable properties of these materials for use as interconnects include electronic conductivity under fuel and oxidant atmospheres, stability in the FC environment and chemical and thermal compatibility with other cell components:

- LSCr has adequate electronic conductivity at 900−1000°C. This perovskite oxide is chemically stable in both reducing and oxidizing atmospheres. The material has insignificant chromium volatilization at the SOFC operating conditions. However, at higher temperatures, LSCr appreciably volatilizes chromium oxides in oxidizing atmospheres. Because of the high chromium volatilization, LSCr is difficult to sinter to high densities under oxidizing atmospheres. LSCr can have excess lanthanum that should be avoided, as it tends to precipitate as La$_2$O$_3$, resulting in hydroxide formation and subsequent disintegration of LSCr at room temperatures.

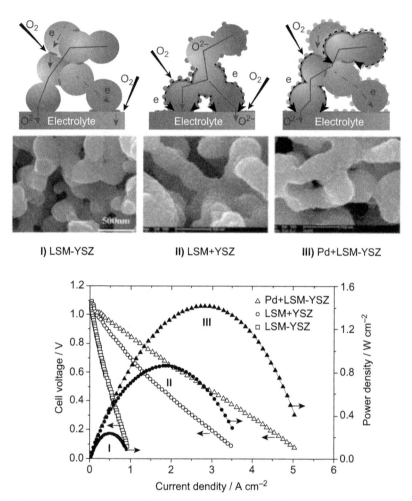

**FIGURE 2.19** Performance of SOFC single cells with LSM-YSZ cathodes at 750°C with hydrogen and air (I: mixed LSM-YSZ, II: YSZ infiltrated with LSM, III: mixed LSM-YSZ infiltrated with Pd) [236]. *LSM*, lanthanum strontium-doped manganite; *SOFC*, solid oxide fuel cell; *YSZ*, yttria-stabilized zirconia.

- SSs are commonly used for planar SOFCs operating at $\leq 800°C$ [238]. SSs form well-adhering surface chromium oxides having an acceptable electronic conductivity at SOFC operating temperatures and possess CTEs acceptably close to other components. Scale growth and adherence, creep strength, and embrittlement are several key factors that require careful consideration for SS interconnects in SOFCs.

An important performance issue with SOFCs with SS interconnects is chromium poisoning of the cathode in long-term operation. Chromium present

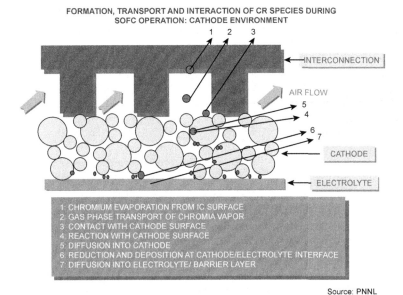

FORMATION, TRANSPORT AND INTERACTION OF CR SPECIES DURING
SOFC OPERATION: CATHODE ENVIRONMENT

1: CHROMIUM EVAPORATION FROM IC SURFACE
2: GAS PHASE TRANSPORT OF CHROMIA VAPOR
3: CONTACT WITH CATHODE SURFACE
4: REACTION WITH CATHODE SURFACE
5: DIFFUSION INTO CATHODE
6: REDUCTION AND DEPOSITION AT CATHODE/ELECTROLYTE INTERFACE
7: DIFFUSION INTO ELECTROLYTE/ BARRIER LAYER

Source: PNNL

**FIGURE 2.20** Potential steps in chromium poisoning of SOFC cathodes with metallic interconnects. *SOFC*, solid oxide fuel cell.

in the metallic interconnect can migrate to cathode reactive sites and interact with the cathode (Fig. 2.20), poisoning the electrode, thereby increasing cathode polarization with time. At present, the most common mitigating approach is to use conductive coatings (e.g., Co−Mn spinel) on the metallic interconnect to minimize the chromium transport and migration, thus reducing degradation rates [239]. Fig. 2.21 shows an example of improved performance degradation of a three-cell stack with Co−Mn spinel−coated interconnects [239,240].

## Single Cell

*Requirement*: An SOFC single cell designed for and fabricated from a selected set of materials must have desired physical, mechanical, chemical, electrical, and electrochemical properties and characteristics set by the operating requirements of the intended application. Table 2.10 summarizes the requirements for SOFC single cells [208]. The requirements given in this table are qualitative, as any specifics very much depend on selected materials and particular cell designs.

*Design*: An SOFC single cell can be configured into various shapes (tube, flat plate, or corrugated structure) because all the components are solid. Each configuration may have several different versions. For example, the tubular cell configuration can be a closed-one-end tube [241], a microtube [242] or a flattened tube [243] (Fig. 2.22). The flat plate or planar cell configuration can be a rectangular plate, a circular disk, or several other shapes [244] (Fig. 2.23).

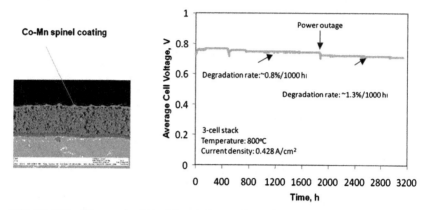

**FIGURE 2.21** Micrograph of Co—Mn spinel—coated metallic interconnect (left) and performance degradation of three-cell stack with anode-supported SOFCs and anode-coated stainless steel interconnects (right). *SOFC*, solid oxide fuel cell.

**TABLE 2.10** Requirements for Solid Oxide Fuel Cell Single Cells

| Requirements | Property Requirements | Targets |
|---|---|---|
| Physical | Shape, dimension, density/porosity | Required shape and dimensional tolerances for cells and cell components<br>Fully dense electrolyte without pinholes and sufficiently porous electrodes |
| Mechanical | Strength and toughness | Adequate strength and toughness for handling<br>Good bonding between components<br>Design to accommodate thermal expansion mismatches |
| Chemical | Chemical stability | Insignificant deleterious chemical interaction, elemental migration/interdiffusion, and phase transformation within cell during fabrication and operation |
| Electrical | Ohmic loss | Cell components as thin as feasible |
| Electrochemical | Open circuit voltage and polarization loss | Insignificant gas cross-leakage and no electrical short in electrolyte<br>Electrode microstructure to provide appropriate and sufficient reactive sites and fast gas diffusion to and from reactive sites |

**Closed-one-end tubular cells**     **Micro-tubular cells**        **Flattened tubular cells**

**FIGURE 2.22**   Tubular cell configurations.

**FIGURE 2.23**   Examples of planar cell configurations.

Various options are available for designing SOFCs. Single cells can be configured using different components for structural support. Thus, cells can be self-supporting or external supporting (built on a separate support material). In the cell-supporting design, one of the cell components, often thickest layer, acts as the cell structural support. Thus, single cells can be designed as electrolyte-supported, anode-supported, or cathode-supported. In the external supporting design, the single cell is configured as thin layers on the interconnect (interconnect-supported) or a substrate (substrate-supported). Fig. 2.24 shows examples of several cell designs [208].

*Fabrication*: Fabrication of SOFC single cells generally involves various process steps to incorporate starting materials into the desired cell configuration.

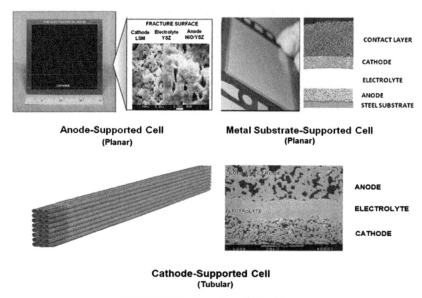

**Anode-Supported Cell**
**(Planar)**

**Metal Substrate-Supported Cell**
**(Planar)**

**Cathode-Supported Cell**
**(Tubular)**

**FIGURE 2.24**   Examples of cell designs.

The selection of processes and process sequences for cell fabrication thus depends on cell materials and cell designs. Any selected process should produce single cells that meet the design specifications and the requirements given in Table 2.10. Considerations should be given to process capability to manage possible mismatches in coefficient of thermal expansion and potential chemical interactions between different cell materials. Other considerations for practical applications include process scalability and fabrication cost. The general issues in cell fabrication mainly relate to cell component microstructures (pinholes in electrolyte, electrode porous structures), thickness uniformity, adhesion/bonding, cell flatness/shape tolerances, and microcracking.

The key step in any selected process sequence is the fabrication of fully dense electrolytes. Fabrication processes used to produce dense oxide electrolytes for the SOFC can be classified into two groups based on the fabrication approach: the particulate approach and the deposition approach. The particulate approach involves compaction of powders into layers and densification at elevated temperatures. Examples of the particulate process include tape casting, tape calendering, screen printing, slurry coating, and electrophoretic deposition. The deposition process involves formation of layers on a substrate by a chemical or physical process. Examples of the deposition approach include chemical vapor deposition/electrochemical vapor deposition, sputtering, pulsed laser deposition, and plasma spraying. A wide range of fabrication processes have been considered and developed for the SOFC [207,208,245]. Fig. 2.25 shows examples of the microstructure of

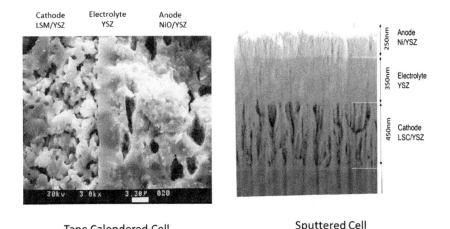

| Cathode | Electrolyte | Anode |
|---------|-------------|-------|
| LSM/YSZ | YSZ | NiO/YSZ |

Anode
Ni/YSZ — 250nm

Electrolyte
YSZ — 350nm

Cathode
LSC/YSZ — 450nm

Tape Calendered Cell

Sputtered Cell

**FIGURE 2.25**    Examples of cell microstructure [246,247].

an anode-supported cell made by tape calendering (a particulate method) [246] and that of a thin-film cell by sputtering [247].

*Performance*: SOFC single cells have exhibited peak power densities as high as 2 W cm$^{-2}$ (with pure hydrogen fuel and air oxidant, low fuel and air utilizations) and operated at temperatures as low as 400–500°C, depending on electrolyte/electrode systems and cell structures and configurations [217,248,249]. An example of SOFC single cell performance is given in Fig. 2.26 [250]. A power density of about 0.4 W cm$^{-2}$ at 0.82 V is achieved at 650°C.

Cell performance generally increases with increasing temperatures and at a specified temperature varies depending on a number of operating parameters such as fuel and oxidant compositions, utilizations, and pressures. Performance of SOFCs is considerably higher when operated in oxygen instead of air [251]. Operation at higher fuel utilizations reduces cell power output. For practical applications, cells must be designed to support high utilizations; fuel utilizations up to 90% have been demonstrated [252]. Pressurized operation, on the other hand, improves cell performance; testing up to 15 atm has been conducted [253]. Significant performance enhancement has been observed as pressure increases from 1 to 3 atm [240] and moderate improvement from 4 to 10 atm. When properly designed, SOFCs have also been shown to have excellent thermal cyclability, but long startup and shutdown times are often required to prevent thermal shock.

SOFC single cells, when properly prepared with conventional high-purity materials and operated on clean fuels and air, show minimal performance degradation for extended periods of time. For example, tubular cells were electrically tested for times as long as 8 years and showed satisfactory performance with less than 0.1% per 1000 h degradation [241]. SOFCs, however,

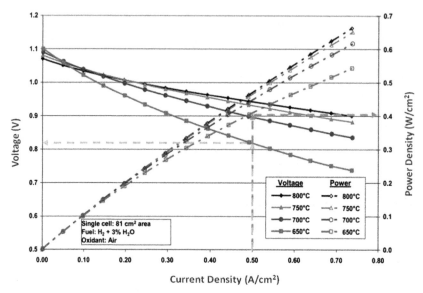

**FIGURE 2.26** Example of SOFC single cell (anode-supported) performance [250]. *SOFC*, solid oxide fuel cell.

can experience significant performance degradation in realistic environments, depending on several factors such as gas input purities and component materials used in the stack/system [254]. Sulfur is the most prevalent fuel impurity in many practical fuels, and its poisoning effects on the Ni/YSZ anode are well known [207]. It has also been shown that silicon impurities present in the fuel (originated from, for example, stack glass sealants or silica-containing insulations in the system) can also poison the Ni/YSZ anode [255]. On the cathode side, the presence of significant amounts of water or carbon dioxide in air can have deleterious effects on cell performance [256–258]. In long-term operation of cells in contact with metallic interconnects, cell performance can degrade due to chromium poisoning as discussed in the Interconnect section.

## Multicell Stack

*Requirement:* An SOFC stack formed from selected single cells and interconnects must be designed to have the desired electrical and electrochemical performance along with required chemical stability, thermal management, gas flow distribution, pressure drop, and mechanical/structural integrity to meet operating requirements of specified power generation applications. Table 2.11 summarizes the requirement for SOFC multicell stacks [208]. The requirements in Table 2.11 are qualitative because the specific requirements depend on specific materials and stack designs.

**TABLE 2.11** Requirements for Solid Oxide Fuel Cell Multicell Stacks

| Requirements | Property Requirement | Targets |
|---|---|---|
| Electrical performance | Minimal ohmic loss | Short current path<br>Good electrical contact and sufficient contact area<br>Current collector design for uniform and short current path |
| Electrochemical performance | Full OCV, minimal cell-to-cell performance variation | Insignificant gas leakage or cross-leakage<br>No electrical short<br>Design for uniform cell-to-cell performance |
| Chemical stability | Required chemical stability | No detrimental chemical interaction and elemental interdiffusion between cell and interconnect and between cell and sealant |
| Thermal management | Required cooling, uniform stack temperature distribution, and highest possible temperature gradient across stack | Simple and efficient means for cooling<br>Appropriate gas flow configuration<br>Design to permit highest temperature gradient |
| Gas flow distribution and pressure drop | Uniform gas distribution and reduced pressure drop | Design for uniform distribution of fuel and oxidant across cell area and to each cell<br>Design to reduce stack pressure drop |
| Mechanical/ structural integrity | Adequate mechanical strength | Design for minimal mechanical and thermal stress |

OCV, open circuit voltage.

*Design*: Four main stack designs have been proposed and developed for the SOFC: the tubular design, the segmented-cells-in-series design, the monolithic design, and the planar (flat plate) design [206–208]. The designs differ in the extent of dissipative losses within the cells, in the manner of fuel and oxidant channels and in the cell-to-cell connection in a stack of cells. Each design may have several different versions. Fig. 2.27 shows, as an example, the schematic diagrams of the various SOFC stack designs. Photographs of several stacks of different designs [253,259–261] are given in Fig. 2.28.

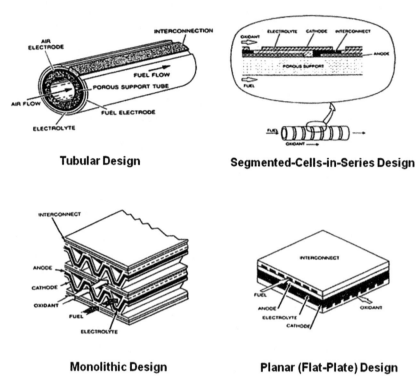

**FIGURE 2.27**    SOFC multicell stack designs performance. *SOFC*, solid oxide fuel cell.

Three important features of the SOFC stack designs relate to gas flow configuration, electrical connection, and gas manifolding, which can be arranged in several ways:

1. *Gas flow configurations*: Fuel and oxidant flows in SOFC stacks can be arranged to be cross-flow, coflow, or counterflow. The selection of a particular flow configuration depends on the stack design and has significant effects on temperature and current distribution within the stack. Various flow patterns can be implemented in the different flow configurations, especially for the planar SOFC, including Z-flow, serpentine, radial, and spiral patterns (Fig. 2.29).

2. *Electrical connection*: In a stack, single cells are normally connected in series electrical connection via interconnects to build voltage. A stack design can also include parallel electrical connection [262–264]. The series and parallel electrical connection can be used to protect stack against complete failure if an individual cell fails.

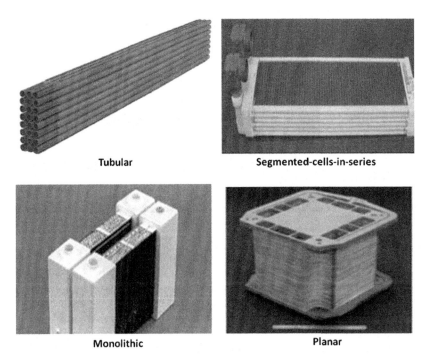

**FIGURE 2.28**  Examples of SOFC multicell stacks of different designs [254,261,263]. *SOFC, solid oxide fuel cell.*

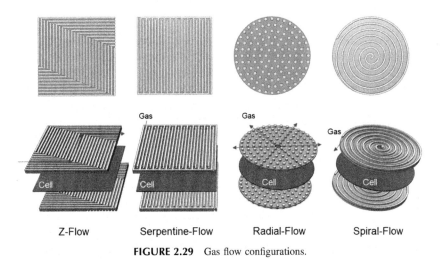

**FIGURE 2.29**  Gas flow configurations.

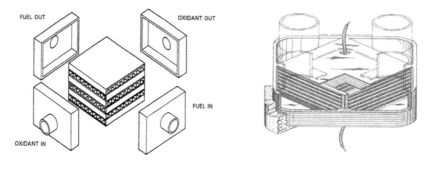

External Gas Manifold    Integral Gas Manifold

**FIGURE 2.30** Gas manifold designs [265,266].

**3.** *Gas manifolding*: Any stack design must include gas manifolds for routing gases from a common supply point to each cell and removing unreacted gases and reaction products. Gas manifolds can be classified as external or integral. External manifolds are constructed separately from the cell or interconnect component of the stack. Examples of external and integral manifold designs [265,266] are given in Fig. 2.30. Depending on the design, gas manifolds often require sealing to prevent gas leakage or crossover (although several sealless designs have been proposed [267,268]). The gas manifold seal is insulating to preclude cell-to-cell electrical shorts. In principle, the manifold must be designed to have low pressure drop (relative to individual cell pressure drop) to provide uniform flow distribution to the stack.

*Performance*: SOFC multicell stacks have demonstrated electrochemical performance under operating conditions appropriate for practical uses, including operation with internal reforming and on gas products directly from external reformers. For example, a 96-cell planar stack shows a power density of about 0.3 W cm$^{-2}$ (voltage of about 0.82 V per cell at 0.364 A cm$^{-2}$), 715°C on air (air utilization Ua of 15%), and fuel containing 14.5% NG (fuel utilization Uf of 68%) (Fig. 2.31) [269].

Multicell stack performance is dependent on performance of individual cells in the stack. Thus, one important characteristic for stable and reliable stack performance is minimal cell-to-cell variation within the stack. Cell-to-cell variations in the stack can be due to cell performance reproducibility, stack design properties, and other conditions during operation. For example, end effects can be seen in multicell stacks because of temperature distributions within the stack. In principle, minimal cell-to-cell variations can be achieved via design and process control.

In SOFC stacks, especially planar stacks with metallic interconnects, contact resistance between the electrodes, especially the cathode, and the

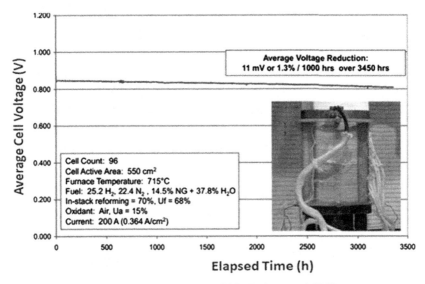

**FIGURE 2.31**    Performance of 96-cell planar stack [269].

metallic interconnect is a major factor in stack performance losses (Fig. 2.32) [270] and long-term performance degradation (discussed below). Conductive contact pastes have been used in planar stacks to minimize contact resistance; however, stability of such contact pastes over long duration is questionable.

An important performance degradation mode is chromium poisoning of the cathode in long-term operation of planar stacks having metallic interconnects as discussed earlier. As mentioned above, another potential degradation mode

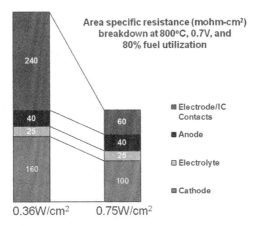

**FIGURE 2.32**    Area specific resistance breakdown (planar stack, anode-supported cells, metallic interconnects) [270].

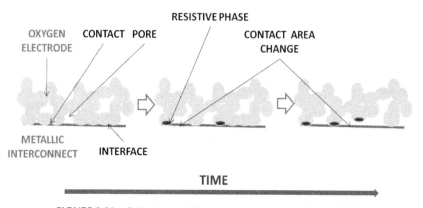

**FIGURE 2.33** Cathode–metallic interconnect contact evolution [271].

in multicell stacks with metallic interconnects is cathode–interconnect contact resistance changes. The contact between the ceramic cathode and the metallic interconnect tends to change because of thermodynamic driving forces (cathode ceramic particles tend to sinter/agglomerate to minimize their surface free energy, thus moving away the metallic interconnect) and other operating characteristics such as temperature distribution and thermal expansion mismatch as operation proceeds. These factors can lead to degradation in long-term operation. It is highly possible that during long-term operation, chemical interaction develops and electrical contact between the cathode and the interconnect evolves, ohmic resistance increases, and contact area reduces, resulting in higher ohmic losses and thus degradation (Fig. 2.33) [271].

SOFC stacks have been operated for tens of thousands of hours, and durability has been demonstrated with low performance degradation rates under specified operating conditions. For example, a short planar SOFC stack (with uncoated metallic interconnects) has been in operation at 700°C for more than 5 years with the overall voltage degradation of about 1% per 1000 h [272]. With coated metallic interconnects, a stack has been tested for more than 14,000 h with a reduced degradation rate of about 0.12% per 1000 h [272]. Fig. 2.31 is an example of performance (with internal reforming) of a 96-cell stack showing 1.3% voltage degradation per 1000 h [269].

## System

An SOFC system is a fully functional unit integrating SOFC stacks and other required components for specified power generation applications [273,274]. The development of an appropriate architecture for an SOFC power system involves design, analysis, and optimization to define system configurations and specify system components and operating parameters that meet the requirements of the target application.

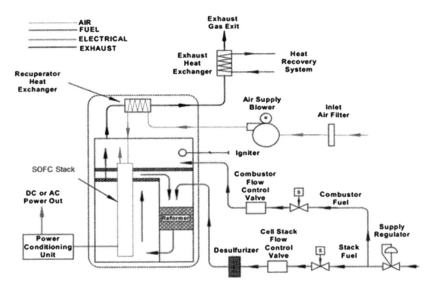

**FIGURE 2.34**   Schematic of a 5-kW SOFC power system [275]. *SOFC*, solid oxide fuel cell.

*System component*: An SOFC power system consists of FC stack(s) and all required components for a fully functional, stand-alone unit. Fig. 2.34 is an example of a simplified schematic depicting the configuration of a complete 5-kW SOFC system [275]. A photograph of an SOFC system showing the stack and other supporting components is given in Fig. 2.35 [276].

In general, system components can be grouped into different subsystems depending on their roles within the system: power generation; fuel processing; fuel, oxidant, and water delivery; thermal management; power conditioning; and control. The components for these subsystems except the power generation are together referred to as BOP.

- *Power generation subsystem*: For the majority of SOFC power systems, the SOFC stack is the only component that produces electricity, thus constituting the power generation subsystem. In this case, the power generation subsystem consists of mainly SOFC stack(s). For certain designs, the power generation subsystem contains additional equipment such as gas turbine (GT) and/or steam turbine (ST) generators.
- *Fuel processing subsystem*: The fuel processing subsystem is used to precondition the fuel (other than hydrogen) before it enters the FC stack. This subsystem consists of a reformer (if the system design includes external reformation) and other equipment for fuel cleanup (e.g., acid gas removal, particulate removal) if any.
- *Fuel, oxidant, and water delivery subsystem*: This subsystem mainly consists of blowers/compressors and valves/orifices to deliver required reactants to the fuel reformer and fuel and oxidant to the SOFC stack.

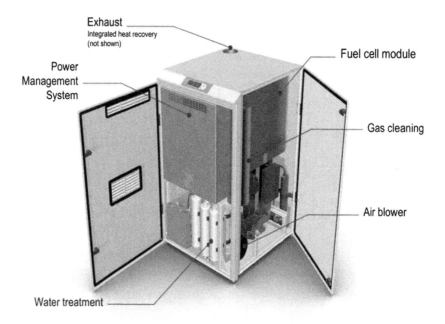

Exhaust
Integrated heat recovery
(not shown)

Power
Management
System

Fuel cell module

Gas cleaning

Air blower

Water treatment

**FIGURE 2.35**  A 2-kW SOFC power system [276]. *SOFC*, solid oxide fuel cell.

- *Thermal management subsystem*: The subsystem consists of a number of heat exchangers/recuperators (including steam generators) and combustors/burners to maintain the SOFC stack temperature at the required level and to control heat supply/removal for efficient operation of the reformer and other cleanup equipment. Insulation is also an important element in thermal management of the system to contain heat losses.
- *Power conditioning subsystem*: The power conditioning subsystem converts variable DC from the FC to regulated DC or AC power appropriate for the application. Depending on the particular application, this subsystem may consist of power electronics (DC−DC converters, DC−AC inverters) and transformers.
- *Control subsystem*: This subsystem is a controller including control software that provides for system startup, shutdown, and normal operation while maintaining the system within its operating constraints when subjected to load changes or disturbances. Sensors are also an important element in the control of the system.

*Type of SOFC power system*: Current SOFC power systems can be divided into two general classes depending on their power generation cycles: simple cycle systems and hybrid cycle systems.

- *Simple cycle SOFC system*: The SOFC is the only power-generating component for a simple cycle system. The simple cycle is the common and preferred configuration for power systems with outputs ranging from tens of watts to hundreds of kilowatts; however, it is also considered for megawatt-size power plants [277]. In simple cycle SOFC systems, the FC typically operates under atmospheric pressure.
- *Hybrid cycle SOFC system*: The SOFC can be combined with another power-generating equipment such as a heat engine [278,279] or a battery [280,281] to form a hybrid power system. Hybridization of the SOFC with a heat engine is the most common hybrid cycle. In a typical hybrid combination with a heat engine, the heat energy of the FC exhaust is used by the heat engine to generate additional electricity. The heat engine can be a GT, a ST, or a combination of heat engines such as a combined cycle and integrated gasification combined cycle [282,283] with SOFC/GT hybrids the most common. An SOFC/GT hybrid system can be based on the direct fired design (in this design, the SOFC operates under pressure) or the indirect fired design (in this design, the SOFC operates under ambient pressure).

*Design*: The focus in designing a FC power system is to develop and optimize the system configuration to meet the specifications of its intended application. These specifications could include the following: cycle efficiency, duty cycle, fuel specification, cost (purchase and installation), reliability, maintenance, size and weight, environmental interfaces, cogeneration, acoustic noise, power quality, and safety. In general, SOFC systems are designed to fulfill the key requirements for practical/commercial products, namely, performance, reliability, and cost. Thus, the design aims at establishing system configurations and defining system components, including component specifications, component performance characteristics, and effects of component and process variables on system performance and reliability. The detailed system design can be used to estimate/determine system costs.

In terms of SOFC stacks, several stack design parameters and operating variables are critical in the design of the system (to meet the performance, reliability, and cost targets of the intended application). These design parameters and operating variables include (1) the number of stacks and stack arrangements, (2) stack voltage and current density, (3) fuel utilization, (4) stack pressure drop, and (5) stack temperature gradient.

*Market and application*: SOFC power systems being developed range from watt-sized devices to multimegawatt power plants, and practical applications being considered cover all market sectors (portable/mobile, transportation, and stationary) (Table 2.12). Many of the applications for SOFC power systems have progressed to hardware demonstration and prototype/precommercial stages, while several applications, especially those with large power outputs, are at the conceptual/design stage (Fig. 2.36).

**TABLE 2.12** Solid Oxide Fuel Cell Power System Applications

| Market Sector | Applications | Power Size | Status |
|---|---|---|---|
| Portable | Power for consumer electronic devices, mobile portable power | 1–100 W | Demonstration, precommercial |
| | Portable power, battery chargers | 200–500 W | Demonstration |
| Transportation | Automobile and truck auxiliary power unit (APU) | 5–50 kW | Demonstration |
| | Aircraft APU | Upto 500 kW | Concept |
| Stationary | Residential/microcombined heat and power (CHP), uninterruptible power | 1–10 kW | Prototype, precommercial |
| | CHP and distributed generation | 100 kW–1 MW | Demonstration and concept |
| | Base load | 100–500 MW | Concept |

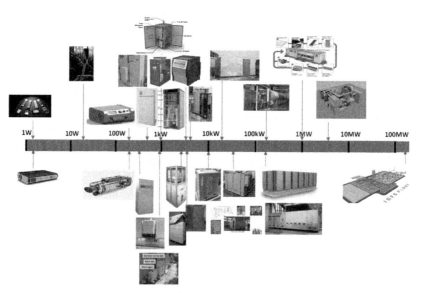

**FIGURE 2.36** Selected SOFC power systems (hardware/prototype and concept). *SOFC*, solid oxide fuel cell.

## Solid Oxide Electrolysis Cell Technology

An SOEC is an SOFC operated in reverse mode. An SOEC thus functions as an electrolyzer, producing, for example, $H_2$ from $H_2O$ inputs when coupled with an energy source (fossil, nuclear, renewable). In addition to hydrogen production from $H_2O$ [284], the SOEC can also be used to produce other chemicals such as syngas from mixtures of $H_2O$ and $CO_2$ [285] and oxygen from $CO_2$ [286]. The SOEC is the only electrolysis cell having this capability. When a cell operates efficiently in both SOFC and SOEC modes, it is referred to as reversible SOFC (RSOFC) [209].

The SOEC operating at high temperatures has the advantage that the electrical energy required for the electrolysis decreases as temperature increases and the unavoidable joule heat is used in the splitting process. Thus, the SOEC can work under the so-called thermoneutral condition; i.e., at the thermoneutral voltage ($V_{tn}$), the electricity input exactly matches the total energy demand of the electrolysis reaction. In this case, the electrical-to-chemical conversion efficiency is 100%. At cell operating voltages $<V_{tn}$, heat must be supplied to the system to maintain the temperature and the electrical-to-chemical conversion efficiency is above 100%. At cell operating voltages $>V_{tn}$, heat must be removed from the system and the efficiency is below 100%.

### *Component*

Because the SOEC is the SOFC operated in reverse mode, the SOEC being developed is typically derived from the more technologically advanced SOFC. Thus, current SOECs, usually based on the planar electrode-supported cell configuration, typically operate at 800°C. The materials for the SOEC are those commonly used in the SOFC, e.g., for single cells, YSZ for the electrolyte, perovskites such as LSM and LSCF for the oxygen electrode, Ni/YSZ cermet for the hydrogen electrode and for stacks, and SSs for the interconnect.

In general, the hydrogen electrode shows performance reversibility (symmetry) between FC and electrolysis modes. The oxygen electrode, on the other hand, may exhibit irreversibility at higher current densities in electrolysis mode. This property, however, depends on a number of factors such as electrode microstructure, material, and operating parameter. Fig. 2.37 shows, as an example, voltage—current density curves of cells with different oxygen electrodes exhibiting irreversibility at higher current densities [287]. This irreversibility becomes more pronounced with predominantly electronic conducting oxides (e.g., LSM) as compared with mixed ionic electronic conducting oxides (e.g., LSCF). The higher applied current density at the irreversibility onset of mixed conducting oxides can be explained by the difference in local current density owing to the spreading of triple phase boundary active sites on the mixed conducting surface (Fig. 2.38). In general, SOEC components derived from those developed for FC operation can operate stably in electrolysis mode with no or minor modifications. However, an

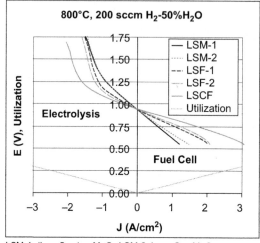

LSM-1: $(La_{0.8}Sr_{0.2})_{0.95}MnO_3$ LSM-2: $La_{0.65}Sr_{0.3}MnO_3$
LSF-1: $(La_{0.8}Sr_{0.2})_{0.95}FeO_{3-\delta}$ ,, LSF-2: $La_{0.65}Sr_{0.3}FeO_{3-\delta}$
LSCF: $(La_{0.6}Sr_{0.4})_{0.98}Fe_{0.8}Co_{0.2}O_{3-\delta}$

**FIGURE 2.37** Voltage–current density curves of cells with different oxygen electrodes [287].

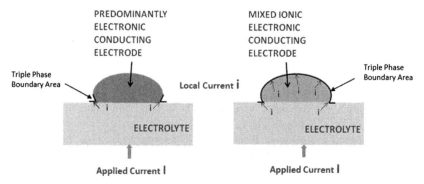

**FIGURE 2.38** Applied and local currents in predominantly electronic conducting and mixed conducting electrodes.

oxygen electrode that works stably in SOFC mode may experience rapid performance decay in electrolysis mode owing to electrode delamination caused by oxygen evolution at the electrode/electrolyte interface [288]. An example of oxygen electrode delamination is given in Fig. 2.39. This type of degradation has been observed for oxygen electrodes based on predominantly electronic conducting oxides (e.g., LSM) when not designed to minimize oxygen pressure built up at the interface during operation.

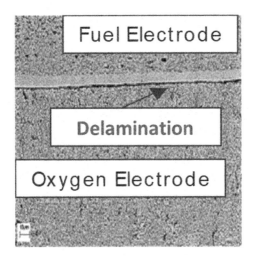

**FIGURE 2.39** Oxygen electrode delamination in SOEC. *SOEC*, solid oxide electrolysis cell.

## Single Cell

To date, SOEC single cells have been shown to have the ability to perform well for hydrogen production from steam. For example, a cell voltage of 1.1 V (below thermoneutral voltage $V_{tn}$) has been obtained for a Ni/YSZ-supported SOEC of 45 $cm^2$ active area at a current density of about 1.4 A $cm^{-2}$, 900°C, 93% $H_2O$ (balance $H_2$) [289]. Fig. 2.40 summarizes data on current densities at

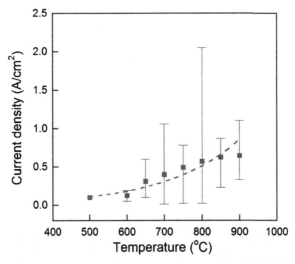

**FIGURE 2.40** Current densities at thermoneutral voltage reported for SOECs [271]. *SOEC*, solid oxide electrolysis cell.

the thermoneutral voltage of about 1.3 V reported in the literature for steam electrolysis between 500 and 900°C [271]. Current density variations (the bars in Fig. 2.40 show the range of values) for the same material systems are due to starting material characteristics, processing, absolute humidity input, and flow rates. At higher temperatures (>900°C), extraordinarily high current densities have been reported, e.g., about 3 A cm$^{-2}$ at 1.3 V, 950°C [285].

In addition to steam electrolysis, SOECs (based on oxygen ion—conducting electrolytes) are also suitable for electrolyzing $CO_2$ and coelectrolyzing $H_2O + CO_2$. Electrolysis of CO and $H_2O + CO_2$ has been considered to produce/recover oxygen for spacecraft life support and propulsion [290–293]. Coelectrolysis of $H_2O + CO_2$ can be used to produce syngas, an intermediate energy carrier that can be converted to a variety of practical chemicals/fuels [294,295]. SOECs have been shown to be capable of syngas production by inputs of $H_2O + CO_2$ at similar current density ranges to steam electrolysis [285,296–298] although the area specific resistance (ASR) for electrolysis of $CO_2$ is generally higher than that of $H_2O$ (Fig. 2.41) [297]. Performance of the $CO/CO_2$ electrode, however, can be improved by modifying electrode materials and structures [299]. The mechanism for the $CO_2$ reaction is not well determined and may depend on electrode microstructures. In the case of $H_2O + CO_2$, it is possible that only $H_2O$ is involved in the electrochemical reaction and the $CO_2$ in the mixture reacts with $H_2$ of the reaction products via a reverse water gas shift reaction.

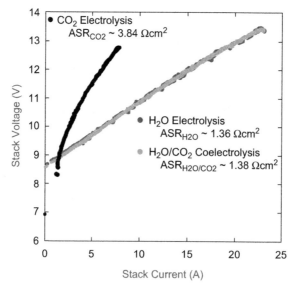

**FIGURE 2.41** Performance curves of 10-cell stack for electrolysis of $H_2O$, $H_2O + CO_2$, and $CO_2$ at 800°C [297]. *ASR*, area specific resistance.

*Stack*

SOEC multicell stacks for hydrogen production from steam have been built and operated in laboratory and system demonstration scales. To date, planar stacks as tall as 60-cell height have been achieved. Reasonable performance has been demonstrated. For example, ASRs of $<0.6\,\Omega\,cm^2$ have been obtained for five-cell planar stacks (100 $cm^2$ cell active area) at 800°C [300]. In general, hydrogen production rates have been found to be in good agreement with dew point measurements and rate predictions based on current (Fig. 2.42) [301].

Fig. 2.43 shows an example of a 10-cell RSOFC (SOFC + SOEC) stack and its performance (in term of ASR) in operation for more than 1000 h (operating alternately between internal reforming FC mode and steam electrolysis mode) [270]. This stack showed an initial power density of 480 mW $cm^{-2}$ at 0.7 V, 800°C with 64% $H_2$−35% $N_2$ as fuel and 80% fuel utilization in FC mode and produced about 6.3 standard liter per minute of hydrogen at 1.26 V cell voltage, 0.62 A $cm^{-2}$ with 30% $H_2$−70% $H_2O$ and steam utilization of about 54%.

SOEC stacks have been tested for thousands of hours, and performance degradation is typically in the order of 5%−10% per 1000 h [302,303]. A recent SOEC work reports low degradation ($<1\%$ per 1000 h) for electrolysis of $H_2O + CO_2$ at current densities below 0.7 A $cm^{-2}$ [304]. Long-term

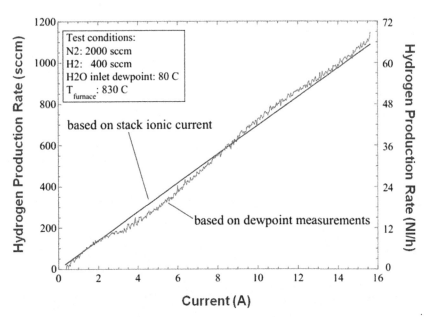

**FIGURE 2.42** Hydrogen production rates of 10-cell SOEC stack [301]. *SOEC*, solid oxide electrolysis cell.

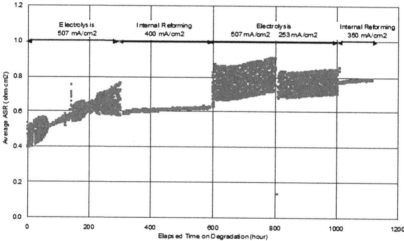

**FIGURE 2.43** Photograph of 10-cell stack (cell active area of 142 cm$^2$) and stack ASR in methane internal reforming fuel cell mode and steam electrolysis mode (fluctuations seen in steam electrolysis due to instability of steam generation and delivery) [270]. *ASR*, area specific resistance.

degradation of SOECs is generally higher than that obtained for similar cells in FC mode [305]. Root causes for this difference, however, are not fully understood.

## System and Application

The SOEC has been considered for hydrogen production from steam [289,306] at distributed plant (e.g., 1500 kg H$_2$/day) and central station (e.g., 150,000 kg H$_2$/day) sizes [288], syngas production for industrial uses [295], and oxygen

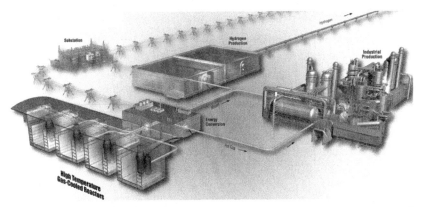

**FIGURE 2.44** Concept for large-scale centralized nuclear hydrogen production with SOECs [300]. *SOEC*, solid oxide electrolysis cell.

generation/recovery for space applications [307,308]. Integration of SOEC systems with nuclear [309] and renewable energy resources such as solar energy [310] has been envisioned. Fig. 2.44 shows a concept of an SOEC-based hydrogen production central system coupled with high-temperature gas-cooled nuclear reactors [300].

In terms of SOEC system integration and demonstration, a 15-kW laboratory facility has been fabricated and hydrogen production rates as high as 5.7 Nm$^3$/h have been achieved [311]. Most recently, a reversible SOFC demonstration system capable of storing 600 kWh of energy, produced in SOEC mode, and of generating 50 kW of power in SOFC mode has been installed and operated [312] (Fig. 2.45).

**FIGURE 2.45** RSOFC system [312]. *RSOFC*, reversible solid oxide fuel cell.

## CONCLUDING REMARKS

Fuel cells are a very promising technology for energy conversion, an area dominated by old technologies, such as the internal combustion engine and the gas turbine, which have to be replaced to allow decarbonization of the energy sector. The advantages of fuel cells over the incumbent technologies are numerous, for example, high energy efficiency, extremely low pollutant emissions, low noise levels, and modularity. During the years, several kinds of fuel cells have been devised, each characterized by peculiar features, in a way that the potential applications of the technology are nowadays extremely widespread, covering the whole range from portable devices to megawatt-sized power plants.

During the last decades, PEMFC technology has made enormous progress to reach commercialization, with the first fuel cell hydrogen-powered vehicles launched in the market in 2015. Performance and durability reached acceptable levels, making PEMFCs one of the most promising and sustainable zero-emission technologies available for energy conversion. However, improvements are still necessary to reduce conveniently the costs for consumers and to reach market competitiveness.

SOFC technology offers a number of attractive features in terms of cell/stack design, fabrication, and operating characteristics, such as design flexibility, multiple cell fabrication options, multifuel capability, and operating temperature choice. On the other hand, the technology also presents several technical challenges, such as limited material selection, increased undesirable chemical interaction/elemental interdiffusion between components and certain operation restrictions. The SOFC has made significant progress toward resolving those challenges, while demonstrating the features suitable for practical applications. To date, baseline cell, stack and system designs, and a set of appropriate materials and manufacturing processes have been developed and specified for the technology. Cell/stack/system performance under realistic operating conditions has reached the level suitable for widespread uses. However, several critical issues such as reliability and reproducibility and long-term performance degradation remain to be addressed. Technological advances in areas such as nanotechnology and digital manufacturing have been considered for incorporation in the SOFC. Improvements in SOFC efficiency and reliability along with cost reduction should move the technology toward widespread applications in the future. As for the SOEC technology, which is fundamentally derived from the SOFC, advancements in SOEC material, design, and operation are expected as SOFC technology progresses.

The diffusion of niche markets, the beginning of mass fabrication and the adoption of specific policies will leverage the industrial and commercial development of fuel cell technologies, leading to a steady growth of engineering applications.

# REFERENCES

[1] U. Bossel, The Birth of the Fuel Cell 1835–1845, European Fuel Cell Forum, 2000.

[2] A. Coralli, H.V. de Miranda, C.F.E. Monteiro, J.F.R. da Silva, P.E.V. de Miranda, Mathematical model for the analysis of structure and optimal operational parameters of a solid oxide fuel cell generator, J. Power Sources 269 (2014) 632–644.

[3] B.J.M. Sarruf, J.-E. Hong, R. Steinberger-Wilckens, P.E.V. de Miranda, $CeO_2$-$Co_3O_4$-CuO anode for direct utilisation of methane or ethanol in solid oxide fuel cells, Int. J. Hydrogen Energy 43 (2018) 6340–6351.

[4] The Fuel Cell Industry Review 2017, E4tech, December 2017.

[5] R. O'Hayre, S.-W. Cha, W.G. Colella, F.B. Prinz, Fuel Cell Fundamentals, Wiley, 2016.

[6] S. Revankar, P. Majmudar, Fuel Cells Principles, Design, and Analysis, CRC Press, 2014.

[7] I. Verhaert, S. Verhelst, H. Huisseune, I. Poels, G. Janssen, G. Mulder, M.D. Paepe, Thermal and electrical performance of an alkaline fuel cell, Appl. Thermal Eng. 40 (2012) 1359–4311.

[8] P.E.V. de Miranda, L.A.C. Bustamante, M. Cerveira, J.C. Bustamante, Pilhas a Combustível, in: Maurício T, Tolmasquim, Ciência Moderna, Rio de Janeiro (Eds.), Fontes Alternativas de Energia no Brasil, 2003.

[9] C. Bae, Status and Challenges of Hydroxide Ion-Conducting Polymers for Anion Exchange Membrane Applications, DOE AMFC Workshop 2016 (04/01/2016).

[10] R. Steinberger-Wilckens, W. Lehnert, Innovations in Fuel Cell Technologies, RSC Publishing, 2011.

[11] C. Wang, A.J. Appleby, D.L. Cocke, Alkaline fuel cell with intrinsic energy storage, J. Electrochem. Soc. 151 (2) (2004) A260–A264.

[12] P.-Y. Olu, N. Job, M. Chatenet, Evaluation of anode (electro)catalytic materials for the direct borohydride fuel cell: methods and benchmarks, J. Power Sources 327 (2016) 235e257.

[13] P. Sapkota, H. Kim, Zinc–air fuel cell, a potential candidate for alternative energy, J. Ind. Eng. Chem. 15 (2009) 445–450.

[14] A.L. Zhua, D.P. Wilkinson, X. Zhang, Y. Xing, A.G. Rozhine, S.A. Kulinich, Zinc regeneration in rechargeable zinc-air fuel cells—a review, J. Energy Storage 8 (2016) 35–50.

[15] H. Hirata, T. Aoki, K. Nakajima, Liquid phase migration effects on the evaporative and condensational dissipation of phosphoric acid in phosphoric acid fuel cell, J. Power Sources 199 (2012) 110–116.

[16] A.L. Dicks, Molten carbonate fuel cells, Curr. Opin. Solid State Mater. Sci. 8 (5) (2004) 379–383.

[17] C.R.I. Chisholm, D.A. Boysen, A.B. Papandrew, S. Zecevic, S. Cha, K.A. Sasaki, Á. Varga, K.P. Giapis, S.M. Haile, From laboratory breakthrough to technological realization: the development path for solid acid fuel cells, Electrochem. Soc. Interface 18 (3) (2009).

[18] N. Mohammad, A.B. Mohamad, A.A.H. Kadhum, K.S. Loh, A review on synthesis and characterization of solid acid materials for fuel cell applications, J. Power Sources 322 (1) (August 2016) 77–92.

[19] UltraCell and SAFCell demo first solid acid fuel cell using propane, Fuel Cells Bull. 2014 (10) (October 2014) 6.

[20] S. Pandit, D. Das, Principles of microbial fuel cell for the power generation, in: Microbial Fuel Cell - A Bioelectrochemical System that Converts Waste to Watts, Springer, 2018.

[21] H.R. Luckarift, P. Atanassov, G.R. Johnson, Enzymatic Fuel Cells - From Fundamentals to Applications, Wiley, 2014.

[22] D. Leech, P. Kavanagh, W. Schuhmann, Enzymatic fuel cells: recent progress, Electrochim. Acta 84 (2012) 223−234.

[23] S. Specchia, C. Francia, P. Spinelli, Polymer electrolyte membrane fuel cells, in: Electrochem. Technol. Energy Storage Convers, 2011, pp. 601−670. https://doi.org/10.1002/9783527639496.ch13.

[24] J.M. Andújar, F. Segura, Fuel cells: history and updating. A walk along two centuries, Renew. Sustain. Energy Rev. 13 (2009) 2309−2322. https://doi.org/10.1016/j.rser.2009.03.015.

[25] W.R. Grove, On a gaseous voltaic battery, Philos. Mag. 92 (2012) 3753−3756. https://doi.org/10.1080/14786435.2012.742293.

[26] G. Cacciola, V. Antonucci, S. Freni, Technology up date and new strategies on fuel cells, J. Power Sources 100 (2001) 67−79. https://doi.org/10.1016/S0378-7753(01)00884-9.

[27] W.G. Colella, M.Z. Jacobson, D.M. Golden, Switching to a U.S. hydrogen fuel cell vehicle fleet: the resultant change in emissions, energy use, and greenhouse gases, J. Power Sources 150 (2005) 150−181. https://doi.org/10.1016/j.jpowsour.2005.05.092.

[28] O. Erdinc, M. Uzunoglu, Recent trends in PEM fuel cell-powered hybrid systems: investigation of application areas, design architectures and energy management approaches, Renew. Sustain. Energy Rev. 14 (2010) 2874−2884. https://doi.org/10.1016/j.rser.2010.07.060.

[29] A.S. Patil, T.G. Dubois, N. Sifer, E. Bostic, K. Gardner, M. Quah, C. Bolton, Portable fuel cell systems for America's army: technology transition to the field, J. Power Sources 136 (2004) 220−225. https://doi.org/10.1016/j.jpowsour.2004.03.009.

[30] M. Oszcipok, M. Zedda, J. Hesselmann, M. Huppmann, M. Wodrich, M. Junghardt, C. Hebling, Portable proton exchange membrane fuel-cell systems for outdoor applications, J. Power Sources 157 (2006) 666−673. https://doi.org/10.1016/j.jpowsour.2006.01.005.

[31] T. Wilberforce, A. Alaswad, A. Palumbo, M. Dassisti, A.G. Olabi, Advances in stationary and portable fuel cell applications, Int. J. Hydrogen Energy 41 (2016) 16509−16522. https://doi.org/10.1016/j.ijhydene.2016.02.057.

[32] I. Verhaert, G. Mulder, M. De Paepe, Evaluation of an alkaline fuel cell system as a micro-CHP, Energy Convers. Manag. 126 (2016) 434−445. https://doi.org/10.1016/j.enconman.2016.07.058.

[33] Y. Liu, W. Lehnert, H. Janßen, R.C. Samsun, D. Stolten, A review of high-temperature polymer electrolyte membrane fuel-cell (HT-PEMFC)-based auxiliary power units for diesel-powered road vehicles, J. Power Sources 311 (2016) 91−102. https://doi.org/10.1016/j.jpowsour.2016.02.033.

[34] D.J. Kim, M.J. Jo, S.Y. Nam, A review of polymer-nanocomposite electrolyte membranes for fuel cell application, J. Ind. Eng. Chem. 21 (2015) 36−52. https://doi.org/10.1016/j.jiec.2014.04.030.

[35] Y. Zhan, Y. Guo, J. Zhu, H. Wang, Intelligent uninterruptible power supply system with back-up fuel cell/battery hybrid power source, J. Power Sources 179 (2008) 745−753. https://doi.org/10.1016/j.jpowsour.2007.12.113.

[36] G. Squadrito, G. Giacoppo, O. Barbera, F. Urbani, E. Passalacqua, L. Borello, A. Musso, I. Rosso, Design and development of a 7 kW polymer electrolyte membrane fuel cell stack for UPS application, Int. J. Hydrogen Energy 35 (2010) 9983−9989. https://doi.org/10.1016/j.ijhydene.2009.11.019.

[37] A.K. Shukla, A.S. Aricò, V. Antonucci, An appraisal of electric automobile power sources, Renew. Sustain. Energy Rev. 5 (2001) 137−155. https://doi.org/10.1016/S1364-0321(00)00011-3.

[38] A. Alaswad, A. Baroutaji, H. Achour, J. Carton, A. Al Makky, A.G. Olabi, Developments in fuel cell technologies in the transport sector, Int. J. Hydrogen Energy 41 (2016) 16499−16508. https://doi.org/10.1016/j.ijhydene.2016.03.164.

[39] L. van Biert, M. Godjevac, K. Visser, P.V. Aravind, A review of fuel cell systems for maritime applications, J. Power Sources 327 (2016) 345−364. https://doi.org/10.1016/j.jpowsour.2016.07.007.

[40] T. Larriba, R. Garde, M. Santarelli, Fuel cell early markets: techno-economic feasibility study of PEMFC-based drivetrains in materials handling vehicles, Int. J. Hydrogen Energy 38 (2013) 2009−2019. https://doi.org/10.1016/j.ijhydene.2012.11.048.

[41] T. Hordé, P. Achard, R. Metkemeijer, PEMFC application for aviation: experimental and numerical study of sensitivity to altitude, Int. J. Hydrogen Energy 37 (2012) 10818−10829. https://doi.org/10.1016/j.ijhydene.2012.04.085.

[42] S. Bégot, F. Harel, D. Candusso, X. François, M.C. Péra, S. Yde-Andersen, Fuel cell climatic tests designed for new configured aircraft application, Energy Convers. Manag. 51 (2010) 1522−1535. https://doi.org/10.1016/j.enconman.2010.02.011.

[43] M.T. Gencoglu, Z. Ural, Design of a PEM fuel cell system for residential application, Int. J. Hydrogen Energy 34 (2009) 5242−5248. https://doi.org/10.1016/j.ijhydene.2008.09.038.

[44] Y. Nagata, Quantitative analysis of $CO_2$ emissions reductions through introduction of stationary-type PEM-FC systems in Japan, Energy 30 (2005) 2636−2653. https://doi.org/10.1016/j.energy.2004.07.009.

[45] L. Barelli, G. Bidini, F. Gallorini, A. Ottaviano, An energetic-exergetic comparison between PEMFC and SOFC-based micro-CHP systems, Int. J. Hydrogen Energy 36 (2011) 3206−3214. https://doi.org/10.1016/j.ijhydene.2010.11.079.

[46] R. Napoli, M. Gandiglio, A. Lanzini, M. Santarelli, Techno-economic analysis of PEMFC and SOFC micro-CHP fuel cell systems for the residential sector, Energy Build 103 (2015) 131−146. https://doi.org/10.1016/j.enbuild.2015.06.052.

[47] Nuvera https://www.nuvera.com/ (accessed February 2018).

[48] Toyota Mirai https://ssl.toyota.com/mirai/fcv.html (accessed February 2018).

[49] Ballard Fuel Cells http://www.ballard.com/markets (accessed February 2018).

[50] Tucson Fuel Cells http://www.hyundaihydrogen.ca/ (accessed February 2018).

[51] Solid Power http://www.solidpower.com/it/ (accessed February 2018).

[52] G. Álvarez, F. Alcaide, P.L. Cabot, M.J. Lázaro, E. Pastor, J. Solla-Gullón, Electrochemical performance of low temperature PEMFC with surface tailored carbon nanofibers as catalyst support, Int. J. Hydrogen Energy 37 (2012) 393−404. https://doi.org/10.1016/j.ijhydene.2011.09.055.

[53] E. Passalacqua, F. Lufrano, G. Squadrito, A. Patti, L. Giorgi, Nafion content in the catalyst layer of polymer electrolyte fuel cells: effects on structure and performance, Electrochim. Acta 46 (2001) 799−805. https://doi.org/10.1016/S0013-4686(00)00679-4.

[54] M.K. Cho, H.-Y. Park, S.Y. Lee, B.-S. Lee, H.-J. Kim, D. Henkensmeier, S.J. Yoo, J.Y. Kim, J. Han, H.S. Park, Y.-E. Sung, J.H. Jang, Effect of catalyst layer ionomer content on performance of intermediate temperature proton exchange membrane fuel cells (IT-PEMFCs) under reduced humidity conditions, Electrochim. Acta 224 (2017) 228−234. https://doi.org/10.1016/j.electacta.2016.12.009.

[55] K. Angjeli, I. Nicotera, M. Baikousi, A. Enotiadis, D. Gournis, A. Saccà, E. Passalacqua, A. Carbone, Investigation of layered double hydroxide (LDH) Nafion-based nanocomposite membranes for high temperature PEFCs, Energy Convers. Manag. 96 (2015) 39−46. https://doi.org/10.1016/j.enconman.2015.02.064.

[56] Q. Li, R. He, J.O. Jensen, N.J. Bjerrum, Approaches and recent development of polymer electrolyte membranes for fuel cells operating above 100°C, Chem. Mater. 15 (2003) 4896−4915. https://doi.org/10.1021/cm0310519.

[57] R.E. Rosli, A.B. Sulong, W.R.W. Daud, M.A. Zulkifley, T. Husaini, M.I. Rosli, E.H. Majlan, M.A. Haque, A review of high-temperature proton exchange membrane fuel cell (HT-PEMFC) system, Int. J. Hydrogen Energy 42 (2017) 9293–9314. https://doi.org/10.1016/j.ijhydene.2016.06.211.

[58] H. Zhang, P.K. Shen, Recent development of polymer electrolyte membranes for fuel cells, Chem. Rev. 112 (2012) 2780–2832. https://doi.org/10.1021/cr200035s.

[59] M.M. Nasef, Radiation-grafted membranes for polymer electrolyte fuel cells: current trends and future directions, Chem. Rev. 114 (2014) 12278–12329. https://doi.org/10.1021/cr4005499.

[60] C. Huang, P.Y.K. Choi, K. Nandakumar, L.W. Kostiuk, Investigation of entrance and exit effects on liquid transport through a cylindrical nanopore, Phys. Chem. Chem. Phys. 10 (2008) 186–192. https://doi.org/10.1039/B709575A.

[61] N.N. Krishnan, D. Joseph, N.M.H. Duong, A. Konovalova, J.H. Jang, H.J. Kim, S.W. Nam, D. Henkensmeier, Phosphoric acid doped crosslinked polybenzimidazole (PBI-OO) blend membranes for high temperature polymer electrolyte fuel cells, J. Memb. Sci. 544 (2017) 416–424. https://doi.org/10.1016/j.memsci.2017.09.049.

[62] Q.F. Li, H.C. Rudbeck, A. Chromik, J.O. Jensen, C. Pan, T. Steenberg, M. Calverley, N.J. Bjerrum, J. Kerres, Properties, degradation and high temperature fuel cell test of different types of PBI and PBI blend membranes, J. Memb. Sci. 347 (2010) 260–270. https://doi.org/10.1016/j.memsci.2009.10.032.

[63] F. Celso, S.D. Mikhailenko, S. Kaliaguine, U.L. Duarte, R.S. Mauler, A.S. Gomes, SPEEK based composite PEMs containing tungstophosphoric acid and modified with benzimidazole derivatives, J. Memb. Sci. 336 (2009) 118–127. https://doi.org/10.1016/j.memsci.2009.03.017.

[64] M. Song, X. Lu, Z. Li, G. Liu, X. Yin, Y. Wang, Compatible ionic crosslinking composite membranes based on SPEEK and PBI for high temperature proton exchange membranes, Int. J. Hydrogen Energy 41 (2016) 12069–12081. https://doi.org/10.1016/j.ijhydene.2016.05.227.

[65] A. Heinzel, F. Mahlendorf, O. Niemzig, C. Kreuz, Injection moulded low cost bipolar plates for PEM fuel cells, J. Power Sources 131 (2004) 35–40. https://doi.org/10.1016/j.jpowsour.2004.01.014.

[66] M. Kim, J. Choe, J.W. Lim, D.G. Lee, Manufacturing of the carbon/phenol composite bipolar plates for PEMFC with continuous hot rolling process, Compos. Struct. 132 (2015) 1122–1128. https://doi.org/10.1016/j.compstruct.2015.07.038.

[67] D. Lee, J.W. Lim, D.G. Lee, Cathode/anode integrated composite bipolar plate for high-temperature PEMFC, Compos. Struct. 167 (2017) 144–151. https://doi.org/10.1016/j.compstruct.2017.01.080.

[68] N.F. Asri, T. Husaini, A.B. Sulong, E.H. Majlan, W.R.W. Daud, Coating of stainless steel and titanium bipolar plates for anticorrosion in PEMFC: a review, Int. J. Hydrogen Energy 42 (2017) 9135–9148. https://doi.org/10.1016/j.ijhydene.2016.06.241.

[69] J. Zhang, Z. Xie, J. Zhang, Y. Tang, C. Song, T. Navessin, Z. Shi, D. Song, H. Wang, D.P. Wilkinson, Z.S. Liu, S. Holdcroft, High temperature PEM fuel cells, J. Power Sources 160 (2006) 872–891. https://doi.org/10.1016/j.jpowsour.2006.05.034.

[70] D.H. Ye, Z.G. Zhan, A review on the sealing structures of membrane electrode assembly of proton exchange membrane fuel cells, J. Power Sources 231 (2013) 285–292. https://doi.org/10.1016/j.jpowsour.2013.01.009.

[71] D. Rohendi, E.H. Majlan, A.B. Mohamad, W.R.W. Daud, A.A.H. Kadhum, L.K. Shyuan, Effects of temperature and backpressure on the performance degradation of MEA in PEMFC, Int. J. Hydrogen Energy 40 (2015) 10960–10968. https://doi.org/10.1016/j.ijhydene.2015.06.161.

[72]  C. Costentin, D.H. Evans, M. Robert, J.-M. Savéant, P.S. Singh, Electrochemical approach to concerted proton and electron transfers. Reduction of the water−superoxide ion complex, J. Am. Chem. Soc. 127 (2005) 12490−12491. https://doi.org/10.1021/ja053911n.

[73]  C. Costentin, M. Robert, J.-M. Savéant, Molecular catalysis of electrochemical reactions, Curr. Opin. Electrochem. 2 (2017) 26−31. https://doi.org/10.1016/j.coelec.2017.02.006.

[74]  N.M. Markovića, S.T. Sarraf, H.A. Gasteiger, P.N. Ross, Hydrogen electrochemistry on platinum low-index single-crystal surfaces in alkaline solution, J. Chem. Soc. Faraday Trans. 92 (1996) 3719−3725. https://doi.org/10.1039/FT9969203719.

[75]  R.M.Q. Mello, E.A. Ticianelli, Kinetic study of the hydrogen oxidation reaction on platinum and Nafion® covered platinum electrodes, Electrochim. Acta 42 (1997) 1031−1039. https://doi.org/10.1016/S0013-4686(96)00282-4.

[76]  S. Lee, S. Mukerjee, E. Ticianelli, J. McBreen, Electrocatalysis of CO tolerance in hydrogen oxidation reaction in PEM fuel cells, Electrochim. Acta 44 (1999) 3283−3293. https://doi.org/10.1016/S0013-4686(99)00052-3.

[77]  K.S. Freitas, P.P. Lopes, E.A. Ticianelli, Electrocatalysis of the hydrogen oxidation in the presence of CO on $RhO_2$/C-supported Pt nanoparticles, Electrochim. Acta 56 (2010) 418−426. https://doi.org/10.1016/j.electacta.2010.08.059.

[78]  S. Mukerjee, Role of structural and electronic properties of Pt and Pt alloys on electrocatalysis of oxygen reduction, J. Electrochem. Soc. 142 (1995) 1409−1422. https://doi.org/10.1149/1.2048590.

[79]  U.A. Paulus, T.J. Schmidt, H.A. Gasteiger, R.J. Behm, Oxygen reduction on a high-surface area Pt/Vulcan carbon catalyst: a thin-film rotating ring-disk electrode study, J. Electroanal. Chem. 495 (2001) 134−145. https://doi.org/10.1016/S0022-0728(00)00407-1.

[80]  H.A. Gasteiger, S.S. Kocha, B. Sompalli, F.T. Wagner, Activity benchmarks and requirements for Pt, Pt-alloy, and non-Pt oxygen reduction catalysts for PEMFCs, Appl. Catal. B Environ. 56 (2005) 9−35. https://doi.org/10.1016/j.apcatb.2004.06.021.

[81]  V.P. Zhdanov, B. Kasemo, Kinetics of electrochemical $O_2$ reduction on Pt, Electrochem. Commun. 8 (2006) 1132−1136. https://doi.org/10.1016/j.elecom.2006.05.003.

[82]  R. Gisbert, G. García, M.T.M. Koper, Oxidation of carbon monoxide on poly-oriented and single-crystalline platinum electrodes over a wide range of pH, Electrochim. Acta 56 (2011) 2443−2449. https://doi.org/10.1016/j.electacta.2010.11.032.

[83]  J. Zhang, PEM Fuel Cell Electrocatalysts and Catalyst Layers - Fundamentals and Applications, Springer, 2008.

[84]  A. Brouzgou, S.Q. Song, P. Tsiakaras, Low and non-platinum electrocatalysts for PEMFCs: current status, challenges and prospects, Appl. Catal. B Environ. 127 (2012) 371−388. https://doi.org/10.1016/j.apcatb.2012.08.031.

[85]  C. Lamy, From hydrogen production by water electrolysis to its utilization in a PEM fuel cell or in a SO fuel cell: some considerations on the energy efficiencies, Int. J. Hydrogen Energy 41 (2016) 15415−15425. https://doi.org/10.1016/j.ijhydene.2016.04.173.

[86]  I. Dedigama, P. Angeli, K. Ayers, J.B. Robinson, P.R. Shearing, D. Tsaoulidis, D.J.L. Brett, In situ diagnostic techniques for characterisation of polymer electrolyte membrane water electrolysers - flow visualisation and electrochemical impedance spectroscopy, Int. J. Hydrogen Energy 39 (2014) 4468−4482. https://doi.org/10.1016/j.ijhydene.2014.01.026.

[87]  R.R. Salem, Theory of the electrolysis of water, Prot. Met. 44 (2008) 120−125. https://doi.org/10.1007/s11124-008-2002-x.

[88]  R. de Levie, The electrolysis of water, J. Electroanal. Chem. 476 (1999) 92−93. https://doi.org/10.1021/ed071p70.

[89] W.D. Wetzels, J.W. Ritter, The Beginnings of Electrochemistry in Germany, in: G. Dubpernell, J.H. Westbrook (Eds.), Sel. Top. Hist. Electrochem., The Electrochemical Society, Inc, Princeton, 1978, pp. 68−73.

[90] M. Carmo, D.L. Fritz, J. Mergel, D. Stolten, A comprehensive review on PEM water electrolysis, Int. J. Hydrogen Energy 38 (2013) 4901−4934. https://doi.org/10.1016/j. ijhydene.2013.01.151.

[91] S.A. Grigoriev, V.I. Porembsky, V.N. Fateev, Pure hydrogen production by PEM electrolysis for hydrogen energy, Int. J. Hydrogen Energy 31 (2006) 171−175. https://doi.org/ 10.1016/j.ijhydene.2005.04.038.

[92] S. Sengodan, R. Lan, J. Humphreys, D. Du, W. Xu, H. Wang, S. Tao, Advances in reforming and partial oxidation of hydrocarbons for hydrogen production and fuel cell applications, Renew. Sustain. Energy Rev. 82 (2018) 761−780. https://doi.org/10.1016/j. rser.2017.09.071.

[93] N.A.K. Aramouni, J.G. Touma, B.A. Tarboush, J. Zeaiter, M.N. Ahmad, Catalyst design for dry reforming of methane: analysis review, Renew. Sustain. Energy Rev. 82 (2017) 2570−2585. https://doi.org/10.1016/j.rser.2017.09.076.

[94] B. Parkinson, M. Tabatabaei, D.C. Upham, B. Ballinger, C. Greig, S. Smart, E. McFarland, Hydrogen production using methane: techno-economics of decarbonizing fuels and chemicals, Int. J. Hydrogen Energy (2018) 1−16. https://doi.org/10.1016/j.ijhydene.2017.12.081.

[95] S. Zhigang, Y. Baolian, H. Ming, Bifunctional electrodes with a thin catalyst layer for 'unitized' proton exchange membrane regenerative fuel cell, J. Power Sources 79 (1999) 82−85. https://doi.org/10.1016/S0378-7753(99)00047-6.

[96] M. Ball, M. Weeda, The hydrogen economy - vision or reality? Int. J. Hydrogen Energy 40 (2015) 7903−7919. https://doi.org/10.1016/j.ijhydene.2015.04.032.

[97] R. Moliner, M.J. Lázaro, I. Suelves, Analysis of the strategies for bridging the gap towards the hydrogen economy, Int. J. Hydrogen Energy 41 (2016) 19500−19508. https://doi.org/ 10.1016/j.ijhydene.2016.06.202.

[98] F. Barbir, T. Molter, L. Dalton, Efficiency and weight trade-off analysis of regenerative fuel cells as energy storage for aerospace applications, Int. J. Hydrogen Energy 30 (2005) 351−357. https://doi.org/10.1016/j.ijhydene.2004.08.004.

[99] B. Paul, J. Andrews, PEM unitised reversible/regenerative hydrogen fuel cell systems: state of the art and technical challenges, Renew. Sustain. Energy Rev. 79 (2017) 585−599. https://doi.org/10.1016/j.rser.2017.05.112.

[100] M. Reuß, T. Grube, M. Robinius, P. Preuster, P. Wasserscheid, D. Stolten, Seasonal storage and alternative carriers: a flexible hydrogen supply chain model, Appl. Energy 200 (2017) 290−302. https://doi.org/10.1016/j.apenergy.2017.05.050.

[101] T. Sinigaglia, F. Lewiski, M.E. Santos Martins, J.C. Mairesse Siluk, Production, storage, fuel stations of hydrogen and its utilization in automotive applications-a review, Int. J. Hydrogen Energy 42 (2017) 24597−24611. https://doi.org/10.1016/j.ijhydene. 2017.08.063.

[102] K. Alanne, S. Cao, Zero-energy hydrogen economy (ZEH2E) for buildings and communities including personal mobility, Renew. Sustain. Energy Rev. 71 (2017) 697−711. https://doi.org/10.1016/j.rser.2016.12.098.

[103] E.S. Hanley, J.P. Deane, B.P.Ó. Gallachóir, The role of hydrogen in low carbon energy futures − a review of existing perspectives, Renew. Sustain. Energy Rev. 82 (2018) 3027−3045. https://doi.org/10.1016/j.rser.2017.10.034.

[104] DOE FCs https://energy.gov/eere/fuelcells/fuel-cells (accessed February 2018).

[105] J. Marcinkoski, J. Spendelow, A. Wilson, D. Papageorgopoulos, DOE PEMFC costs (2015). https://www.hydrogen.energy.gov/pdfs/15015_fuel_cell_system_cost_2015.pdf (accessed February 2018).

[106] JM Prices http://www.platinum.matthey.com/prices (accessed February 2018).

[107] M. Shao, Q. Chang, J.-P. Dodelet, R. Chenitz, Recent advances in electrocatalysts for oxygen reduction reaction, Chem. Rev. 116 (2016) 3594–3657. https://doi.org/10.1021/acs.chemrev.5b00462.

[108] B. Li, D.C. Higgins, Q. Xiao, D. Yang, C. Zhng, M. Cai, Z. Chen, J. Ma, The durability of carbon supported Pt nanowire as novel cathode catalyst for a 1.5 kW PEMFC stack, Appl. Catal. B Environ. 162 (2015) 133–140. https://doi.org/10.1016/j.apcatb.2014.06.040.

[109] S. Hong, M. Hou, H. Zhang, Y. Jiang, Z. Shao, B. Yi, A high-performance PEM fuel cell with ultralow platinum electrode via electrospinning and underpotential deposition, Electrochim. Acta 245 (2017) 403–409. https://doi.org/10.1016/j.electacta.2017.05.066.

[110] W.S. Jung, B.N. Popov, Hybrid cathode catalyst with synergistic effect between carbon composite catalyst and Pt for ultra-low Pt loading in PEMFCs, Catal. Today 295 (2017) 65–74. https://doi.org/10.1016/j.cattod.2017.06.019.

[111] K.C. Wang, H.C. Huang, C.H. Wang, Synthesis of Pd@Pt$_3$Co/C core–shell structure as catalyst for oxygen reduction reaction in proton exchange membrane fuel cell, Int. J. Hydrogen Energy 42 (2017) 11771–11778. https://doi.org/10.1016/j.ijhydene.2017.03.084.

[112] G. Wu, K.L. More, C.M. Johnston, P. Zelenay, High-performance electrocatalysts for oxygen reduction derived from polyaniline, iron, and cobalt, Science 332 (2011) 443–447. https://doi.org/10.1126/science.1200832.

[113] E. Proietti, F. Jaouen, M. Lefèvre, N. Larouche, J. Tian, J. Herranz, J.P. Dodelet, Iron-based cathode catalyst with enhanced power density in polymer electrolyte membrane fuel cells, Nat. Commun. 2 (2011). https://doi.org/10.1038/ncomms1427.

[114] L. Xiong, A. Manthiram, High performance membrane-electrode assemblies with ultra-low Pt loading for proton exchange membrane fuel cells, Electrochim. Acta 50 (2005) 3200–3204. https://doi.org/10.1016/j.electacta.2004.11.049.

[115] G.Y. Chen, C. Wang, Y.J. Lei, J. Zhang, Z. Mao, Z.Q. Mao, J.W. Guo, J. Li, M. Ouyang, Gradient design of Pt/C ratio and Nafion content in cathode catalyst layer of PEMFCs, Int. J. Hydrogen Energy 42 (2017) 29960–29965. https://doi.org/10.1016/j.ijhydene.2017.06.229.

[116] M. Shao, B. Merzougui, K. Shoemaker, L. Stolar, L. Protsailo, Z.J. Mellinger, I.J. Hsu, J.G. Chen, Tungsten carbide modified high surface area carbon as fuel cell catalyst support, J. Power Sources 196 (2011) 7426–7434. https://doi.org/10.1016/j.jpowsour.2011.04.026.

[117] C.V. Rao, A.L.M. Reddy, Y. Ishikawa, P.M. Ajayan, Synthesis and electrocatalytic oxygen reduction activity of graphene-supported Pt$_3$Co and Pt$_3$Cr alloy nanoparticles, Carbon 49 (2011) 931–936. https://doi.org/10.1016/j.carbon.2010.10.056.

[118] M. Shao, Palladium-based electrocatalysts for hydrogen oxidation and oxygen reduction reactions, J. Power Sources 196 (2011) 2433–2444. https://doi.org/10.1016/j.jpowsour.2010.10.093.

[119] Y.H. Cho, B. Choi, Y.H. Cho, H.S. Park, Y.E. Sung, Pd-based PdPt(19:1)/C electrocatalyst as an electrode in PEM fuel cell, Electrochem. Commun. 9 (2007) 378–381. https://doi.org/10.1016/j.elecom.2006.10.007.

[120] R. Jasinski, A new fuel cell cathode catalyst, Nature 201 (1964) 1212–1213.

[121] A. Serov, C. Kwak, Review of non-platinum anode catalysts for DMFC and PEMFC application, Appl. Catal. B Environ. 90 (2009) 313–320. https://doi.org/10.1016/j.apcatb.2009.03.030.

[122] E.F. Holby, P. Zelenay, Linking structure to function: the search for active sites in non-platinum group metal oxygen reduction reaction catalysts, Nano Energy 29 (2016) 54–64. https://doi.org/10.1016/j.nanoen.2016.05.025.

[123] A.H.A. Monteverde Videla, L. Osmieri, S. Specchia, Non-noble metal (NNM) catalysts for fuel cells: tuning the activity by a rational step-by-step single variable evolution, in: F.B. José, H. Zagal (Eds.), Electrochem. N4 Macrocycl. Met. Complexes, II, 2016, pp. 69–101. https://doi.org/10.1007/978-3-319-31172-2.

[124] L. Zhang, J. Zhang, D.P. Wilkinson, H. Wang, Progress in preparation of non-noble electrocatalysts for PEM fuel cell reactions, J. Power Sources 156 (2006) 171–182. https://doi.org/10.1016/j.jpowsour.2005.05.069.

[125] C.W.B. Bezerra, L. Zhang, K. Lee, H. Liu, A.L.B. Marques, E.P. Marques, H. Wang, J. Zhang, A review of Fe-N/C and Co-N/C catalysts for the oxygen reduction reaction, Electrochim. Acta 53 (2008) 4937–4951. https://doi.org/10.1016/j.electacta.2008.02.012.

[126] L. Osmieri, A.H.A. Monteverde Videla, P. Ocón, S. Specchia, Kinetics of oxygen electroreduction on Me-N-C (Me = Fe, Co, Cu) catalysts in acidic medium: insights on the effect of the transition metal, J. Phys. Chem. C 121 (2017) 17796–17817. https://doi.org/10.1021/acs.jpcc.7b02455.

[127] A.H.A. Monteverde Videla, D. Sebastian, N.S. Vasile, L. Osmieri, A.S. Aricò, V. Baglio, S. Specchia, Performance analysis of Fe-N-C catalyst for DMFC cathodes: effect of water saturation in the cathodic catalyst layer, Int. J. Hydrogen Energy 41 (2016) 22605–22618. https://doi.org/10.1016/j.ijhydene.2016.06.060.

[128] D. Sebastián, A. Serov, I. Matanovic, K. Artyushkova, P. Atanassov, A.S. Aricò, V. Baglio, Insights on the extraordinary tolerance to alcohols of Fe-N-C cathode catalysts in highly performing direct alcohol fuel cells, Nano Energy 34 (2017) 195–204. https://doi.org/10.1016/j.nanoen.2017.02.039.

[129] Y.-C. Wang, Y.-J. Lai, L. Song, Z.-Y. Zhou, J.-G. Liu, Q. Wang, X.-D. Yang, C. Chen, W. Shi, Y.-P. Zheng, M. Rauf, S.-G. Sun, S-doping of an Fe/N/C ORR catalyst for polymer electrolyte membrane fuel cells with high power density, Angew. Chem. Int. Ed. 54 (2015) 9907–9910. https://doi.org/10.1002/anie.201503159.

[130] I. Martinaiou, T. Wolker, A. Shahraei, G.R. Zhang, A. Janßen, S. Wagner, N. Weidler, R.W. Stark, B.J.M. Etzold, U.I. Kramm, Improved electrochemical performance of Fe-N-C catalysts through ionic liquid modification in alkaline media, J. Power Sources 375 (2018) 222–232. https://doi.org/10.1016/j.jpowsour.2017.07.028.

[131] J. Tian, A. Morozan, M.T. Sougrati, R. Lefèvre, R. Chenitz, J.P. Dodelet, D. Jones, F. Jaouen, Optimized synthesis of Fe/N/C cathode catalysts for PEM fuel cells: a matter of iron-ligand coordination strength, Angew. Chem. Int. Ed. 52 (2013) 6867–6870. https://doi.org/10.1002/anie.201303025.

[132] H. Barkholtz, L. Chong, Z. Kaiser, T. Xu, D.-J. Liu, Highly active non-PGM catalysts prepared from metal organic frameworks, Catalysts 5 (2015) 955–965. https://doi.org/10.3390/catal5020955.

[133] S. Yuan, J.L. Shui, L. Grabstanowicz, C. Chen, S. Commet, B. Reprogle, T. Xu, L. Yu, D.J. Liu, A highly active and support-free oxygen reduction catalyst prepared from ultrahigh-surface-area porous polyporphyrin, Angew. Chem. Int. Ed. 52 (2013) 8349–8353. https://doi.org/10.1002/anie.201302924.

[134] A. Serov, K. Artyushkova, P. Atanassov, Fe-N-C oxygen reduction fuel cell catalyst derived from carbendazim: synthesis, structure, and reactivity, Adv. Energy Mater. 4 (2014) 1−7. https://doi.org/10.1002/aenm.201301735.

[135] A. Collier, H. Wang, X. Zi Yuan, J. Zhang, D.P. Wilkinson, Degradation of polymer electrolyte membranes, Int. J. Hydrogen Energy 31 (2006) 1838−1854. https://doi.org/10. 1016/j.ijhydene.2006.05.006.

[136] K.A. Mauritz, R.B. Moore, State of understanding of Nafion, Chem. Rev. 104 (2004) 4535−4585. https://doi.org/10.1021/cr0207123.

[137] N. Yoshida, T. Ishisaki, A. Watakabe, M. Yoshitake, Characterization of Flemion® membranes for PEFC, Electrochim. Acta 43 (1998) 3749−3754. https://doi.org/10.1016/ S0013-4686(98)00133-9.

[138] J.T. Hinatsu, Water uptake of perfluorosulfonic acid membranes from liquid water and water vapor, J. Electrochem. Soc. 141 (1994) 1493. https://doi.org/10.1149/1.2054951.

[139] Y. Wang, K.S. Chen, J. Mishler, S.C. Cho, X.C. Adroher, A review of polymer electrolyte membrane fuel cells: technology, applications, and needs on fundamental research, Appl. Energy 88 (2011) 981−1007. https://doi.org/10.1016/j.apenergy.2010.09.030.

[140] M.A. Hickner, H. Ghassemi, Y.S. Kim, B.R. Einsla, J.E. McGrath, Alternative polymer systems for proton exchange membranes (PEMs), Chem. Rev. 104 (2004) 4587−4611. https://doi.org/10.1021/cr020711a.

[141] Y. You, S.Y. Park, Inter-ligand energy transfer and related emission change in the cyclometalated heteroleptic iridium complex: facile and efficient color tuning over the whole visible range by the ancillary ligand structure, J. Am. Chem. Soc. 127 (2005) 12438−12439. https://doi.org/10.1021/ja052880t.

[142] M.L. Di Vona, D. Marani, C. D'Ottavi, M. Trombetta, E. Traversa, I. Beurroies, P. Knauth, S. Licoccia, A simple new route to covalent organic/inorganic hybrid proton exchange polymeric membranes, Chem. Mater. 18 (2006) 69−75. https://doi.org/10.1021/ cm051546t.

[143] M.M. Hasani-Sadrabadi, E. Dashtimoghadam, N. Mokarram, F.S. Majedi, K.I. Jacob, Triple-layer proton exchange membranes based on chitosan biopolymer with reduced methanol crossover for high-performance direct methanol fuel cells application, Polymer 53 (2012) 2643−2651. https://doi.org/10.1016/j.polymer.2012.03.052.

[144] N.S. Vasile, A.H.A. Monteverde Videla, C. Simari, I. Nicotera, S. Specchia, Influence of membrane-type and flow field design on methanol crossover on a single-cell DMFC: an experimental and multi-physics modeling study, Int. J. Hydrogen Energy 42 (2017) 27995−28010. https://doi.org/10.1016/j.ijhydene.2017.06.214.

[145] V. Parthiban, S. Akula, A.K. Sahu, Surfactant templated nanoporous carbon-Nafion hybrid membranes for direct methanol fuel cells with reduced methanol crossover, J. Memb. Sci. 541 (2017) 127−136. https://doi.org/10.1016/j.memsci.2017.06.081.

[146] S.D. Knights, K.M. Colbow, J. St-Pierre, D.P. Wilkinson, Aging mechanisms and lifetime of PEFC and DMFC, J. Power Sources 127 (2004) 127−134. https://doi.org/10.1016/j. jpowsour.2003.09.033.

[147] A. Pozio, R.F. Silva, M. De Francesco, L. Giorgi, Nafion degradation in PEFCs from end plate iron contamination, Electrochim. Acta 48 (2003) 1543−1549. https://doi.org/10. 1016/S0013-4686(03)00026-4.

[148] H.J. Lee, M.K. Cho, Y.Y. Jo, K.S. Lee, H.J. Kim, E. Cho, S.K. Kim, D. Henkensmeier, T.H. Lim, J.H. Jang, Application of TGA techniques to analyze the compositional and structural degradation of PEMFC MEAs, Polym. Degrad. Stab 97 (2012) 1010−1016. https://doi.org/10.1016/j.polymdegradstab.2012.03.016.

[149] G. Gavello, J. Zeng, C. Francia, U.A. Icardi, A. Graizzaro, S. Specchia, Experimental studies on Nafion® 112 single PEM-FCs exposed to freezing conditions, Int. J. Hydrogen Energy 36 (2011). https://doi.org/10.1016/j.ijhydene.2011.01.182.

[150] P. Ferreira-Aparicio, A.M. Chaparro, M.A. Folgado, J.J. Conde, E. Brightman, G. Hinds, Degradation study by start-up/shut-down cycling of superhydrophobic electrosprayed catalyst layers using a localized reference electrode technique, ACS Appl. Mater. Interfaces 9 (2017) 10626−10636. https://doi.org/10.1021/acsami.6b15581.

[151] J. Kang, D.W. Jung, S. Park, J.H. Lee, J. Ko, J. Kim, Accelerated test analysis of reversal potential caused by fuel starvation during PEMFCs operation, Int. J. Hydrogen Energy 35 (2010) 3727−3735. https://doi.org/10.1016/j.ijhydene.2010.01.071.

[152] A. Taniguchi, T. Akita, K. Yasuda, Y. Miyazaki, Analysis of degradation in PEMFC caused by cell reversal during air starvation, Int. J. Hydrogen Energy 33 (2008) 2323−2329. https://doi.org/10.1016/j.ijhydene.2008.02.049.

[153] I. Jang, I. Hwang, Y. Tak, Attenuated degradation of a PEMFC cathode during fuel starvation by using carbon-supported IrO₂, Electrochim. Acta 90 (2013) 148−156. https://doi.org/10.1016/j.electacta.2012.12.034.

[154] A. Amirfazli, S. Asghari, M. Sarraf, An investigation into the effect of manifold geometry on uniformity of temperature distribution in a PEMFC stack, Energy 145 (2018) 141−151. https://doi.org/10.1016/j.energy.2017.12.124.

[155] C. Francia, V.S. Ijeri, S. Specchia, P. Spinelli, Estimation of hydrogen crossover through Nafion® membranes in PEMFCs, J. Power Sources 196 (2011). https://doi.org/10.1016/j.jpowsour.2010.09.058.

[156] S. Galbiati, A. Baricci, A. Casalegno, R. Marchesi, Gas crossover leakage in high temperature polymer electrolyte fuel cells: in situ quantification and effect on performance, J. Power Sources 205 (2012) 350−353. https://doi.org/10.1016/j.jpowsour.2012.01.055.

[157] F.N. Büchi, B. Gupta, O. Haas, G.G. Scherer, Study of radiation-grafted FEP-G-polystyrene membranes as polymer electrolytes in fuel cells, Electrochim. Acta 40 (1995) 345−353. https://doi.org/10.1016/0013-4686(94)00274-5.

[158] C. Zhou, M.A. Guerra, Z.M. Qiu, T.A. Zawodzinski, D.A. Schiraldi, Chemical durability studies of perfluorinated sulfonic acid polymers and model compounds under mimic fuel cell conditions, Macromolecules 40 (2007) 8695−8707. https://doi.org/10.1021/ma071603z.

[159] L. Ghassemzadeh, S. Holdcroft, Quantifying the structural changes of perfluorosulfonated acid ionomer upon reaction with hydroxyl radicals, J. Am. Chem. Soc. 135 (2013) 8181−8184. https://doi.org/10.1021/ja4037466.

[160] C. D'Urso, C. Oldani, V. Baglio, L. Merlo, A.S. Aricò, Towards fuel cell membranes with improved lifetime: aquivion® perfluorosulfonic acid membranes containing immobilized radical scavengers, J. Power Sources 272 (2014) 753−758. https://doi.org/10.1016/j.jpowsour.2014.09.045.

[161] J. Hao, Y. Jiang, X. Gao, F. Xie, Z. Shao, B. Yi, Degradation reduction of poly-benzimidazole membrane blended with CeO₂ as a regenerative free radical scavenger, J. Memb. Sci. 522 (2017) 23−30. https://doi.org/10.1016/j.memsci.2016.09.010.

[162] D.E. Curtin, R.D. Lousenberg, T.J. Henry, P.C. Tangeman, M.E. Tisack, Advanced materials for improved PEMFC performance and life, J. Power Sources 131 (2004) 41−48. https://doi.org/10.1016/j.jpowsour.2004.01.023.

[163] L. Gubler, H. Kuhn, T.J. Schmidt, G.G. Scherer, H.P. Brack, K. Simbeck, Performance and durability of membrane electrode assemblies based on radiation-grafted FEP-g-polystyrene membranes, Fuel Cells 4 (2004) 196−207. https://doi.org/10.1002/fuce.200400019.

[164] A. Arvay, J. French, J.C. Wang, X.H. Peng, A.M. Kannan, Nature inspired flow field designs for proton exchange membrane fuel cell, Int. J. Hydrogen Energy 38 (2013) 3717−3726. https://doi.org/10.1016/j.ijhydene.2012.12.149.

[165] M. Kim, D.G. Lee, Optimum design of the carbon composite bipolar plate (BP) for the open cathode of an air breathing PEMFC, Compos. Struct. 140 (2016) 675−683. https://doi.org/10.1016/j.compstruct.2015.12.061.

[166] D. Lee, D.G. Lee, Carbon composite bipolar plate for high-temperature proton exchange membrane fuel cells (HT-PEMFCs), J. Power Sources 327 (2016) 119−126. https://doi.org/10.1016/j.jpowsour.2016.07.045.

[167] C.H. Lin, S.Y. Tsai, An investigation of coated aluminium bipolar plates for PEMFC, Appl. Energy 100 (2012) 87−92. https://doi.org/10.1016/j.apenergy.2012.06.045.

[168] W.L. Wang, S.M. He, C.H. Lan, Protective graphite coating on metallic bipolar plates for PEMFC applications, Electrochim. Acta 62 (2012) 30−35. https://doi.org/10.1016/j.electacta.2011.11.026.

[169] Y. Zhao, L. Wei, P. Yi, L. Peng, Influence of Cr-C film composition on electrical and corrosion properties of 316L stainless steel as bipolar plates for PEMFCs, Int. J. Hydrogen Energy 41 (2016) 1142−1150. https://doi.org/10.1016/j.ijhydene.2015.10.047.

[170] Z. Ren, D. Zhang, Z. Wang, Stacks with TiN/titanium as the bipolar plate for PEMFCs, Energy 48 (2012) 577−581. https://doi.org/10.1016/j.energy.2012.10.020.

[171] F.G. Boyacı San, O. Okur, The effect of compression molding parameters on the electrical and physical properties of polymer composite bipolar plates, Int. J. Hydrogen Energy 42 (2017) 23054−23069. https://doi.org/10.1016/j.ijhydene.2017.07.175.

[172] A.P. Manso, F.F. Marzo, J. Barranco, X. Garikano, M. Garmendia Mujika, Influence of geometric parameters of the flow fields on the performance of a PEM fuel cell. A review, Int. J. Hydrogen Energy 37 (2012) 15256−15287. https://doi.org/10.1016/j.ijhydene.2012.07.076.

[173] J.P. Kloess, X. Wang, J. Liu, Z. Shi, L. Guessous, Investigation of bio-inspired flow channel designs for bipolar plates in proton exchange membrane fuel cells, J. Power Sources 188 (2009) 132−140. https://doi.org/10.1016/j.jpowsour.2008.11.123.

[174] R. Roshandel, F. Arbabi, G.K. Moghaddam, Simulation of an innovative flow-field design based on a bio inspired pattern for PEM fuel cells, Renew. Energy 41 (2012) 86−95. https://doi.org/10.1016/j.renene.2011.10.008.

[175] J.P. Owejan, T.A. Trabold, D.L. Jacobson, M. Arif, S.G. Kandlikar, Effects of flow field and diffusion layer properties on water accumulation in a PEM fuel cell, Int. J. Hydrogen Energy 32 (2007) 4489−4502. https://doi.org/10.1016/j.ijhydene.2007.05.044.

[176] D.H. Ahmed, H.J. Sung, Effects of channel geometrical configuration and shoulder width on PEMFC performance at high current density, J. Power Sources 162 (2006) 327−339. https://doi.org/10.1016/j.jpowsour.2006.06.083.

[177] J.H. Nam, K.J. Lee, S. Sohn, C.J. Kim, Multi-pass serpentine flow-fields to enhance under-rib convection in polymer electrolyte membrane fuel cells: design and geometrical character-ization, J. Power Sources 188 (2009) 14−23. https://doi.org/10.1016/j.jpowsour.2008.11.093.

[178] T. Yoshida, K. Kojima, Toyota MIRAI fuel cell vehicle and progress toward a future hydrogen society, Interface Mag. 24 (2015) 45−49. https://doi.org/10.1149/2.F03152if.

[179] Linde http://www.linde.com/ (accessed February 2018).

[180] Kier http://www.kier.re.kr (accessed February 2018).

[181] Hyundai www.hyundai.new (accessed February 2018).

[182] Honda http://world.honda.com/FuelCell/ (accessed February 2018).

[183] Plugpower http://www.plugpower.com (accessed February 2018).

[184] BMW news www.nydailynews.com/autos/news (accessed February 2018).

[185] P. Agnolucci, Hydrogen infrastructure for the transport sector, Int. J. Hydrogen Energy 32 (2007) 3526–3544. https://doi.org/10.1016/j.ijhydene.2007.02.016.

[186] J. Alazemi, J. Andrews, Automotive hydrogen fuelling stations: an international review, Renew. Sustain. Energy Rev. 48 (2015) 483–499. https://doi.org/10.1016/j.rser.2015.03.085.

[187] K. Reddi, A. Elgowainy, N. Rustagi, E. Gupta, ScienceDirect: Impact of hydrogen refueling configurations and market parameters on the refueling cost of hydrogen, Int. J. Hydrogen Energy 42 (2017) 21855–21865. https://doi.org/10.1016/j.ijhydene.2017.05.122.

[188] M. Iordache, D. Schitea, I. Iordache, ScienceDirect: Hydrogen refuelling station infrastructure roll-up, an indicative assessment of the commercial viability and profitability in the Member States of Europe Union, Int. J. Hydrogen Energy 42 (2017) 29629–29647. https://doi.org/10.1016/j.ijhydene.2017.09.146.

[189] W. Mcdowall, Technology roadmaps for transition management: the case of hydrogen energy, Technol. Forecast. Soc. Chang 79 (2012) 530–542. https://doi.org/10.1016/j.techfore.2011.10.002.

[190] K. Kendall, B.G. Pollet, Hydrogen and Fuel Cells in Transport, Elsevier Ltd., 2012. https://doi.org/10.1016/B978-0-08-087872-0.00419-4.

[191] NewBusFuel http://newbusfuel.eu/ (accessed February 2018).

[192] P.E.V. de Miranda, E.S. Carreira, U.A. Icardi, G.S. Nunes, Brazilian hybrid electric-hydrogen fuel cell bus: improved on-board energy management system, Int. J. Hydrogen Energy 42 (2017) 13949–13959. https://doi.org/10.1016/j.ijhydene.2016.12.155.

[193] N. Briguglio, M. Ferraro, L. Andaloro, V. Antonucci, New simulation tool helping a feasibility study for renewable hydrogen bus fleet in Messina, Int. J. Hydrogen Energy 33 (2008) 3077–3084. https://doi.org/10.1016/j.ijhydene.2008.03.059.

[194] F. Sergi, L. Andaloro, G. Napoli, N. Randazzo, V. Antonucci, Development and realization of a hydrogen range extender hybrid city bus, J. Power Sources 250 (2014) 286–295. https://doi.org/10.1016/j.jpowsour.2013.11.006.

[195] G. Napoli, S. Micari, G. Dispenza, S. Di Novo, V. Antonucci, L. Andaloro, Development of a fuel cell hybrid electric powertrain: a real case study on a Minibus application, Int. J. Hydrogen Energy 42 (2017) 28034–28047. https://doi.org/10.1016/j.ijhydene.2017.07.239.

[196] T. Hua, R. Ahluwalia, L. Eudy, G. Singer, B. Jermer, N. Asselin-Miller, S. Wessel, T. Patterson, J. Marcinkoski, Status of hydrogen fuel cell electric buses worldwide, J. Power Sources 269 (2014) 975–993. https://doi.org/10.1016/j.jpowsour.2014.06.055.

[197] DOE reports highlight fuel cell bus economy, hydrogen fuel quality, Fuel Cells Bull. 2016 (2016) 12. https://doi.org/10.1016/S1464-2859(16)30362-5.

[198] G.E. Haslam, J. Jupesta, G. Parayil, Assessing fuel cell vehicle innovation and the role of policy in Japan, Korea, and China, Int. J. Hydrogen Energy 37 (2012) 14612–14623. https://doi.org/10.1016/j.ijhydene.2012.06.112.

[199] J. Du, M. Ouyang, J. Chen, Prospects for Chinese electric vehicle technologies in 2016–2020: ambition and rationality, Energy 120 (2017) 584–596. https://doi.org/10.1016/j.energy.2016.11.114.

[200] L. Lucas, Toyota delivers first fuel cell bus in Tokyo, recalls all Mirai cars, Fuel Cells Bull. 2017 (2017) 2–3. https://doi.org/10.1016/S1464-2859(17)30102-5.

[201] L. Lucas, Fuel cell bus with Ballard power passes 25,000 h in service, Fuel Cells Bull. 2017 (2017) 2. https://doi.org/10.1016/S1464-2859(17)30339-5.

[202] M. Moreno-Benito, P. Agnolucci, L.G. Papageorgiou, Towards a sustainable hydrogen economy: optimisation-based framework for hydrogen infrastructure development, Comput. Chem. Eng. 102 (2017) 110–127. https://doi.org/10.1016/j.compchemeng.2016.08.005.

[203]  J.C. González Palencia, M. Araki, S. Shiga, Energy, environmental and economic impact of mini-sized and zero-emission vehicle diffusion on a light-duty vehicle fleet, Appl. Energy 181 (2016) 96–109. https://doi.org/10.1016/j.apenergy.2016.08.045.

[204]  M. Sykes, J. Axsen, No free ride to zero-emissions: simulating a region's need to implement its own zero-emissions vehicle (ZEV) mandate to achieve 2050 GHG targets, Energy Policy 110 (2017) 447–460. https://doi.org/10.1016/j.enpol.2017.08.031.

[205]  D.L. Greene, S. Park, C. Liu, Public policy and the transition to electric drive vehicles in the U.S.: the role of the zero emission vehicles mandates, Energy Strateg. Rev. 5 (2014) 66–77. https://doi.org/10.1016/j.esr.2014.10.005.

[206]  N.Q. Minh, J. Am. Ceram. Soc. 76 (1993) 563.

[207]  N.Q. Minh, T. Takahashi, Science and Technology of Ceramic Fuel Cells, Elsevier, Amsterdam, The Netherlands, 1995.

[208]  N.Q. Minh, in: K. Kendall, M. Kendall (Eds.), High-temperature Solid Oxide Fuel Cells for the 21st Century, Academic Press, London, UK, 2016, p. 255.

[209]  N.Q. Minh, in: D. Stolten, B. Emonts (Eds.), Hydrogen Science and Engineering, Materials, Processes, Systems and Technology, Wiley-VCH, Weinheim, Germany, 2016, p. 359.

[210]  J.A. Kilner, J. Druce, T. Ishihara, in: K. Kendall, M. Kendall (Eds.), High-temperature Solid Oxide Fuel Cells for the 21st Century, Academic Press, London, UK, 2016, p. 85.

[211]  S.P. Jiang, S.H. Chan, J. Mater. Sci. 39 (2004) 4405.

[212]  T. Kawada, T. Horita, in: K. Kendall, M. Kendall (Eds.), High temperature Solid Oxide Fuel Cells for the 21st Century, Academic Press, London, UK, 2016, p. 161.

[213]  J.W. Fergus, Solid State Ionics 171 (2004) 1.

[214]  T. Akbay, in: T. Ishihara (Ed.), Perovskite Oxide for Solid Oxide Fuel Cells, Springer, New York, NY, 2009, p. 183.

[215]  M.C. Tucker, J. Power Sources 195 (2010) 4570.

[216]  H. Huang, M. Makamura, P. Su, R. Fasching, Y. Sato, F.B. Prinz, J. Electrochem. Soc. 154 (2007) B20.

[217]  E.D. Wachsman, K.T. Lee, Science 334 (2011) 935.

[218]  H. Yahiro, Y. Baba, K. Eguchi, H. Arai, J. Electrochem. Soc. 135 (1988) 2077.

[219]  A.N. Virkar, G. Tao, Int. J. Hydrogen Energy 40 (2015) 5561.

[220]  N.Q. Minh, ECS Trans. 57 (1) (2013) 197.

[221]  S. Park, J.M. Vohs, R.J. Gorte, Nature 404 (2000) 265.

[222]  S. McIntosh, R.J. Gorte, Chem. Rev. 104 (2004) 4845.

[223]  E.N. Armstrong, J.-W. Park, N.Q. Minh, ECS Trans. 45 (1) (2012) 499.

[224]  T. Klemensø, C. Chung, P.H. Larsen, M. Mogensen, in: S.C. Singhal, J. Mizusaki (Eds.), SOFC-IX, vol. 2, The Electrochemical Society, Pennington, NJ, 2005, p. 1226.

[225]  T. Klemensø, M. Mogensen, J. Am. Ceram. Soc. 90 (2007) 3582.

[226]  K.J. Haltiner Jr, S. Mukerjee, J. Tachtler, B. Edlinger, D.M. England, M.T. Faville, S.M. Kelly, WO Patent 2004001875, 2003.

[227]  A. Faes, J.M. Fuerbringer, D. Mohamedi, A. Hessler-Wyser, G. Cabochi, J. van Herle, J. Power Sources 196 (2011) 7058.

[228]  O.A. Marina, N.L. Canfield, J.W. Stevenson, Solid State Ionics 149 (2002) 21.

[229]  H. Yokokawa, N. Sakai, T. Kawada, M. Dokiya, J. Electrochem. Soc. 138 (1991) 2719.

[230]  S.P. Simner, J.F. Bonnett, N.L. Canfield, K.D. Meinhardt, J.P. Shelton, V.L. Sprenkle, J.W. Stevenson, J. Power Sources 113 (2003) 1.

[231]  S. Bebelis, N. Kotsionopoulos, A. Mai, F. Tietz, J. Appl. Electrochem. 37 (2007) 15.

[232]  N.Q. Minh, in: S.C. Singhal (Ed.), High Temperature Materials, The Electrochemical Society, Pennington, NJ, 2002, p. 54.

[233] National Energy Technology Laboratory, Recent Solid Oxide Fuel Cell Cathode Studies, Report DOE/NETL-2013/1618, 2013.

[234] T.Z. Sholklapper, C.P. Jacobson, S.J. Visco, L.C. De Jonghe, Fuel Cells 5 (2008) 303.

[235] J.M. Vohs, R.J. Gorte, Adv. Mater. 21 (2009) 943.

[236] F.L. Liang, J. Chen, S.P. Jiang, B. Chi, J. Pu, L. Jian, Electrochem. Commun. 11 (2009) 1048.

[237] Y. Gong, D. Palacio, X. Song, R.L. Patel, X. Liang, X. Zhao, J.B. Goodenough, K. Huang, Nano Lett. 13 (2013) 4340.

[238] L. Niewolak, F. Tietz, W.J. Quadakkers, in: K. Kendall, M. Kendall (Eds.), High-temperature Solid Oxide Fuel Cells for the 21st Century, Academic Press, London, UK, 2016, p. 195.

[239] N.Q. Minh, S.C. Singhal, M.C. Williams, ECS Trans. 17 (1) (2009) 211.

[240] N.Q. Minh, ECS Trans. 7 (1) (2007) 45.

[241] K. Kendall, N.Q. Minh, S.C. Singhal, in: S.C. Singhal, K. Kendall (Eds.), High Temperature Solid Oxide Fuel Cells: Fundamentals, Design and Applications, Elsevier, Oxford, UK, 2003, p. 197.

[242] S. Küln, T. Pessara, D. Rückemann, A. Weber, K. Paciejewska, A. Stoeck, S. Mnich, L. Winkler, Presentation at Hannover Messe 2014, www.h2fc.com/hm14/images/tech-forum-presentations/2014-0409-1030.pdf.

[243] S.D. Vora, in: 8th Annual SECA Workshop, 2007, in: www.netl.doe.gov/events/conference-proceedings/2007/seca-worksop.

[244] www.elcogen.com (accessed February 2018).

[245] J. Will, A. Mitterdorfer, C. Kleindogel, D. Perednis, L.J. Gauckler, Solid State Ionics 131 (2000) 79.

[246] N.Q. Minh, C.R. Horne, in: F.W. Poulsen (Ed.), High-temperature Electrochemical Behaviour of Fast Ion and Mixed Conductors, Risø National Laboratory, Roskilde, Denmark, 1993, p. 337.

[247] N.Q. Minh, in: 18th Annual SOFC Project Review Meeting, 2017, in: www.netl.doe.gov/events/conference-proceedings/2017/sofc.

[248] J.-W. Kim, A.V. Virkar, K.-Z. Fung, K. Mehta, S.C. Singhal, J. Electrochem. Soc. 146 (1999) 69.

[249] C.-C. Chao, C.-M. Hsu, Y. Cui, F.B. Prinz, ACS Nano 5 (2011) 5692.

[250] H. Ghezel-Ayagh, in: 12th Annual SECA Workshop, 2011, in: www.netl.doe.gov/events/conference-proceedings/2011/seca-workshop.

[251] Z. Zhang, D.M. Bierschenk, J.C. Cronin, S.A. Barnett, Energy Environ. Sci. 4 (2011) 3951.

[252] M. Mogensen, hydrail.org/sites/hydrail.org/files/2_Mogensen.pdf.

[253] S.C. Singhal, MRS Bull. 25 (3) (2000) 16.

[254] H. Yokokawa, K. Yamaji, M.E. Brito, H. Kishimoto, T. Horita, J. Power Sources 196 (2011) 7070.

[255] Y.L. Liu, S. Primdahl, M. Mogensen, Solid State Ionics 161 (2003) 1.

[256] S.H. Kim, T. Ohshima, Y. Shiratori, K. Itoh, K. Sasaki, in: V. Pthenakis, A. Dillon, N. Savage (Eds.), MRS Proceedings, Volume 1041, Materials Research Society, Warrendale, PA, 2007. 1041-R03-10.

[257] A. Hagen, K. Neufeld, Y.L. Liu, J. Electrochem. Soc. 157 (2010) B1343.

[258] W. Zhou, F. Liang, Z. Shao, Z. Zhu, Sci. Rep. 2 (2012) 327. https://doi.org/10.1038/srep00327.

[259] G. Agnew, R. Goettler, in: 13th Annual SECA Workshop, 2012, in: www.netl.doe.gov/events/conference-proceedings/2012/seca-workshop.

[260] A. Nakanishi, M. Hattori, Y. Sasaki, K. Kimura, Y. Ando, H. Miyamoto, T. Onodera, S. Kanehira, K. Takenobu, M. Nishiura, H. Oozawa, in: S.C. Singhal, J. Mizusaki (Eds.), SOFC-IX, vol. 1, The Electrochemical Society, Pennington, NJ, 2005, p. 82.

[261] J. Doyon, in: 8th Annual SECA Workshop, 2007, in: www.netl.doe.gov/events/conference-proceedings/2007/seca-workshop.

[262] Westinghouse Electric Corporation, Solid Oxide Fuel Cell Power Generation System, The Status of the Cell Technology — A Topical Report, U.S. Department of Energy, Washington, DC, 1984. Report No. DOE/ET/17089—15.

[263] E. Ivers-Tiffée, W. Wersing, B. Reichelt, in: 1990 Fuel Cell Seminar Abstracts, Courtesy Associates, Washington, DC, 1990, p. 137.

[264] R. Goettler, T. Ohm, in: 11th Annual SECA Workshop, 2010, in: www.netl.doe.gov/events/conference-proceedings/2010/seca-workshop.

[265] N.Q. Minh, C.R. Horne, U.S. Patent 5,162,167, 1992.

[266] N.Q. Minh, T.L. Stillwagon, U.S. Patent 5,256,499, 1993.

[267] S.C. Singhal, in: F. Grosz, P. Zegers, S.C. Singhal, O. Yamamoto (Eds.), Proceedings of the Second International Symposium on Solid Oxide Fuel Cells, Commission of European Communities, Luxembourg, 1991, p. 25.

[268] J. Piascik, D. Dalfonzo, J. Yamanis, E. Ong, U.S. Patent US 6677069 B1, 2004.

[269] H. Ghezel-Ayagh, in: 13th Annual SECA Workshop, 2012, in: www.netl.doe.gov/events/conference-proceedings/2012/seca-workshop.

[270] N.Q. Minh, J. Korean Ceram. Soc. 47 (2010) 1.

[271] N.Q. Minh, M.B. Mogensen, ECS Interface 22 (4) (2013) 55.

[272] L. Blum, U. Packbier, I.C. Vinke, L.G.J. de Haart, Fuel Cells 13 (2013) 646.

[273] N.Q. Minh, in: D. Stolten, B. Emonts (Eds.), Fuel Cell Science and Engineering, vol. 2, Wiley-VCH, Weinheim, Germany, 2012, p. 963.

[274] N.Q. Minh, in: K. Kendall, M. Kendall (Eds.), High-temperature Solid Oxide Fuel Cells for the 21st Century, Academic Press, London, UK, 2016, p. 283.

[275] R.A. George, in: Proceedings of 3rd Annual SECA Workshop, U.S. Department of Energy, National Energy Technology Laboratory, Morgantown, WV, 2002, in: www.alrc.doe.gov/publications/proceedings/02/SECA/seca02.html.

[276] Ceramic Fuel Cells Limited, Introducing BlueGen Modular Generator: Power + Heat, 2009. www.cfcl.com.au/Assets/Files/BlueGen_Launch_information_(Web)_May-2009.pdf.

[277] N.Q. Minh, Solid State Ionics 174 (2004) 271.

[278] W. Winkler, in: U. Bossel (Ed.), Proceedings of First European Solid Oxide Fuel Cell Forum, vol. 2, Lucerne, Switzerland, 1994, p. 821.

[279] S. Samuelsen, Fuel Cell/Gas Turbine Hybrid Systems, ASME International Gas Turbine Institute, Norcross, GA, 2004.

[280] P. Aguiar, D.J.L. Brett, N.P. Brandon, J. Power Sources 171 (2007) 186.

[281] G. Napoli, M. Ferraro, F. Sergi, G. Brunaccini, G. Dispenza, L. Andaloro, A. Yasin, V. Antonucci, ECS Trans. 42 (1) (2012) 209.

[282] F. Zabihian, A. Fung, Int. J. Eng. 3 (2009) 85.

[283] W. Winkler, H. Lorenz, in: U. Bossel (Ed.), Proceedings of Fourth European Solid Oxide Fuel Cell Forum, vol. 1, Lucerne, Switzerland, 2000, p. 413.

[284] W. Dönitz, E. Erdle, Int. J. Hydrogen Energy 10 (1985) 291.

[285] S.H. Jensen, P.H. Larsen, M. Mogensen, Int. J. Hydrogen Energy 32 (2007) 3253.

[286] J. Guan, R. Doshi, G. Lear, K. Montgomery, E. Ong, N. Minh, J. Am. Ceram. Soc. 85 (2002) 2651.

[287] N.Q. Minh, M.C. Williams, ECS Trans. 68 (1) (2015) 3301.

[288] GE Hybrid Power Generation Systems, High Performance Flexible Reversible Solid Oxide Fuel Cell, Final Technical Report, 2008.

[289] A. Brisse, J. Schefold, M. Zahid, J. Hydrogen Energy 33 (2008) 5375.

[290] N. Minh, B. Chung, R. Doshi, K. Montgomery, E. Ong, M. Reddig, A. MacKnight, S. Fuhs, Zirconia Electrolysis Cells for Oxygen Generation from Carbon Dioxide for Mars In-Situ Resource Utilization, SAE Technical Paper 981655, SAE International, Warrendale, PA, 1998.

[291] K.R. Sridhar, B.T. Vaniman, Solid State Ionics 93 (1997) 321.

[292] A.O. Isenberg, C.E. Verostko, Carbon Dioxide and Water Vapor High Temperature Electrolysis, SAE Technical Paper 891506, SAE International, Warrendale, PA, 1989.

[293] M.G. McKellar, C.M. Stoots, M.S. Sohal, L.M. Mulloh, B. Luna, M.B. Abney, The Concept and Analytical Investigation of Carbon Dioxide and Steam Co-Electrolysis for Resource Utilization in Space Exploration, AIAA Paper 2010-6273, American Institute of Aeronautics and Astronautics, Reston, VA, 2010.

[294] C. Graves, S.D. Ebbesen, M. Mogensen, Solid State Ionics 192 (2011) 398.

[295] Q. Fu, in: A. Indarto, J. Palgunadi (Eds.), Syngas: Production, Applications and Environmental Impact, Nova Science Publishers, Hauppauge, NY, 2011, p. 209.

[296] J. Hartvigsen, L. Frost, S. Elangovan, 2012. www.fuelcellseminar.com/media/51332/sta44-4.pdf.

[297] C.M. Stoots, J.E. O'Brien, J.S. Herring, K.G. Condie, J.J. Hartvigsen, Idaho National Laboratory Experimental Research in High Temperature Electrolysis for Hydrogen and Syngas Production, INL/CON-08—14622, 2008.

[298] P. Kim-Lohsoontorn, J. Bae, J. Power Sources 196 (2011) 7161.

[299] F. Bidrawn, G. Kim, G. Corre, J.T.S. Irvine, J.M. Vohs, R.J. Gorte, Electrochem. Solid-State Lett. 11 (2008) B167.

[300] J.E. O'Brien, X. Zhang, C.R. O'Brien, G. Tao, 2011. www.fuelcellseminar.com/media/9066/hrd34-5.pdf.

[301] S. Herring, in: 2005 DOE Hydrogen, Fuel Cells & Infrastructure Technologies Program Review, 2005.

[302] J. Schefold, A. Brisse, M. Zahid, J.P. Ouweltjes, J.U. Nielsen, ECS Trans. 35 (2011) 2915.

[303] Y. Zheng, Q. Li, W. Guan, C. Xu, W. Wu, W.G. Wang, Ceram. Int. 40 (2014) 5901.

[304] S.D. Ebbesen, J. Høgh, K.A. Nielsen, J.U. Nielsen, M. Mogensen, Int. J. Hydrogen Energy 36 (2011) 7363.

[305] A. Hauch, S.H. Jensen, S.D. Ebbesen, M. Mogensen, in: L.S. Petersen, H. Larsen (Eds.), Energy Solutions for Sustainable Development Proceedings, Risø Energy International Conference 2007, Risø-R-1608 (EN), Risø National Laboratory Technical University of Denmark, Roskilde, Denmark, 2007, p. 327.

[306] B. Yu, W.Q. Zhang, J. Chen, J.M. Xu, S.R. Wang, Sci. China Ser. B-Chem. 51 (2008) 289.

[307] K.R. Sridhar, R. Foertner, J. Propul. Power 16 (2000) 1105.

[308] D. Weng, S. Yates, AIAA Paper 2010-8675, American Institute of Aeronautics and Astronautics, Reston, VA, 2010.

[309] R. Rivera-Tinoco, C. Mansilla, C. Bouallou, F.F. Werkoff, Int. J. Nucl. Hydrogen Production Appl. 1 (2008) 249.

[310] J. Padin, T.N. Veziroglu, A. Shahin, Int. J. Hydrogen Energy 25 (2000) 295.

[311] C.M. Stoots, J.E. O'Brien, K.G. Condie, J.J. Hartvigsen, Int. J. Hydrogen Energy 35 (2010) 4861.

[312] http://www.navy.mil/submit/display.asp?story_id=92948 (accessed in February 2018).

Chapter 3

# Potential of Hydrogen Production From Biomass

Vaishali Singh, Debabrata Das
*Indian Institute of Technology, Kharagpur, West Bengal, India*

## INTRODUCTION

Rapid increase in population leads to radical increment in food and energy requirement of humanity [1]. In addition to population growth, the standard of living of human being is also increasing, which exerts tremendous pressure on the earth to regenerate its resources. So far a huge share of energy supply is coming from fossil fuel and its adverse effects such as climate change, ozone layer depletion, and greenhouse gas emission are well known. Moreover, fossil fuels are also running off at a quicker rate, as they are nonrenewable in nature. In case of developing countries, such as India, nearly about 80% of total crude oil is imported from other countries to fulfill the energy demand. In addition to this, around 54% of electricity production comes from thermal power plant that uses coal as energy source. Negative impacts of utilization of fossil fuels are leading toward application of renewable energy sources [2]. A potent alternative to fossil fuel—based energy system can be renewable energy—based system, as it can be replenished gradually. There are various forms of nonconventional energy sources such as hydro, solar, wind, nuclear power, geothermal, and biomass-derived energy. All of the renewable energy sources have great potential to supply the energy demand of the 21st century and future. Shift toward a carbon-neutral economy appears to be very crucial depending on present environmental scenario. As far as carbon-neutral energy source is concern, hydrogen gas seems to be more suitable in comparison with other sources. Hydrogen is a major source of energy behind solar power and most abundant element on the earth, nearly about three-fourth of all matter [3]. Hydrogen fuel has capability of supplying the energy demand without exerting negative impacts on environment. Supremacy of hydrogen as a fuel can be proved by its highest energy content of about 141 MJ/kg among all known fuels. There are several applications of hydrogen such as for electricity generation through fuel cell and as a fuel in internal combustion engines [4], in sea

*Science and Engineering of Hydrogen-Based Energy Technologies.* https://doi.org/10.1016/B978-0-12-814251-6.00003-4

transportation vehicles [5], and in chemical industries. Nowadays, various countries such as North America, Japan, Korea, India, China, and Europe are taking steps for the implementation of hydrogen as a fuel for the fuel cell busses [6]. Hyundai has displayed its hydrogen-fueled SUV (sports-utility vehicle) car "Nexo" at 201 consumer electronics show [7], and Toyota has launched "Mirai," a fuel cell car [8]. In addition, "Hydrogen Council," comprising originally 13 leading industries related to energy and transport, plans to increase the investment in hydrogen-related products up to 10 billion euros for the next 5 years to stimulate hydrogen as a key player of the future energy systems [9]. Advancements in hydrogen-based technologies can lead us to the conclusion of the beginning of hydrogen era. There are various methods of hydrogen production such as electrolysis of water, steam reforming of natural gas, partial oxidation of hydrocarbons, plasma arc decomposition, thermolysis, thermocatalysis [10], autothermal reforming, pyrolysis, coal gasification, and biomass gasification [11]. Percentage distribution of hydrogen production through different methods has shown in Fig. 3.1.

This chapter focuses on hydrogen production from biomass through biological routes. Various types of fermentation process for hydrogen production with special consideration of dark fermentation process are considered. Metabolic engineering of strains, mathematical modeling, and integration of different energy generation processes have been taken into consideration.

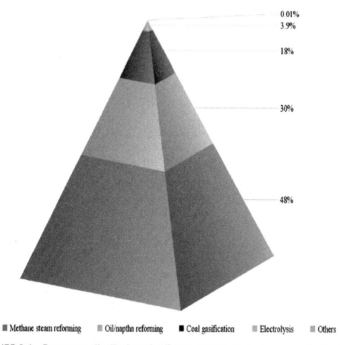

0.01%
3.9%
18%
30%
48%

■ Methane steam reforming    ▩ Oil/naptha reforming    ■ Coal gasification    ▩ Electrolysis    ▩ Others

**FIGURE 3.1** Percentage distribution of different methods for hydrogen production [12].

## HYDROGEN PRODUCTION FROM BIOMASS

Biomass is one of the major renewable sources for energy generation. It can be derived from several sources such as agricultural residues, forest residues, waste generated from various industries, and domestic household and municipal waste. Transformation of biomass into energy can be performed in various ways, e.g., biogas generation, hydrogen production, ethanol production, and biodiesel production. Biomass acts as natural battery, which stores the light energy of sun in the form of chemical bonds unless it is not harnessed. The two major ways of hydrogen production through biomass are thermochemical and biological routes. Fig. 3.2 shows different routes of hydrogen production from biomass.

### Hydrogen Production Through Thermochemical Process

Thermochemical process includes pyrolysis, gasification, and liquefaction. Pyrolysis is a process of production of liquid product, solid charcoal, and gaseous compounds from biomass, i.e., bio-oil. Bio-oil is a source for production of various products further. During pyrolysis, application of heat of about 650–800 K leads to dissociation of complex molecule into simpler compounds. If bio-oil is used for catalytic steam reforming over nickel catalyst at 750–850°C, then maximum hydrogen yield will be 0.172 g $H_2$/g bio-oil [11]. Gasification is a process involving a series of chemical reactions that convert biomass into mixture of gases at high temperature. This gas is

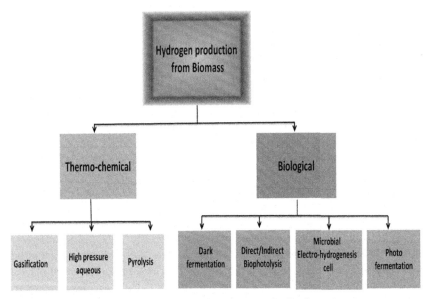

**FIGURE 3.2** Different routes of hydrogen production from biomass.

combustible in nature and contains hydrogen along with carbon dioxide, carbon monoxide, and methane [13]. To modulate the composition of gases, different parameters such as configuration of reactor and types of biomass need to be optimized. Low hydrogen content is one of the disadvantages of biomass gasification at larger scale. Liquefaction is the process of heating biomass to 525–600 K in water by applying pressure of 50–200 bars in the absence of air [14]. Addition of solvents or catalyst to the process can be performed for enhanced hydrogen production. Less hydrogen yield is the major limitation of biomass liquefaction.

## Hydrogen Production Through Biological Process

Hydrogen production through biological routes is renewable and eco-friendly in nature. Numerous kinds of biomass can be utilized for hydrogen generation, e.g., agricultural residues, organic wastewater, municipal solid waste. Dark fermentation, photofermentation, direct photolysis, indirect photolysis, and microbial electrohydrogenesis cells (MECs) are the various methods of conversion of biomass into hydrogen [15]. Fermentative bacteria are capable of fermenting complex substrate present in organic wastes into $H_2$ and $CO_2$. The photofermentation and direct/indirect photolysis of water have major problem of light dependency and lower yield of hydrogen production. Among all the processes, the most reliable method is dark fermentation by virtue of fact of higher rate and yield of hydrogen production. Moreover, it has process simplicity and moderate operational conditions.

## BIOMASS AS A FEEDSTOCK FOR HYDROGEN PRODUCTION

Biomass is the potent renewable sources for energy generation in developing countries mostly but recently in first world also. Presently, major focus is on identification of suitable feedstock based on the high-energy output. Since ancient times, biomass has been an important source of energy for humanity and it is evaluated that it contributes 10%–14% of total world's energy supply [16]. Of this, 25% biomass is used in developed countries and 75% in developing countries. There is an estimation that $3.87 \times 10^9$ ha of forest is present throughout the world [17]. Biomass feedstock can be of different types such as agricultural residues, forest residues, energy crops, agroindustrial waste, and municipal waste. Fig. 3.3 depicts various types of organic residues and wastes for hydrogen production. Besides biomass, organic wastes are also an important source of energy. Organic wastes can be converted into fuels using different technologies. In India, about 500 million metric tons of organic wastes per year are generated. This has a potential of about 17,500 MW power generation [18]. India has around 580 odd sugar mills, which generates bagasse. These bagasses have potential of generating about 5000 MW by cogeneration technique.

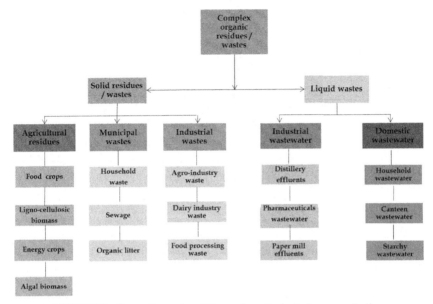

**FIGURE 3.3**  Types of organic residues and wastes for hydrogen production.

## Agricultural Crops

Globally, about 11.4 billion tons of biomass production from forest and agricultural land has been estimated per year, in which 40% is from agricultural production, 30% is from pasture, 18% stems from wood, and 12% are by-products [19]. The amount of agricultural residues generation is very high. Although the amount is large, there is huge agroclimatic diversity across globe, which leads to the need of development of different approaches for the utilization of different types of biomass-based energy generation system.

## Lignocellulosic and Agroforestry-Based Biomass

In plants, starch and cellulose are the two abundant polymer molecules whose basic unit is glucose. Plants store energy in the form of starch, whereas cellulose forms the structural skeleton of cell walls for the stalks, leaves, stems, and woody parts [20]. Moreover, cell walls of plants and some algal cultures also contain xylans. Xylans are polymeric sugars having xylose (a pentose sugar) as a basic unit. Similar to cellulose, xylans also have predominantly $\beta$-glycosidic bonds between xylose subunits [21]. Thus because of large abundance of lignocellulosic raw materials, it might be considered as suitable feedstock for the future biofuel industries. Lignocelluloses are biopolymers consisting of tightly bound lignin, cellulose, and hemicellulose. Sugar-rich plants such as sugarcane [22], *Miscanthus*, corn, and beetroot could be considered as a source of fermentable sugars [23]. Cultivation of such crops for biofuel purpose has evoked the debate on food versus fuel issues. Use of switch grass, rice husk, fodder straw, etc. for biofuel production has gained importance in recent years. Lignocellulosic crops need a systematic pretreatment step to remove the lignin content. After lignin removal, the crystalline cellulose is still not accessible to microbes [24]. Further saccharification of this crystalline cellulose yields simple sugars that could be used for biohydrogen production. Requirement of pretreatment and saccharification process increases the operational cost of the process. Moreover, many growth inhibitors such as furfurals are produced during pretreatment and saccharification processes. Lignocellulosic biomass might be in high abundance, but requirement of harsh pretreatment and saccharification processes limits their use as feedstock [25].

## Food Industry Wastes

According to a study conducted by the European Union in 2010, about 90 million tons of food waste was generated from food industry annually throughout the European Union and the United States [26].

Food processing industry is one of the booming businesses that contribute a chunk toward India's GDP. The processed food industries generate a lot of organic-rich wastewater. Disposal of food wastes possesses a great environmental threat. Food processing wastes contain high organic materials that are suitable substrates for dark fermentation [27].

## Dairy Industry Wastewater

Dairy industry is one of the rising industrial sectors across the globe. With increase in production of dairy products, dairy wastewater has also increased; for example, a typical dairy industry in the European Union produces approximately 180,000 m$^3$ of waste effluent per year [28]. These effluents have high biological oxygen demand (BOD) and chemical oxygen demand (COD), which make them harmful for environment if disposed untreated. The organic content of these wastewaters makes them ideal substrate for fermentative bacteria. Usage of such wastewater has been successful in case of biogas production. Hydrothermal pyrolysis could also be used to produce a variety of liquid fuels.

## Distillery Effluent

There is a huge production of alcohol throughout the world, which also produces a large amount of distillery waste; for example, annually, India produces 40 billion liter of distillery effluent. Approximately 12 L of stillage is generated per liter of alcohol production [29]. Distillery or alcoholic beverage industry wastewaters contain high BOD in the form of sugars, organic acids, dextrin, hemicelluloses, and resins. Distilleries based on molasses produce

8−15 L of wastewater per liter of ethanol. This wastewater has high COD of approximately 80−160 g/L [30]. This wastewater is also suitable for biogas production. The other technology of energy generation is incineration of distillery stillage. As there are various nutrients present in the effluent, it has the capability of sustaining the growth of microbes. Utilizing distillery effluent for hydrogen production not only helps in energy generation but also in wastewater treatment.

## Municipal Wastewater

In wastewater treatment process, a huge amount of activated sludge produces due to the production of biomass in a range of 0.3−0.5 g/g soluble COD reduction; for example, in 2010, about 10 million tons of dry sludge in the European Union, 8 million tons in the United States, and about 11 million tons in China have been generated during wastewater treatment [31].

In terms of availability, India produces more than 38,254 million liters of sewage every day [32]. The figure would further increase in the near future by virtue of development of the country's economy. Organic fractions of municipal solid wastes are commonly existing renewable supply, which can be utilized as a substrate for biohydrogen production. It is rich in polysaccharides and proteins. The dried sewage sludge generally incinerated to generate heat energy. Anaerobic digestion of these wastes generates good amount of methane that could be used to produce electricity [33].

# HYDROGEN PRODUCTION FROM BIOMASS USING BIOLOGICAL ROUTE

## Dark Fermentation

Anaerobic digestion process of organic matter is dark fermentation. It does not require light and occurs at normal temperature and pressure. For biohydrogen production, dark fermentation is the most potent and simplest way. The yield and rate of hydrogen production is higher in case of dark fermentation as compared with other biological hydrogen producing methods, viz., photosynthetic methods [3]. Dark fermentation is environmentally friendly method of hydrogen generation and lucrative alternative to fossil fuel−based systems. Dark fermentation involves two types of microorganisms either facultative or obligate anaerobes. Facultative anaerobes can grow in the presence of small amount of oxygen but only produce hydrogen in anaerobic condition. Obligate anaerobes are unable to grow even in trace amount of oxygen, which is highly toxic for growth of obligate anaerobes. Mixed acid metabolic pathway occurs in dark fermentation, which leads to production of different types of volatile fatty acids and alcohols such as acetate, butyrate, propionate, lactate, ethanol, acetone, and butanol. If organism follows acetate pathway, then total 4 mol of hydrogen will be produced from 1 mol of hexose, and in case of butyrate, 2 mol of hydrogen will be produced. If ethanol or lactate pathway occurs, then no hydrogen will be produced [34].

## Microorganism Involved in Dark Fermentation

Various organisms, growing in anaerobic environment, can produce hydrogen naturally. There are various fermentative organisms, which produce hydrogen via dark fermentation. The major types of fermentative bacteria can be classified based on the temperature requirement and other growth requirements such as the presence or absence of oxygen. Both facultative and obligate anaerobes can produce hydrogen via dark fermentation. Based on the temperature requirement, they can be further classified as mesophiles, which grow at ambient temperature, and thermophiles, which require high temperature for growth. Facultative mesophilic organisms are always more advantageous for performing experiments, as they are easier to handle. On the other hand, handling of obligate or thermophilic anaerobes is very tedious job. However, the yield of hydrogen is close to the theoretical value in case of thermophiles, as the thermodynamic barrier is less [35].

### Facultative Anaerobes

The major facultative species involved in hydrogen production is *Enterobacter* sp. This species mostly possesses formate hydrogen lyase (FHL) or [FeFe] hydrogenase enzymes, which are responsible for higher yield and high rate of hydrogen production [36]. Most studied strains of *Enterobacteriaceae* are *Enterobacter aerogenes* E.82005 isolated from leaves of *Mirabilis jalapa* by Tanisho et al. and *Enterobacter cloacae* IIT-BT 08 strain isolated from leaf extracts by Kumar and Das [37,38].

### Obligate Anaerobes

Obligate anaerobes are a more potent hydrogen producer in comparison with facultative anaerobes. In addition, they can utilize numerous varieties of carbohydrates along with various types of wastewater. The well-known obligate anaerobe for hydrogen production is *Clostridium* species. The metabolic pathway of hydrogen production is different. Hydrogen production occurs in exponential phase and in stationary phase, and metabolic shift occurs toward solvent production [39]. *Clostridium thermolacticum, Clostridium acetobutylicum, Clostridium thermocellum, Clostridium tyrobutyricum, Clostridium paraputrificum,* and *Clostridium saccharoperbutylacetonicum* are various strains mostly studied for hydrogen production. Zhang et al. achieved 0.680–1.270 mL $H_2$/g glucose from *C. acetobutylicum* ATCC 824 [40].

Among obligate anaerobes, some organisms require very high temperature for growth, known as thermophile. These organisms are found in deep-sea vent, hot springs etc. Thermophiles have different temperature as well as nutritional requirement based on their source of isolation [41,42]. One major component of medium is L-cysteine hydrochloride, which is a reducing agent for removing extra traces of oxygen from medium. Because the growth temperature is high, utilizing the complex substrate is easier for thermophiles as compared with mesophiles; for example, the genus *Caldicellulosiruptor* is a well-studied thermophile and cellulolytic in nature (Onyenwoke et al., 2006). Other examples of thermophilic genera are *Thermoanaerobacter, Ethanoligenens*, and *Thermotoga* [43,44].

## Coculture and Mixed Culture

Apart from pure culture, coculture and mixed culture have also been used in dark fermentation for hydrogen production. Coculture and mixed culture are preferably required for metabolizing complex substrate such as agroindustry waste and sewage sludge. Use of microbial consortium leads to efficient utilization of complex wastes because of the presence of various types of organisms. Another benefit of using mixed consortium is that it is less prone to contamination as compared with pure culture. On the other hand, one disadvantage of using mixed consortium at larger scale or industrial scale can be the stability of culture because the population dynamics keep on changing with respect to time and condition.

Similarly, coculture is beneficial when the characteristics of both organisms are required such as culturing facultative organisms with obligate anaerobes, which will eliminate the requirement of addition of expensive reducing agent [45].

## Metabolic Pathway Involved in Dark Fermentation

Dark fermentation is the metabolism of carbohydrates under anaerobic environment for producing energy for cell in the form of ATP. In this condition, kerb cycle is blocked. As a result, metabolism transfers the extra electrons for formation of reduced end metabolites such as acid and alcohol. Hydrogen is also produced in a similar way for maintenance of redox potential of cell. The major metabolites produced from glucose as substrate are acetic acid, butyric acid, and hydrogen. Glucose is first converted into pyruvate through glycolysis, and then pyruvate is converted into different subsequent metabolites. In case of organic polymers such as agroindustrial waste, it is first hydrolyzed into simple sugars and then converted into pyruvate and further metabolites. Pyruvate can be involved in two biochemical pathways either acetate producing or butyrate producing depending on the conditions: 4 mol of hydrogen is produced from 1 mol of glucose, whereas 2 mol of hydrogen is produced from 1 mol of glucose by butyrate pathway (Fig. 3.4).

In the metabolic pathway of anaerobes, glucose is first converted into pyruvate by Embden-Meyerhof pathway, and then pyruvate is oxidized to acetyl CoA. This results in generation of ATP along with ferredoxin (Fd(red)) or formate. Facultative anaerobes derive hydrogen through formate by utilizing

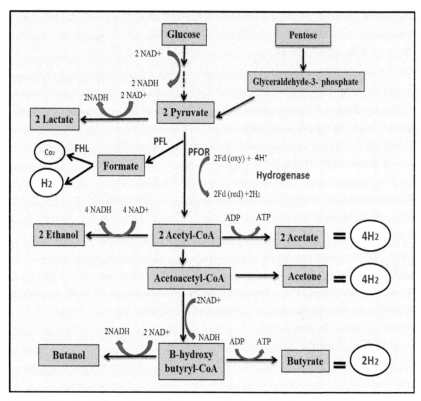

**FIGURE 3.4** Metabolic pathway of hydrogen production. *FHL*, formate hydrogen lyase; *PFL*, pyruvate formate lyase.

enzyme FHL, and obligate anaerobes derive hydrogen through Fd (red) by utilizing enzyme [FeFe] hydrogenase. Reduced ferredoxin is oxidized by hydrogenase, which leads to formation of Fd and electrons in the form of molecular hydrogen [46]. In case of obligate anaerobes, pyruvate is converted into acetyl CoA by enzyme pyruvate ferredoxin oxidoreductase, which produces reduced Fd and subsequently hydrogen. The overall reaction is depicted in Eqs. (3.1) and (3.2)

$$\text{Pyruvate} + \text{CoA} + 2\text{Fd(ox)} \rightarrow \text{acetyl CoA} + 2\text{Fd(red)} + \text{CO}_2 \quad (3.1)$$

$$2\text{H}^+ + \text{Fd(red)} \rightarrow \text{H}_2 + \text{Fd(ox)} \quad (3.2)$$

If an organism follows mixed acid pathway, then the ratio of acetate and butyrate is very crucial for hydrogen production. Eqs. (3.3) and (3.4) depict the overall biochemical reaction of production of acetic acid and butyric acid as end metabolites.

$$\text{C}_6\text{H}_{12}\text{O}_6 + 2\text{H}_2\text{O} \rightarrow 2\text{CH}_3\text{COOH} + 2\text{CO}_2 + 4\text{H}_2 \quad (3.3)$$

$$C_6H_{12}O_6 \rightarrow CH_3CH_2CH_2COOH + 2CO_2 + 2H_2 \qquad (3.4)$$

In facultative anaerobes such as *Escherichia coli*, the pyruvate is oxidized to acetyl CoA and formate by the action of pyruvate formate lyase (PFL) [47,48]. This reaction is shown in Eq. (3.5).

$$\text{Pyruvate} + \text{CoA} \rightarrow \text{acetyl CoA} + \text{formate} \qquad (3.5)$$

Carbon dioxide and hydrogen are produced due to the cleavage of formate by the help of FHL (Eq. 3.6). It was first explained by Stephenson and Stickland in the 1930s [49].

$$HCOOH \rightarrow CO_2 + H_2 \qquad (3.6)$$

However, if lactic acid, propionic acid, and ethanol are produced, no hydrogen will be produced.

## Metabolic Engineering for the Improvement of Hydrogen Production

Biological processes of hydrogen production are more eco-friendly as compared with chemical processes, but the rate and yield of hydrogen production is low in case of biological processes. There is a need of application of various strategies for the improvement in production of hydrogen. Metabolic engineering is one of the most significant methods to improve the overall process. Currently, many advances such as expression analysis, genome sequencing, and gene engineering have been developed in metabolic engineering. By application of metabolic engineering, production of native or nonnative products can be improved [50]. Metabolic engineering is a set of steps for the improvement of cellular characteristics with the help of modification in a specific metabolic reaction or introduction of new reactions, by applying recombinant DNA technology. Cellular properties depend on the metabolic fluxes. Thus, by controlling metabolic flux, desired metabolic reactions or pathway can be directed in a particular organism. For the modification of metabolic pathways, first complete knowledge about the pathway, such as what are the intermediate metabolites, end metabolites, cofactors, and enzymes, is required. This is followed by desired modification to produce required product. In these methods, molecular approaches for controlling the enzymes or gene products are very essential [51]. Various metabolic modification strategies, such as increment in flux through deletion of genes of competing pathways, upregulation of hydrogen evolving enzymes, and knockout of hydrogenase uptake enzymes, can be applied for the enhancement of hydrogen production. Metabolic pathways for facultative and obligate anaerobes are different, so different strategies are required for modification of the involved pathways [52]. The theoretical yield of hydrogen can be achieved only when end metabolite is acetate or acetone. If the end metabolite is butyric acid, then 2 mol of hydrogen can be produced. If the end metabolite of the pathway is ethanol, lactic acid, or propionate, then no hydrogen production takes place. Key enzyme for hydrogen production is [Fe–Fe] hydrogenase that is encoded by *hyd* gene in *C. acetobutylicum* [53]. However, the enzyme needs additional accessory proteins for functional hydrogenase to be expressed. The *hyd E*, *hyd F*, and *hyd G* genes of *C. acetobutylicum* are known to encode accessory proteins. There are various studies conducted on genetic modification for the enhancement of hydrogen production. *Clostridium paraputrificum* M-21 has been mutated by overexpression of *hydA* gene that encodes [Fe–Fe] hydrogenase. This leads to 1.7-fold increment in hydrogen production from *N*-acetylglucosamine (GlcNAc). Improvement in hydrogen production is due to increased acetate production and decreased lactic acid production [54]. It was observed that 1.5-fold increase in production of hydrogen and 1.4-fold increment in hydrogenase activity by knockout of *ack* encoding acetate kinase for production of acetate in *C. tyrobutyricum* [55]. There is another approach of addition of cellulose degrading pathway in *Clostridium* by heterologous expression of alcohol dehydrogenase and pyruvate decarboxylase from *Zymomonas mobilis*; this approach will reduce the toxic effect of pyruvate buildup. Ren et al. showed that *C. acetobutylicum* and *Clostridium populeti* give highest yield of hydrogen by metabolizing cellulose. *C. acetobutylicum* gave 2.3 mol/mol glucose from cellobiose [56]. In case of facultative fermentative anaerobes, FHL system is a key player for hydrogen production. It is a complex containing multienzymes, which leads to molecular hydrogen production from formate. FHL complex has been reported to be present in many facultative bacteria such as *Klebsiella pneumoniae, Rhodospirillum rubrum* [57], *Salmonella typhimurium* [58], and *E. coli* [59]. In facultative anaerobes, such as *E. coli*, pyruvate is converted into acetyl CoA and formate by action of enzyme PFL. Bisaillon et al. have studied the effect of mutation in uptake hydrogenase enzyme, i.e., Hyd-1, Hyd-2 along with *ldhA* and *fhlA* in *E. coli* BW 545. This mutation of *ldhA* and *fhlA* resulted in 18% and 11% increase in hydrogen production, and double mutant leads to 47% increase in hydrogen production [60]. Another study was conducted by Maeda et al. [61,62] by creating mutation in *E. coli* for the improvement of hydrogen production. In this study, *hyaB* and *hyaC* genes were deleted to remove hydrogen uptake activity of *Hyd-1* and *Hyd-2*. Along with this, *fhlA* gene was overexpressed; that is, the activator of FHL complex and *hycA* gene was deleted, that is, FHL repressor. Additionally, they deleted the alpha subunit of formate dehydrogenase-N and formate dehydrogenase-O encoded by *fdnG* and *fdoG*, respectively, and the alpha subunit of nitrate reductase A encoded by *narG* to inactivate formate consumption. Export of formate has been prevented by *focA* and *focB* deletions. This mutant strain with four modifications of BW25113 showed 141-fold higher hydrogen production in comparison with the wild-type strain BW25113. Various metabolic modifications performed in different microbes such as bacteria and cyanobacteria for the improvement in hydrogen production are shown in Table 3.1.

**TABLE 3.1** Metabolic Engineering Approaches for Enhanced Biohydrogen Production Using Anaerobes

| Microorganisms | Genetic Modifications | Outcomes | References |
|---|---|---|---|
| *Caldicellulosiruptor bescii* | Knockout of lactate dehydrogenase gene (*ldh*) utilizing dual selection marker around *pyrF* | Increment in acetate and hydrogen production | [63] |
| *Clostridium paraputrificum M-21* | Homologous overexpression of *hydA* gene | Hydrogen production increased by 1.7-fold | [54] |
| *Clostridium perfringens* strain W11 | Lactate dehydrogenase gene (*ldh*) deleted | Production of hydrogen increased by 51% | [64] |
| *Clostridium saccharoperbutylacetonicum* strain N1-4 | Down regulation of *hupCBA* gene cluster; by antisense RNA strategy | Production of hydrogen increased by 3.1-fold | [65] |
| *Clostridium tyrobutyricum* | Integrational mutagenesis to inactivate acetate kinase gene (*ack*) | Production of hydrogen increased by 50% | [55] |
| *Clostridium tyrobutyricum JM1* | Homologous overexpression of *hydA* gene | Hydrogen production increased by 1.5-fold | [66] |
| *Enterobacter aerogenes IAM 1183-O* | Knockout of uptake hydrogenase | Hydrogen yield increased by 20% | [67] |
| *Escherichia coli* | Overexpression of hydrogenase | Hydrogen production increased by 41-fold | [68] |
| *E. coli BW25113* | Overexpression of (*fhlA*) and knockout of uptake hydrogenase | Hydrogen production increased by 4.6-fold | [61,62] |
| *E. coli K-12 BW25113* | Hydrogenase 3 protein engineering | Hydrogen yield increased by 23-fold | [69] |

*Continued*

**TABLE 3.1** Metabolic Engineering Approaches for Enhanced Biohydrogen Production Using Anaerobes—cont'd

| Microorganisms | Genetic Modifications | Outcomes | References |
|---|---|---|---|
| E. coli strain MC4100 | Knockout of twin-arginine translocation (tat) system genes | Rate of hydrogen production doubled in mutant | [70] |
| E. coli strain S13 | Deletion of FHL repressor (hycA) and overexpression of FHL activator (fhlA) | Hydrogen productivity increased by 2.8-fold | [71] |
| Enterobacter cloacae CICC10017 | Heterologous overexpression of hydrogen promoting protein of Enterobacter cloacae IIT-BT 08 | Hydrogen yield increased by 2-fold | [72] |
| Synechococcus sp. strain PCC 7002 | Mutation in ldhA | Hydrogen production increased by 5-fold | [73] |
| Thermoanaerobacterium aotearoense | Deletion of lactate dehydrogenase gene (ldh) | Hydrogen production increased by 2-fold | [74] |

FHL, formate hydrogen lyase.

## *Effect of Physicochemical Parameters on Hydrogen Production*

Various factors, such as temperature, pH, medium composition, hydrogen partial pressure, volatile fatty acid, hydraulic retention time (HRT), affect hydrogen production via dark fermentation. Each factor has its own importance on the overall yield and rate of hydrogen production.

### Temperature

Dark fermentation can be performed at different range of temperatures, 25−80°C. There are different microorganisms, such as mesophiles (25−40°C), thermophiles (40−65°C), hyperthermophiles (65−80°C), and extreme thermophiles (>80°C), carrying dark fermentation [75]. Temperature affects the growth rate of organisms. Hydrogen is a growth-associated product; thus if growth is affected, then it will influence the hydrogen production. Most of the studies in laboratory scale have been conducted in mesophilic conditions, as it is easy to maintain. Zhang et al. (2005) have studied the effect of temperature on hydrogen, and they found increasing fermentation temperature that favored hydrogen production in the range of 25−40°C, but after increasing the temperature beyond 40°C, hydrogen production was inhibited. The probable reason can be inhibition of cellular enzymes. The suitable temperature for hydrogen production with yield of 185 mL $H_2$/g of sucrose was found to be 35°C. Their study demonstrated that temperature affects biohydrogen production by shifting the microorganism metabolic pathway [76]. Thermodynamically higher temperature is favorable for hydrogen production because of increment in the entropy of the system. However, from the economic perspective, extreme thermophilic conditions may not be viable because of intensive energy requirements to maintain the high temperatures [77].

### pH

pH of fermentation medium is one of the most influencing parameters for hydrogen production, as it affects the functioning of different cellular enzymes, ultimately affecting metabolic pathway of organisms. The activity of hydrogenase enzyme also gets affected with change in pH that is responsible for hydrogen production. The hydrogen yield depends on the metabolic pathway of the fermentative bacteria [78]. If acid concentration increases in fermentation medium, it leads to reduction in medium pH in response to this metabolism shift toward conversion of acids into alcohols to reduce the toxicity offered by acid accumulation [79].

### Medium Composition

Influence of medium composition affects the hydrogen production, as hydrogen is a growth-associated product. Nutrient is necessary in proper concentration for optimum growth of fermentative organisms. Appropriate concentration of carbon source, nitrogen source, phosphate source, vitamins, and trace metals is very essential to support desired growth and maintain maximum specific growth rate. Nitrogen is required for formation of amino acids, subsequently proteins and enzymes; hence, it is very essential for growth of microbes. It has been observed that the addition of nitrogen supplement to culture medium leads to significant increase in hydrogen production. Moreover, the C/N ratio also plays a significant role in stabilizing the dark fermentation process and affecting the hydrogen productivity and the specific hydrogen production rate [80]. For efficient production of hydrogen, supplementation of trace metals is very necessary, e.g., iron is required for functioning of [FeFe] hydrogenase, that is, the essential enzyme involved in hydrogen production. Several reports on the effect of iron on hydrogen production are available. In another study, Lin and Lay (2004) reported the effect of various trace elements such as Mg, Na, Zn, Fe, K, I, Co, Mn, Ni, Cu, Mo, and Ca for hydrogen production using *Clostridium pasteurianum*. The study showed that suitable concentrations of Mg, Na, Zn, and Fe were necessary for higher hydrogen yield. Based on these results, they proposed an optimal nutrient formulation containing (mg/L): $MgCl_2 \cdot 6H_2O$ 120, NaCl 1000, $ZnCl_2$ 0.5, and $FeSO_4 \cdot 7H_2O$ 3 [81].

## Partial Pressure

Partial pressure of hydrogen is an important factor that affects the hydrogen production. According to Le Chatelier's principle (Eq. 3.7),

$$\text{Equilibrium constant, } K_c = \frac{\text{Product concentration}}{\text{Substrate concentration}} \qquad (3.7)$$

Removal of the product from the reaction mixture can increase the substrate conversion efficiency. Therefore, the partial pressure and hydrogen production are inversely proportional. As hydrogen production will reduce, then metabolic pathway will also shift toward the more reduced end metabolites such as ethanol, lactic acid, butanol, and acetone. Many studies have been conducted to find out optimal partial pressure for hydrogen production. Junghare et al. (2011) studied that decreased partial pressure increases the hydrogen yield from 2.67 to 3.1 mol of $H_2$/mol of glucose. Thus, maintaining reduced partial pressure inside the system is very important to achieve maximum production [82].

## Soluble End Metabolites

Generation of number of moles of hydrogen depends on the type of end metabolite produced inside the system; hence the concentration of each metabolite as well as its ratio is the deciding factor in hydrogen production process. Major volatile fatty acids produced in dark fermentation are acetic acid, butyric acid, ethanol, propionic acid, lactic acid, etc. [83]. At higher concentration of acids, the ionic strength increase inside the cell leads to permeation of proton from cell membrane, which can disturb the physiological balance of cell. Owing to this disruption, extra maintenance energy is required for the growth of organisms and hydrogen production [84]. Increasing concentration of acids can have inhibitory effect on hydrogen production [83].

### Hydraulic Retention Time

In case of continuous hydrogen production, HRT is a very important factor for hydrogen production. It is directly related to the maximum specific growth rate of the organisms considered for the hydrogen production. If the growth rate of organisms is high, then HRT should be less and vice versa to get the maximum output. HRT in continuous stirred tank reactor can be controlled by controlling the dilution rate. HRT and dilution rate are inversely proportional in nature. Dilution rate can be controlled by controlling the feed flow rate. Eq. (3.8) can express dilution rate

$$D = \frac{F}{V} \tag{3.8}$$

where $D$ is dilution rate, $F$ is volumetric flow rate, and $V$ is reactor volume.

As the volume of reactor is constant, by changing the flow rate, variation in dilution rate can be attained. HRT is very important mode for selection of microbial flora, based on growth rate. In the chemostat, if HRT is less than the generation time of the slow growing microbes, these cells will be washed out from the system and only fast-growing organisms will survive. For example, methanogens are slow growing in nature, and their specific growth rate is very low, approximately $0.0167-0.02$ h$^{-1}$, as compared with acidogens, $0.172$ h$^{-1}$ [85]. Thus, in case of mixed culture studies, lower HRT will support the growth of acidogens and limit the growth of methanogens. Hydrogen production from different pure substrates has been shown in Table 3.2 and from different types of biomass has been shown in Table 3.3.

## Mathematical Modeling of Biohydrogen Production Processes

Mathematical model deals with the representation of a process. It helps in the explanation of behavior of the system. Validation of the model using experimental results is crucial to prove the significance of application of the model. Theoretical prediction provided by the model has to be compared with the repeated observation obtained from the experiment. The deviation between the experimental value and predicted value should be less than 5%. If the deviation is more than 5%, then the model requires some modifications.

**TABLE 3.2** Hydrogen Production From Pure Organic Substrate by Dark Fermentation

| Microorganisms | Fermentation Modes | Substrates | Substrate Concentration | Hydrogen Yield | References |
|---|---|---|---|---|---|
| Bacillus coagulans IIT-BT S1 | Batch | Glucose | 20 g/L | 2.28 mol/mol glucose | [86] |
| Clostridium acetobutylicum | Continuous | Glucose | – | 2 mol/mol glucose | [87] |
| C. acetobutylicum | Batch | Glucose | 15 g/L | 1.57 mol/mol glucose | [88] |
| C. saccharoperbutylacetonicum | Batch | Glucose | 15 g/L | 2.48 mol/mol glucose | [88] |
| C. beijerinckii | Batch | Glucose | – | 2.81 mol/mol hexose | [89] |
| C. cellobioparum | Batch | Glucose | – | 2.73 mol/mol hexose | [90] |
| Clostridium sp. | Continuous | Glucose | 20 g COD/L | 1.7 mol/mol hexose | [91] |
| Clostridium CGS2 | Batch | Starch | – | 2.03 mol/mol glucose | [92] |
| Clostridium sp. | Batch | Glucose | – | 2.8 mol/mol hexose | [93] |
| E. cloacae BL-21 | Batch | Glucose | 10 g/L | 3.12 mol/mol hexose | [94] |
| E. cloacae DM11 | Batch | Glucose | 10 g/L | 3.31 mol/mol glucose | [95] |
| E. cloacae IIT-BT 08 | Batch | Glucose | 10 g/L | 2.2 mol/mol glucose | [38] |
| E. cloacae IIT-BT 08 | Batch | Cellobiose | 10 g/L | 5.4 mol/mol cellobiose | [38] |
| E. cloacae IIT-BT 08 | Continuous | Glucose | 10 g/L | 75.6 mmol/L/h | [96] |
| E. coli | Batch | Glucose | 20 g/L | $4.73 \times 10^{-8}$ mol/mol glucose | [97] |
| E. coli | Batch | Glucose | – | 2.0 mol/mol hexose | [60] |

| | | | | | |
|---|---|---|---|---|---|
| Klebsiella oxytoca HP1 | Batch | Glucose | 50 mM | 1 mol/mol glucose | [98] |
| Klebsiella oxytoca HP1 | Batch | Sucrose | 50 mM | 1.5 mol/mol sucrose | [98] |
| Mixed culture | Batch | Glucose | 1 g COD/L | 0.9 mol/mol glucose | [99] |
| Mixed culture | Batch | Sucrose | 1 g COD/L | 1.8 mol/mol sucrose | [99] |
| Mixed culture | Continuous | Starch | 10 g/L | 0.83 mol/mol starch d | [100] |
| Mixed culture | Batch | Starch | 1 g COD/L | 0.59 mol/mol starch | [99] |
| Mixed culture | Continuous | Glucose | 13.7 g/L | 1.2 mol/mol glucose | [101] |
| Mixed culture | Continuous | Sucrose | 20 g COD/L | 3.47 mol/mol sucrose | [102] |
| Mixed culture | Continuous | Sucrose | — | 2.1 mol/mol sucrose | [103] |
| Mixed culture | Batch | Glucose | 10 g/L | 44% hydrogen fraction | [22] |
| Mixed culture | Continuous | Sucrose | 20 g COD/L | 2.6 mol/mol glucose | [80] |
| Mixed culture | Continuous | Sucrose | 20 g COD/L | 1.5 mol/mol sucrose | [104] |
| Mixed culture | Continuous | Sucrose | 20 g COD/L | 1.48 mol/mol sucrose | [105] |
| Mixed culture | Continuous | Starch | 6 kg starch/m$^3$ | 1.29 L/g starch COD | [106] |
| Mixed microflora | Batch | Starch | — | 1.5 mol/mol hexose | [107] |
| Sludge compost | Batch | Glucose | 10 g/L | 2.1 mol/mol glucose | [108] |
| Thermoanaerobacterium | Batch | Glucose | — | 2.67 mol/mol hexose | [109] |
| Thermoanaerobacterium | Batch | Starch | 4.6 g/L | 92 mL/g starch | [110] |

COD, chemical oxygen demand.

**TABLE 3.3** Hydrogen Production From the Different Organic Wastes by Dark Fermentation

| Substrates | Microorganisms | Types of Fermentation | Substrate Concentration | Hydrogen Yield | References |
|---|---|---|---|---|---|
| Algal biomass (*Scenedesmus obliquus*) | *Enterobacter aerogenes* | Batch | 2.5 g algae/L | 40.9 mL $H_2$/g alga | [111] |
| Brewery wastewater | Mixed culture | Batch | 19.2 g VSS/L | 6.11 mmol/g COD | [112] |
| Cattle wastewater | Mixed culture | Batch | 2.648 g COD/L | 12.41 mmol/g COD | [113] |
| Cheese processing wastewater | Mixed culture | Continuous | – | 3.21 mmol/g COD | [114] |
| Cheese whey | *C. saccharoperbutylacetonicum* ATCC27021 | Batch | 102 g COD/L | 2.7 mol $H_2$/mol lactose | [115] |
| Cheese whey wastewater | Mixed culture | Continuous | – | 22.00 mmol/g COD | [116] |
| Citric acid wastewater | Mixed culture | Continuous | – | 0.84 mol $H_2$/mol hexose | [117] |
| Coffee drink manufacturing wastewater | Mixed culture | Continuous | – | 0.20 mol $H_2$/mol hexose | [118] |
| Condensed molasses fermentation solubles | Mixed culture | Batch | 40 g COD/L | 1.5 mol $H_2$/mol hexose | [39] |
| Condensed molasses fermentation solubles | Mixed culture | Continuous | 80 g COD/L-d* | 0.9 mol $H_2$/mol hexose | [119] |
| Dairy waste | Mixed culture | Continuous | – | 1.105 mmol $H_2$/m$^3$/min | [120] |

| Substrate | Culture | Mode | Concentration | Yield | Reference |
|---|---|---|---|---|---|
| Depackaging wastes (DWs) | Mixed consortium | Batch | 25 g/L | 0.4 mol/mole lactate | [121] |
| Distillery effluent | Coculture of C. freundii 01, E. aerogenes E10 and R. palustris P2 | Batch | 6.95 g/L | 2.76 mol H$_2$/mol hexose | [122] |
| Distillery wastewater | Mixed culture | Batch | 9.6 kg COD/m$^3$-day | 6.98 mol H$_2$/kg COD$_R$-day | [123] |
| Domestic sewage | Mixed culture | Batch | 305 mg VSS/L | 6.01 mmol/g COD | [124] |
| Food industry waste | Mixed culture | Continuous | – | 101.75 ± 3.71 L H$_2$/kg FIW | [125] |
| Food waste | Mixed culture | Continuous | – | 0.39 L H$_2$/g COD | [39] |
| Fruit and vegetable waste | T. maritima | Batch | 91 g COD/L | 3.46 mol/mol | [126] |
| Glycerin wastewater | Mixed culture | Batch | 255.0 mg VSS/L | 6.03 mmol/g COD | [124] |
| Lagoon wastewater | Mixed culture | Batch | 3 g COD/L | 0.5 mol H$_2$/mol hexose | [127] |
| Microcrystalline cellulose | C. acetobutylicum X9 + Ethanoligenens harbinense B49 | Batch | 10 g/L | 1.8 L H$_2$/L-POME | [128] |
| Molasses | Mixed culture | Continuous | 3.11–85.57 kg COD/m$^3$ | 5.57 m$^3$ H$_2$/m$^3$ reactor.d | [125] |
| Olive mill wastewater | Mixed culture | Batch | – | 0.54 mmol/g COD | [129] |
| Olive pulp water | Mixed culture | Continuous | – | 2.8 mol H$_2$/mol hexose | [130] |
| POME | Thermoanaerobacterium-rich sludge | Batch | 70–90 g COD/L | 6.33 L H$_2$/L-POME | [131] |

Continued

**TABLE 3.3** Hydrogen Production From the Different Organic Wastes by Dark Fermentation—cont'd

| Substrates | Microorganisms | Types of Fermentation | Substrate Concentration | Hydrogen Yield | References |
|---|---|---|---|---|---|
| POME | Mixed culture | Fed-Batch | — | 4.7 L $H_2$/L-POME | [132] |
| POME | Mixed culture | Batch | — | 2.3 L $H_2$/L-POME | [132] |
| POME | Mixed culture | Continuous | 59 g COD/L | 0.42 L/g $COD_{reduced}$ | [133] |
| Probiotic wastewater | Mixed culture | Batch | 9.48 g COD/L | 1.8 mol $H_2$/mol hexose | [134] |
| Purified terephthalic acid | Mixed culture | Continuous | 16 kg COD/$m^3$ d | 19.29 mmol/g COD | [135] |
| Rice slurry | Mixed culture | Batch | 5.5 g carbohydrates/L | 346 mL $H_2$/g carbohydrate | [136] |
| Rice spent wash | Mixed culture | Batch | 60 ± 2.6 g COD/L | 464 mL/g reducing sugar | [137] |
| Rice winery wastewater | Mixed culture | Continuous | 14–36 g COD/L | 2.14 mol $H_2$/mol hexose | [138] |
| Starch wastewater | *Thermoanaerobacterium* sp. mixed culture | Batch | 9.2–36.6 g/L | 92 mL $H_2$/g starch | [110] |
| Sugar beet wastewater | Mixed culture | Continuous | 16 Kg total sugars/$m^3$.d | 1.7 mol $H_2$/mol hexose | [139] |
| Sugary wastewater | Sludge compost | Continuous | — | 2.52 mol $H_2$/mol hexose | [140] |

COD, chemical oxygen demand

### Monod Growth Model for Cell Growth Kinetics

Monod growth model explains the cell growth kinetics. Jacques Monod [141] proposed an equation showing the relation between limiting substrate concentration and specific growth rate of microbes (Eq. 3.9).

$$\mu = \frac{\mu_{max}S}{K_S + S} \tag{3.9}$$

where $\mu$ is the specific growth rate of microorganisms ($h^{-1}$), $\mu_{max}$ is the maximum specific growth rate of microorganisms ($h^{-1}$), $S$ is the limiting substrate concentration (g/L), and $K_S$ is the saturation constant (g/L).

The Monod equation can also be written in the Lineweaver–Burk plot (Eq. 3.10):

$$\frac{1}{\mu} = \frac{K_S}{\mu_{max}S} + \frac{1}{S} \tag{3.10}$$

Regression analysis can be used to find the best fit for a straight line of $1/\mu$ versus $1/S$. The $K_S$ and $\mu_{max}$ values can be calculated from the intercept and the slope of the straight line, respectively.

### Modeling of Biohydrogen Production Using Modified Gompertz Equation

The cumulative production of biohydrogen, acetate, butyrate, etc. in a batch process can be nonlinearly modeled by the modified Gompertz equation (Eq. 3.11), which can be written as

$$P_i = P_{max,i} \exp\left\{ -\exp\left[ \frac{R_{max,i}e}{P_{max,i}} (\gamma - t) + 1 \right] \right\} \tag{3.11}$$

where $P_i$ is the product $i$ produced per liter of the reactor volume at fermentation time $t$, $P_{max,i}$ is the potential maximum product formed per liter of the reactor volume, and $R_{max,i}$ is the maximum rate of the product formed [142].

### Luedeking–Piret Model for Product Formation Kinetics

The relationship of cell growth and product formation can be modeled using Luedeking–Piret model as shown in Eq. (3.12):

$$\frac{dp}{dt} = \frac{dx}{dt}\alpha + \beta X \tag{3.12}$$

where $\alpha$ is the growth-associated formation coefficient of the product and $\beta$ is the non–growth-associated formation coefficient of product $i$ [142].

## Photobiological Processes

### Photofermentation

Photosynthetic bacteria can convert organic compounds such as volatile fatty acids into hydrogen and carbon dioxide by utilizing light energy of sun under anaerobic conditions; this process is known as photofermentation. Example of such bacteria is purple nonsulfur (PNS) bacteria [3,143]. PNS bacteria can grow photoheterotrophically to produce hydrogen. They can also utilize glucose and sucrose as a carbon source other than volatile fatty acid (VFA) for production of hydrogen [144]. There are several other options of metabolism for PNS, such as photoautotrophic metabolism, aerobic anaerobic respiration, and fermentation, which are non−hydrogen-producing pathways. Hence, it is necessary to maintain the conditions in such a way that organism grows photoheterotrophically [145]. There are several PNS bacteria, such as *Rhodobacter sphaeroides*, *Rhodopseudomonas palustris*, *Rhodobacter capsulatus*, and *Rhodospirillum rubrum*, used for hydrogen production [146−148].

The main enzymes involved in hydrogen production through photofermentation are nitrogenase and uptake hydrogenase. Nitrogenase enzyme promotes hydrogen production, whereas uptake hydrogenase consumes hydrogen. The growth medium of PNS should contain less nitrogen that means high C/N ratio, which will help the organism in dumping the excess energy and reducing power for hydrogen production [145].

The hydrogen production takes place by the action of nitrogenase enzyme, which works under the absence of molecular nitrogen. Eq. (3.13) depicts the formation of hydrogen by nitrogenase enzyme.

$$2H^+ + 2e^- + 4ATP \xrightarrow[\text{Nitrogenase}]{\text{Light}} H_2 + 4ADP + 4P_i \qquad (3.13)$$

For proper functioning of nitrogenase enzyme, huge number of ATP is required; thus strict control of environmental conditions is required [145]. Optimum parameters, such as pH, temperature, light intensity, and substrate concentration, are very necessary for the desired growth of PNS. The optimum values of these parameters vary with strain to strain and substrate to substrate. Thitirut et al. (2015) have found that for *Rhodobacter* sp. KKU-PS, the optimum initial pH, temperature, and light intensity are 7.0, 25.6°C, and 2500 lux, respectively. Under these conditions, they achieve maximum hydrogen production of 1353 mL/L of medium [149].

Photofermentation can be used in integration with dark fermentation, as the huge amount of spent is produced after dark fermentation, which is rich in volatile fatty acids such as acetate, butyrate, and propionate. If photofermentation is followed by dark fermentation, the overall energy recovery from the substrate can be maximized. Photofermentative bacteria can utilize the VFAs of dark fermentative spent for production of hydrogen and carbon dioxide. This is a two-stage process of hydrogen production. Total 12 mol of hydrogen can be produced from 1 mol of glucose theoretically from the two-stage process [150]. Eq. (3.14) depicts the reaction that takes place in photofermentation. 8 mol of $H_2$ can be produced from photofermentation, and 4 mol, from dark fermentation.

$$2CH_3COOH + 4H_2O \rightarrow 8H_2 + 4CO_2 \qquad (3.14)$$

## Algal Fermentation

Algae are the feedstock of third-generation biofuel, and they are used to minimize the food versus fuel conflict [151]. Algae have appeared as sustainable biomass for bioenergy generation in recent years. They have several benefits compared with other biomass such as it can grow in nonarable land, it can grow in wastewater and marine water, its growth rate is faster than terrestrial plants, and their photosynthetic efficiency is higher than that of higher plants. Hydrogen can be produced from algae through various routes such as direct photolysis, indirect photolysis, photofermentation, and dark fermentation as substrate for the acidogenic bacteria.

### Direct Biophotolysis

It is the process of hydrogen generation by utilizing the reductive equivalents, produced through photolysis of water. Algae use sunlight through the process of photosynthesis for splitting of water into hydrogen and oxygen. Hydrogen production takes place at the reducing site of photosystem I with simultaneous oxygen generation at the oxidizing site of photosystem II. The electrons generated through the oxidation of water molecules flow to Fd, which donates the electrons to hydrogenase for production of hydrogen.

Various green microalgae such as *Chlamydomonas reinhardtii*, *Chlorococcum littorale*, *Chlorella fusca*, *Platymonas subcordiformis*, and *Scenedesmus obliquus* have [FeFe]-hydrogenase enzyme for hydrogen production [152].

$$2H_2O \xrightarrow{\text{Light energy}} 2H_2 + O_2 \qquad (3.15)$$

### Indirect Biophotolysis

Indirect biophotolysis is the process of hydrogen production in which electrons are derived from stored organic compounds such as starch in algae and glycogen in cyanobacteria. It is a two-phase process; in first phase, photosynthesis and carbohydrate storage take place. In second phase, fermentation of stored carbohydrates leads to production of hydrogen. Hydrogen evolution has sensitivity problem by oxygen evolution; hence separation of two stages is necessary to achieve hydrogen production. Separation can be temporal and/or spatial. In these two stages, $CO_2$ fixation and evolution take place. Process of indirect biophotolysis is shown in Eqs. (3.16) and (3.17):

$$12H_2O + 6CO_2 + \text{light} \rightarrow C_6H_{12}O_6 + 6O_2 \qquad (3.16)$$

$$C_6H_{12}O_6 + 12H_2O + \text{light} \rightarrow 12H_2 + 6CO_2 \qquad (3.17)$$

Apart from biophotolysis, fermentation is another way of hydrogen production through microalgae. By the action of nitrogenase enzyme, hydrogen can be produced through photofermentation [153].

## Microalgae as Substrate for Dark Fermentation

Microalgae have been used for dark fermentation and hydrogen generation from years, as it has potential to provide various nutrients required for the bioenergy generation [154−156]. Unlike second-generation feedstock, i.e., lignocellulosic biomass, it does not have lignin in the cell wall. As a result, rigorous pretreatment is not required for exposure of stored carbohydrates inside the cell of microalgae. There are numerous studies available for hydrogen production using microalgae as substrate such as Roy et al. 2014 found $958 \, dm^3$ $H_2$ per kg of volatile suspended solids using *Chlorella sorokiniana* microalgal biomass by mixed culture [155]. Another study conducted by Tamanh et al. (2010) got hydrogen yield of 2.5 mol $H_2$/mol glucose using algal biomass of *Chlamydomonas reinhardtii* by thermophilic organism *Thermotoga neapolitana* [157]. Efficient hydrogen production can be achieved by utilizing algae biomass in place of pure substrate or other waste. Naturally grown algae utilization will lead to huge cost reduction of the process along with the benefit of $CO_2$ sequestration. Such technology can be environmentally friendly. However, in case of algae also, pretreatment of biomass is needed, as the carbohydrates are present in complex form. These complex carbohydrates are needed to convert into simple sugars, so that bacteria can utilize them properly. Before going for the production of hydrogen from algal biomass or any other waste, it is very necessary to acclimatize the desired culture for particular substrate, then only desired production could be achieved.

## Microbial Electrolysis Cell

Hydrogen production through dark fermentation is limited up to 4 mol of hydrogen from 1 mol of glucose as compared with stoichiometric potential of 12 mol of hydrogen per mole of glucose [157,158]. Microbial fuel cells (MFCs) provide direct method of conversion of organic substrate to electricity by the help of exoelectrogens [159]. Exoelectrogens are microorganisms that are capable of producing electrons by degradation of organic substrate. Hydrogen can be produced by modifying MFCs to MECs, in which small voltage has been added to the voltage produced by the bacteria and anaerobic conditions are maintained. According to thermodynamic analysis, $>0.11 \, V$ needs to be added to the generated voltage by bacteria, i.e., $-0.3 \, V$. By using electrohydrogenesis process, endothermic barrier of 4 mol hydrogen through dark fermentation can be overcome [160]. Here the microbes present in anodic chamber will ferment the organic compound, and anode will act as a terminal electron acceptor. In the first chamber, microbes will produce hydrogen through dark fermentation; then in the second stage by electrohydrogenesis,

12 mol/mol of glucose hydrogen can be achieved by addition of small amount of electricity [47]. Still the rate of hydrogen production is less in MECs as compared with dark fermentation. MECs have immense potential for the future large-scale hydrogen generation technology. There are various hurdles, such as replacement of expensive electrodes, increasing current density, and reduction in voltage input, that need to be overcome [47].

## SCALE-UP OF BIOHYDROGEN PRODUCTION PROCESS

Hydrogen production through biomass has huge potential to sustain or cater the energy need of future generation. However, its potential cannot be analyzed properly if its production has not been tested at pilot scale or commercial scale. At present, the biohydrogen production is not cost-effective compared with conventional fuels. Therefore, there is immense scope of improvement in terms of development of cost-effective reactors, high-yielding strains, and use of cheaper feedstock for sustainable biohydrogen production. A comprehensive technoeconomic analysis is required to compare the prospect of biohydrogen with other conventional fossil fuels. An economic survey based on fuel production cost, transport system, and storage makes the technoeconomic studies complicated. The acceptance of a fuel for anthropogenic activity at relevant costs depends on many factors such as the emission profile in terms of pollution, short-term/long-term environmental effects, direct/indirect health costs, and socioeconomical acceptance. Because hydrogen is considered a truly carbon-neutral fuel, it is the most logical choice worldwide as a future energy source.

The suitability of any process lies in its capacity to be scaled up to the industrial level. When it comes to scaling up of dark fermentative hydrogen production, the following are the parameters that could be considered:

- Geometric similarity in scale-up
- Scale-up based on volumetric power consumption
- Constant impeller tip speed
- Reynolds number
- Constant mixing time

There are few studies on the pilot-scale production of hydrogen by utilizing biomass. At pilot-scale study, various operational and technical difficulties come to the picture, which cannot be studied at laboratory scale. Hence, for conversion of any process into technology, its realistic pilot-scale study is very crucial. So far, biggest scale for production of $H_2$ is 100,000 L done by Vatsala et al. In this study, distillery effluent has been used as nutritional source for the coculture of *Citrobacter freundii* 01, *E. aerogenes* E10, and *Rhodopseudomonas palustris* P2. They got hydrogen yield of 2.76 mol of hydrogen per mole of glucose [122]. Ren et al. have studied hydrogen production at a scale of 1480 L using cane molasses as a substrate. The organic loading rate for reactor was 3.11–85.57 kg COD/1000L. They achieved 8240 L $H_2$/day with yield of 26.13 mol/kg $COD_{removed}$ [125]. The feasibility study on scaling up of the biohydrogen production process by *Klebsiella pneumoniae* IIT-BT 08 using cane molasses supplemented with groundnut deoiled cake as a substrate was explored in 10 $m^3$ bioreactor at biohydrogen pilot plant facility, Indian Institute of Technology Kharagpur, India. Cumulative gas production ($H_2$ and $CO_2$) of 158.9 $m^3$ was observed from 10 $m^3$ of production media. Total cumulative hydrogen production of 76.2 $m^3$ was achieved from the 10-$m^3$ reactor. COD removal and energy conversion efficiency from the substrate were 18.1 kg/$m^3$ and 26.68%, respectively. Ethanol, acetate, and butyrate are the most dominant end metabolites that were obtained from the spent media. Groundnut deoiled cake appears to be most promising as a nutritional supplement in the hydrogen production process. This pilot-scale study shows that use of organic wastes/residues could become a promising way for economical and sustainable clean energy generation, which also leads to waste management. Scale-up studies of biohydrogen production have been shown in Table 3.4.

## MATERIAL AND ENERGY ANALYSIS OF BIOHYDROGEN PRODUCTION PROCESS

### Material Analysis

For the assessment of potential and feasibility of any new technology, there is a need of technoeconomic evaluation. It can be performed on various grounds; one of them is material and energy analysis of process [167]. Material analysis is important factor for tracking of different material throughout the fermentation process. Input and output of the system along with accumulation can be estimated. Amount of substrate utilized for product formation and other cellular activities can be estimated using material balance. It can be performed in various ways: for substrates with known composition, elemental balance can be performed, and for substrates with unknown composition such as organic waste or residues, COD balance can be performed [168].

### Energy Analysis

Energy analysis can be performed on the basis of gaseous energy output in term of hydrogen with respect to total energy input in terms of substrate and process energy requirement. For a system, net energy ratio is the ratio of energy produced in terms of hydrogen and input energy in the form of substrate along with other operations such as pumping and mixing. Based on the substrate energy, recovery can be calculated according to the following equation (Eq. 3.18) [16]:

$$\text{Energy recovery} = \frac{\text{Lower heating value of hydrogen} \times \text{Hydrogen yield}}{\text{Lower heating value of substrate and overall process energy requirement}}$$

$$(3.18)$$

**TABLE 3.4** Scale-Up Studies of Biohydrogen Production Processes

| Volume (m³) | Reactor Types | Inoculum | Substrates | Hydrogen Yield | Hydrogen Production Rate (L/L-d) | References |
|---|---|---|---|---|---|---|
| 0.4 | Agitated granular sludge bed reactor | Mixed consortia | Molasses | 1.04 mol $H_2$/mol sucrose | 15.59 | [161] |
| 0.03 | CSTR | Clostridium sp. | Waste sugar | 2.93 mol $H_2$/mol hexose | – | [162] |
| 2.0 | CSTR | Mixed consortia | Molasses | 26.13 mol/kg $COD_{removed}$ | 5.57 | [125] |
| 100 | CSTR | Enterobacter aerogenes E10 and Citrobacter freundii 01 | Distillery effluent | 2.76 mol $H_2$/mol glucose | – | [122] |
| 4 | Continuous-flow photoreactor | Photofermentative bacterial consortium | Glucose | – | 2.13 | [163] |
| 0.008 | Draft tube fluidized bed reactor | Thermophilic mixed culture | Glucose | – | 4.80 | [164] |
| 0.008 | Draft tube fluidized bed reactor | Clostridium sp. | Sucrose | 2.5 mol $H_2$/mol glucose | 54.48 | [165] |
| 0.005 | Fluidized bed reactor | Thermophilic mixed culture | Glucose | – | 6.00 | [164] |
| 0.006 | Granule-based CSTR | Mixed consortia | Glucose | 1.84 mol $H_2$/mol glucose | 78.24 | [166] |

COD, chemical oxygen demand; CSTR, continuous stirred tank reactor; FHL, formate hydrogen lyase.

Feng et al. (2004) have performed energy analysis by using sucrose as a substrate for hydrogen production, and energy recovery from the substrate was 7.6%, which is 26.7% of the theoretical value [104].

## IMPROVEMENT OF ENERGY RECOVERY BY TWO-STAGE PROCESSES

The major drawback of biohydrogen production via dark fermentation is the less overall energy recovery from the substrate. This limitation can be overcome by the integration of dark fermentation with other processes such as biomethanation, photofermentation, ABE fermentation, and bioelectrochemical systems. Process integration can lead to enhancement in substrate energy recovery via production of various products from single substrate.

### Improvement of Gaseous Energy Generation by Biohythane Process

Biogas production from biomass is a very well-established technique and has wider application in energy production. Hydrogen is an intermediate product of process that is rapidly consumed by methane-producing bacteria. However, in case of biohythane process, separation of two stages occurs or the production of hydrogen and methane is decoupled. Biohythane, mixture of hydrogen and methane, is known as Hythane and it is trademarked by Eden [169,170]. Biohythane has various advantages as compared with single product as for vehicular application is concerned such as

- addition of hydrogen to methane leads to acceleration of methane combustion in engine;
- lean flammability improvement occurs by mixing hydrogen to methane;
- hydrogen has eightfold more flame speed as compared with methane at lean air/fuel mixtures.

Biological production of hydrogen and methane holds huge potential to be fuel for the future. It can be produced through various types of waste. Biohythane production from waste solves dual purpose of improvement of gaseous energy recovery from it and waste remediation. In regular dark fermentation, theoretical possible energy recovery is only 30% [3]. After dark fermentation a huge amount of spent is produced, which is rich in volatile fatty acids, ethanol, etc. This spent can be directly used for methane production just by increasing the pH. This will result in enhancement of overall energy recovery from a particular substrate [171].

### Improvement of Gaseous Energy Generation by Integration of Photofermentation

Hydrogen generation can be increased by integrating photofermentation to dark fermentation from same amount of substrate. Theoretical yield of 12 mol of hydrogen from 1 mol of glucose is achievable by photofermentation followed by dark fermentation. There are various studies conducted on biohydrogen production by dark fermentation followed by photofermentation. Tatyana et al. (2010) used potato hydrolysate as a substrate for biohydrogen production; they achieved 5.6 mol of hydrogen per mole of glucose by two-stage process, which is much higher than single-stage process [172]. Another study used *K. pneumoniae* DM11 for dark fermentation and *Rhodobacter sphaeroides* O.U. 001 for photofermentation of the spent of dark fermentation, which led to the hydrogen yield of 5.14 mol of hydrogen per mole of glucose [173].

## Integration of Dark Fermentation and Bioelectrochemical System

Dark fermentation can be linked to bioelectrochemical systems such as MFCs and microbial electrolysis cells. High organic content of the dark fermentation spent makes it ideal substrate for MFCs and MECs. Volatile fatty acids present in spent can be converted into hydrogen by MECs or electricity by MFCs. Overall energy recovery can be significantly improved by application of bioelectrochemical systems in integration with dark fermentation. MFCs can also be used as wastewater treatment process of waste generated after dark fermentation. It will help in COD removal from dark fermentative effluent. In a study conducted by Xiao-Hu Li et al. (2014), they reported hydrogen yield of 0.129 L $H_2$/g-corn stalk at substrate concentration of 20 g/L of corn stalk. After that, the dark fermentative spent was subjected to MECs, and hydrogen yield of 0.2573 L $H^2$/g-corn stalk was achieved with an energy efficiency of 166%, which was obtained under applied voltage of 0.8 V [174]. Another study showed that, in an integrated system using a single-chambered MEC and 0.6 V voltage, the overall hydrogen recovery was 96%, with a production rate of 2.11 L/L-d [175].

Similarly, dark fermentative spent can also be subjected to microbial fuel cell for generation of bioelectricity. A study conducted by Jhansi et al. (2015) showed thermophilic hydrogen production of 3.79 L/L and yielded 2.92 mol/mol of glucose with energy recovery of 28% from single stage; then spent of this process was subjected to two-chambered MFCs and maximum power density of 85.05 $mW/m^2$ was obtained, along with energy recovery of 2.49%. This two-stage process achieved overall substrate energy of 30.49% [176]. Application of integrated approach can help in energy recovery improvement as well as wastewater treatment.

## CONCLUSION

Hydrogen production from biomass is one of the potent clean energy generation processes. The challenge of serving the dual purpose of supporting future energy demand as well as reducing greenhouse gas emission could be achieved by biomass-based hydrogen production. A proper planning, execution, and maintenance of technologies is the need of the hour. By application of process optimization and metabolic engineering, industrial microbial strain can be developed, which is capable of producing hydrogen at a higher rate and yield by utilizing various types of organic wastes. Centralized energy generation facilities should be promoted because of variation in quantity and quality of biomass availability. Considering the numerous advancements in the field of hydrogen technologies, it can be concluded that hydrogen era has already been started. Despite the fact that there are various renewable energy sources, the single type of energy source cannot replace fossil fuel completely. Integration of processes, technologies, and energy sources is required for meeting the energy needs of humanity.

## REFERENCES

[1] F. Macdonald, We Just Used Up all of Earth's Resources for the Year, and It's Only August - ScienceAlert, Science Alert, 2016 [Online]. Available: https://www.sciencealert.com/we-just-used-up-all-of-earth-s-resources-for-the-year.

[2] M. Hiloidhari, D. Das, D.C. Baruah, Bioenergy potential from crop residue biomass in India, Renew. Sustain. Energy Rev. 32 (2014) 504–512.

[3] D. Das, T.N. Veziroğlu, Hydrogen production by biological processes: a survey of literature, Int. J. Hydrogen Energy 26 (1) (2001) 13–28.

[4] R. Cracknell, H. Walmsley, S. Global, Hydrogen for Internal Combustion Engines? Springer, Singapore, 2018, pp. 39–54.

[5] Y. Bicer, I. Dincer, Clean fuel options with hydrogen for sea transportation: a life cycle approach, Int. J. Hydrogen Energy 43 (2) (2018) 1179–1193.

[6] T. Hua, R. Ahluwalia, L. Eudy, G. Singer, B. Jermer, N. Asselin-Miller, S. Wessel, T. Patterson, J. Marcinkoski, Status of hydrogen fuel cell electric buses worldwide, J. Power Sources 269 (2014) 975–993.

[7] Move over electric vehicles!, Hyundai unveils hydrogen fuel cell-powered SUV at CES [Online]. Available: https://economictimes.indiatimes.com/magazines/panache/move-over-electric-vehicles-hyundai-bets-on-hydrogen-fuel-cell-powered-car-at-ces-2018/articleshow/62423790.cms, 2018.

[8] Toyota Company, Mirai | New Cars | Toyota UK. [Online]. Available: https://www.toyota.co.uk/new-cars/new-mirai/landing.json.

[9] Major hydrogen investment mooted as OEMs and energy companies form Hyd - Autovista Group Market Reports. [Online]. Available: https://reports.autovistagroup.com/blogs/news/major-hydrogen-investment-mooted-as-oems-and-energy-companies-form-hydrogen-council.

[10] I. Dincer, Green methods for hydrogen production, Int. J. Hydrogen Energy 37 (2) (2012) 1954–1971.

[11] S. Mohapatra, Hydrogen Production Technologies with Specific Reference to Biomass, 2(3) Gazi Univ., Fac. of Technology, Dep. of Electrical et Electronics Eng, 2012.

[12] P. Nikolaidis, A. Poullikkas, A comparative overview of hydrogen production processes, Renew. Sustain. Energy Rev. 67 (Jan. 2017) 597–611.

[13] T. Wakayama, J. Miyake, J. Miyake, T. Matsunaga, A. San Pietro, Hydrogen from biomass, Curr. Sci. 85 (3) (2001) 41–51.

[14] J.D. Holladay, J. Hu, D.L. King, Y. Wang, An overview of hydrogen production technologies, Catal. Today 139 (4) (2009) 244–260.

[15] S. Subudhi, Hydrogen Production Through Biological Route, Springer, Singapore, 2018, pp. 23–38.

[16] P. McKendry, Energy production from biomass (part 1): overview of biomass, Bioresour. Technol. 83 (1) (2002) 37–46.

[17] M. Parikka, Global biomass fuel resources, Biomass Bioenergy 27 (6) (2004) 613–620.

[18] A. Kumar, N. Kumar, P. Baredar, A. Shukla, A review on biomass energy resources, potential, conversion and policy in India, Renew. Sustain. Energy Rev. 45 (2005) 530–539.

[19] I. Lewandowski, Biobased resources and value chains, in: Bioeconomy, Springer International Publishing, Cham, 2018, pp. 75–94.

[20] I. Kögel-Knabner, The macromolecular organic composition of plant and microbial residues as inputs to soil organic matter: fourteen years on, Soil Biol. Biochem. 105 (2) (2017) A3–A8.

[21] M. Kačuráková, N. Wellner, A. Ebringerová, Z. Hromádková, R. Wilson, P. Belton, Characterisation of xylan-type polysaccharides and associated cell wall components by FT-IR and FT-Raman spectroscopies, Food Hydrocoll. 13 (1) (1999) 35–41.

[22] J. Penniston, E.B. Gueguim Kana, Impact of medium pH regulation on biohydrogen production in dark fermentation process using suspended and immobilized microbial cells, Biotechnol. Biotechnol. Equip. 32 (1) (2018) 204–212.

[23] D. Rebaque, R. Martínez-Rubio, S. Fornalé, P. García-Angulo, A. Alonso-Simón, J.M. Álvarez, D. Caparros-Ruiz, J.L. Acebes, A. Encina, Characterization of structural cell wall polysaccharides in cattail (Typha latifolia): evaluation as potential biofuel feedstock, Carbohydr. Polym. (2017) 679–688.

[24] N. Mosier, C. Wyman, B. Dale, R. Elander, Y.Y. Lee, M. Holtzapple, M. Ladisch, Features of promising technologies for pretreatment of lignocellulosic biomass, Bioresour. Technol. 96 (6) (2005) 673–686.

[25] P. Kumar, D.M. Barrett, M.J. Delwiche, P. Stroeve, Methods for pretreatment of lignocellulosic biomass for efficient hydrolysis and biofuel production, Ind. Eng. Chem. Res. 48 (8) (2009), 3713–3729. 2.

[26] R. Ravindran, A.K. Jaiswal, Exploitation of food industry waste for high-value products, Trends Biotechnol. 34 (1) (2016) 58–69.

[27] N.H.M. Yasin, T. Mumtaz, M.A. Hassan, N. Abd Rahman, Food waste and food processing waste for biohydrogen production: a review, J. Environ. Manage. 130 (2013) 375–385.

[28] R. Kothari, V. Kumar, V.V. Pathak, V.V. Tyagi, Sequential hydrogen and methane production with simultaneous treatment of dairy industry wastewater: bioenergy profit approach, Int. J. Hydrogen Energy 42 (8) (2017) 4870–4879.

[29] P. Chowdhary, A. Raj, R.N. Bharagava, Environmental pollution and health hazards from distillery wastewater and treatment approaches to combat the environmental threats: a review, Chemosphere 194 (2018) 229–246.

[30] S.K. Nataraj, K.M. Hosamani, T.M. Aminabhavi, Distillery wastewater treatment by the membrane-based nanofiltration and reverse osmosis processes, Water Res. 40 (12) (2006) 2349–2356.

[31] J. Gu, G. Xu, Y. Liu, An integrated AMBBR and IFAS-SBR process for municipal wastewater treatment towards enhanced energy recovery, reduced energy consumption and sludge production, Water Res. 110 (2017) 262–269.

[32] B. Subedi, K. Balakrishna, R.K. Sinha, N. Yamashita, V.G. Balasubramanian, K. Kannan, Mass loading and removal of pharmaceuticals and personal care products, including psychoactive and illicit drugs and artificial sweeteners, in five sewage treatment plants in India, J. Environ. Chem. Eng. 3 (4A) (2015) 2882–2891.

[33] J. Mata-Alvarez, S. Macé, P. Llabrés, Anaerobic digestion of organic solid wastes. An overview of research achievements and perspectives, Bioresour. Technol. 74 (1) (2000) 3–16.

[34] X.M. Guo, E. Trably, E. Latrille, H. Carrre, J.P. Steyer, Hydrogen production from agricultural waste by dark fermentation: a review, Int. J. Hydrogen Energy 35 (19) (2010) 10660–10673.

[35] D. Das, Advances in biohydrogen production processes: an approach towards commercialization, Int. J. Hydrogen Energy 34 (17) (2009) 7349–7357.

[36] M.A. Rachman, Y. Nakashimada, T. Kakizono, N. Nishio, Hydrogen production with high yield and high evolution rate by self-flocculated cells of *Enterobacter aerogenes* in a packed-bed reactor, Appl. Microbiol. Biotechnol. 49 (4) (1998) 450–454.

[37] S. Tanisho, Y. Suzuki, N. Wakao, Fermentative hydrogen evolution by *Enterobacter aerogenes* strain E.82005, Int. J. Hydrogen Energy 12 (9) (1987) 623–627.

[38] N. Kumar, D. Das, Enhancement of hydrogen production by *Enterobacter cloacae* IIT-BT 08, Process Biochem. 35 (6) (2000) 589–593.

[39] S.K. Han, H.S. Shin, Biohydrogen production by anaerobic fermentation of food waste, Int. J. Hydrogen Energy 29 (6) (2004) 569–577.

[40] H. Zhang, M.A. Bruns, B.E. Logan, Biological hydrogen production by *Clostridium acetobutylicum* in an unsaturated flow reactor, Water Res. 40 (4) (2006) 728–734.

[41] C. Schröder, M. Selig, P. Schönheit, Glucose fermentation to acetate, $CO_2$ and $H_2$ in the anaerobic hyperthermophilic eubacterium *Thermotoga maritima*: involvement of the Embden-Meyerhof pathway, Arch. Microbiol. 161 (6) (1994) 460–470.

[42] E.W.J. Van Niel, M.A.W. Budde, G. De Haas, F.J. Van der Wal, P.A.M. Claassen, A.J.M. Stams, Distinctive properties of high hydrogen producing extreme thermophiles, *Caldicellulosiruptor saccharolyticus* and *Thermotoga elfii*, Int. J. Hydrogen Energy 27 (11–12) (2002) 1391–1398.

[43] A.I. Slobodkin, T.P. Tourova, B.B. Kuznetsov, N.A. Kostrikina, N.A. Chernyh, E.A. Bonch-Osmolovskaya, *Thermoanaerobacter siderophilus* sp. nov., a novel dissimilatory Fe(III)-reducing, anaerobic, thermophilic bacterium, Int. J. Syst. Bacteriol. 49 (4) (1999) 1471–1478.

[44] N. Pradhan, L. Dipasquale, G. D'Ippolito, A. Panico, P.N.L. Lens, G. Esposito, A. Fontana, Hydrogen production by the thermophilic bacterium *Thermotoga neapolitana*, Int. J. Mol. Sci. 16 (6) (2015) 12578–12600.

[45] H. Yokoi, R. Maki, J. Hirose, S. Hayashi, Microbial production of hydrogen from starch-manufacturing wastes, Biomass Bioenergy 22 (5) (2002) 389–395.

[46] P.C. Hallenbeck, J.R. Benemann, Biological hydrogen production; fundamentals and limiting processes, Int. J. Hydrogen Energy 27 (11–12) (2002) 1185–1193.

[47] P.C. Hallenbeck, D. Ghosh, Advances in fermentative biohydrogen production: the way forward? Trends Biotechnol. 27 (5) (2009) 287–297.

[48] J. Knappe, G. Sawers, A radical-chemical route to acetyl-CoA: the anaerobically induced pyruvate formate-lyase system of *Escherichia coli*, FEMS Microbiol. Lett. 75 (4) (1990) 383–398.

[49] M. Stephenson, L. H. Stickland, Hydrogenlyases: Bacterial enzymes liberating.

[50] P. Sinha, S. Roy, D. Das, Genomic and proteomic approaches for dark fermentative bio-hydrogen production, Renew. Sustain. Energy Rev. 56 (2016) 1308–1321.

[51] G. Stephanopoulos, Metabolic fluxes and metabolic engineering, Metab. Eng. 1 (1) (1999) 1–11.

[52] P. Sinha, A. Pandey, An evaluative report and challenges for fermentative biohydrogen production, Int. J. Hydrogen Energy 36 (13) (2011) 7460–7478.

[53] P. Soucaille, Molecular characterization and transcriptional analysis of the putative hydrogenase gene of *Clostridium acetobutylicum* ATCC 824. Microbiology 178 (9) (1996) 2668–2675.

[54] K. Morimoto, T. Kimura, K. Sakka, K. Ohmiya, Overexpression of a hydrogenase gene in *Clostridium paraputrificum* to enhance hydrogen gas production, FEMS Microbiol. Lett. 246 (2) (2005) 229–234.

[55] X. Liu, Y. Zhu, S.T. Yang, Construction and characterization of *ack* deleted mutant of *Clostridium tyrobutyricum* for enhanced butyric acid and hydrogen production, Bio-technol. Prog. 22 (5) (2006) 1265–1275.

[56] Z. Ren, T.E. Ward, B.E. Logan, J.M. Regan, Characterization of the cellulolytic and hydrogen-producing activities of six mesophilic *Clostridium* species, J. Appl. Microbiol. 103 (6) (2007) 2258–2266.

[57] H. Voelskow, G. Schön, $H_2$ production of *Rhodospirillum rubrum* during adaptation to anaerobic dark conditions, Arch. Microbiol. 125 (3) (1980) 245–249.

[58] M. Chippaux, M.C. Pascal, F. Casse, Formate hydrogenlyase system in *Salmonella typhimurium* LT2, Eur. J. Biochem. 72 (1) (Jan. 1977) 149–155.

[59] H.D. Peck, H. Gest, Formic dehydrogenase and the hydrogenlyase enzyme complex in coli-aerogenes bacteria, J. Bacteriol. 73 (6) (1957) 706–721.

[60] J. Turcot, A. Bisaillon, P.C. Hallenbeck, Hydrogen production by continuous cultures of *Escherichia coli* under different nutrient regimes, Int. J. Hydrogen Energy 33 (5) (2008) 1465–1470.

[61] T. Maeda, V. Sanchez-Torres, T.K. Wood, Enhanced hydrogen production from glucose by metabolically engineered *Escherichia coli*, Appl. Microbiol. Biotechnol. 77 (4) (2007) 879–890.

[62] T. Maeda, V. Sanchez-Torres, T.K. Wood, Metabolic engineering to enhance bacterial hydrogen production, Microb. Biotechnol. 1 (1) (2008) 30–39.

[63] M. Cha, D. Chung, J.G. Elkins, A.M. Guss, J. Westpheling, Metabolic engineering of *Caldicellulosiruptor bescii* yields increased hydrogen production from lignocellulosic biomass, Biotechnol. Biofuels 6 (1) (2013) 85.

[64] R. Wang, W. Zong, C. Qian, Y. Wei, R. Yu, Z. Zhou, Isolation of *Clostridium perfringens* strain W11 and optimization of its biohydrogen production by genetic modification, Int. J. Hydrogen Energy 36 (19) (2011) 12159–12167.

[65] S.I. Nakayama, T. Kosaka, H. Hirakawa, K. Matsuura, S. Yoshino, K. Furukawa, Meta-bolic engineering for solvent productivity by downregulation of the hydrogenase gene cluster hupCBA in *Clostridium saccharoperbutylacetonicum* strain N1-4, Appl. Microbiol. Biotechnol. 78 (3) (2008) 483–493.

[66] J.H. Jo, C.O. Jeon, S.Y. Lee, D.S. Lee, J.M. Park, Molecular characterization and homologous overexpression of [FeFe]-hydrogenase in *Clostridium tyrobutyricum* JM1, Int. J. Hydrogen Energy 35 (3) (2010) 1065−1073.

[67] H. Zhao, K. Ma, Y. Lu, C. Zhang, L. Wang, X.H. Xing, Cloning and knockout of formate hydrogen lyase and $H_2$-uptake hydrogenase genes in *Enterobacter aerogenes* for enhanced hydrogen production, Int. J. Hydrogen Energy 34 (1) (2009) 186−194.

[68] T. Maeda, G. Vardar, W.T. Self, T.K. Wood, Inhibition of hydrogen uptake in *Escherichia coli* by expressing the hydrogenase from the cyanobacterium Synechocystis sp. PCC 6803, BMC Biotechnol. 7 (1) (2007) 25.

[69] T. Maeda, V. Sanchez-Torres, T.K. Wood, Protein engineering of hydrogenase 3 to enhance hydrogen production, Appl. Microbiol. Biotechnol. 79 (1) (2008) 77−86.

[70] D.W. Penfold, C.F. Forster, L.E. Macaskie, Increased hydrogen production by *Escherichia coli* strain HD701 in comparison with the wild-type parent strain MC4100, Enzyme Microb. Technol. 33 (2−3) (2003) 185−189.

[71] A. Yoshida, T. Nishimura, H. Kawaguchi, M. Inui, H. Yukawa, Enhanced hydrogen production from formic acid by formate hydrogen lyase-overexpressing *Escherichia coli* strains, Appl. Environ. Microbiol. 71 (11) (2005) 6762−6768.

[72] W. Song, J. Cheng, J. Zhao, D. Carrieri, C. Zhang, J. Zhou, K. Cen, Improvement of hydrogen production by over-expression of a hydrogen-promoting protein gene in *Enterobacter cloacae*, Int. J. Hydrogen Energy 36 (11) (2011) 6609−6615.

[73] K. McNeely, Y. Xu, N. Bennette, D.A. Bryant, G.C. Dismukes, Redirecting reductant flux into hydrogen production via metabolic engineering of fermentative carbon metabolism in a cyanobacterium, Appl. Environ. Microbiol. 76 (15) (2010) 5032−5038.

[74] S. Li, C. Lai, Y. Cai, X. Yang, S. Yang, M. Zhu, J. Wang, X. Wang, High efficiency hydrogen production from glucose/xylose by the *ldh*-deleted *Thermoanaerobacterium* strain, Bioresour. Technol. 101 (22) (2010) 8718−8724.

[75] D.B. Levin, L. Pitt, M. Love, Biohydrogen production: prospects and limitations to practical application, Int. J. Hydrogen Energy 29 (2) (2004) 173−185.

[76] Y. Zhang, J. Shen, Effect of temperature and iron concentration on the growth and hydrogen production of mixed bacteria, Int. J. Hydrogen Energy 31 (4) (2006) 441−446.

[77] P.C. Hallenbeck, Fundamentals of the fermentative production of hydrogen, Water Sci. Technol. 52 (1−2) (2005) 21−29.

[78] G. Antonopoulou, K. Stamatelatou, N. Venetsaneas, M. Kornaros, G. Lyberatos, Biohydrogen and methane production from cheese whey in a two-stage anaerobic process, Ind. Eng. Chem. Res. 47 (15) (2008) 5227−5233.

[79] N. Khanna, S.M. Kotay, J.J. Gilbert, D. Das, Improvement of biohydrogen production by *Enterobacter cloacae* IIT-BT 08 under regulated pH, J. Biotechnol. 152 (1−2) (2011) 9−15.

[80] C.Y. Lin, C.H. Lay, Carbon/nitrogen-ratio effect on fermentative hydrogen production by mixed microflora, Int. J. Hydrogen Energy 29 (1) (2004) 41−45.

[81] J. Wang, W. Wan, Factors influencing fermentative hydrogen production: a review, Int. J. Hydrogen Energy 34 (2) (2009) 799−811.

[82] M. Junghare, S. Subudhi, B. Lal, Improvement of hydrogen production under decreased partial pressure by newly isolated alkaline tolerant anaerobe, *Clostridium butyricum* TM-9A: optimization of process parameters, Int. J. Hydrogen Energy 37 (4) (2012) 3160−3168.

[83] Y.J. Lee, T. Miyahara, T. Noike, Effect of pH on microbial hydrogen fermentation, J. Chem. Technol. Biotechnol. 77 (6) (2002) 694−698.

[84] D.T. Jones, D.R. Woods, Acetone-butanol fermentation revisited, Microbiol. Rev. 50 (4) (1986) 484–524.

[85] Y.C. Lo, Y.C. Su, C.Y. Chen, W.M. Chen, K.S. Lee, J.S. Chang, Biohydrogen production from cellulosic hydrolysate produced via temperature-shift-enhanced bacterial cellulose hydrolysis, Bioresour. Technol. 100 (23) (2009) 5802–5807.

[86] S.M. Kotay, D. Das, Microbial hydrogen production with *Bacillus coagulans* IIT-BT S1 isolated from anaerobic sewage sludge, Bioresour. Technol. 98 (6) (2007) 1183–1190.

[87] H.L. Chin, Z.S. Chen, C.P. Chou, Fedbatch operation using *Clostridium acetobutylicum* suspension culture as biocatalyst for enhancing hydrogen production, Biotechnol. Prog. 19 (2) (2003) 383–388.

[88] R. Mitra, G. Balachandar, V. Singh, P. Sinha, D. Das, Improvement in energy recovery by dark fermentative biohydrogen followed by biobutanol production process using obligate anaerobes, Int. J. Hydrogen Energy 42 (8) (2017), 4880–489.

[89] P.Y. Lin, L.M. Whang, Y.R. Wu, W.J. Ren, C.J. Hsiao, S.L. Li, J.S. Chang, Biological hydrogen production of the genus *Clostridium*: metabolic study and mathematical model simulation, Int. J. Hydrogen Energy 32 (12) (2007) 1728–1735.

[90] K.T. Chung, Inhibitory effects of $H_2$ on growth of *Clostridium cellobioparum*, Appl. Environ. Microbiol. 31 (3) (1976) 342–348.

[91] H. Liu, T. Zhang, H.H.P. Fang, Thermophilic $H_2$ production from a cellulose-containing wastewater, Biotechnol. Lett. 25 (4) (2003) 365–369.

[92] C.Y. Chen, M.H. Yang, K.L. Yeh, C.H. Liu, J.S. Chang, Biohydrogen production using sequential two-stage dark and photo fermentation processes, Int. J. Hydrogen Energy 33 (18) (2008) 4755–4762.

[93] S.W. Van Ginkel, B. Logan, Increased biological hydrogen production with reduced organic loading, Water Res. 39 (16) (2005) 3819–3826.

[94] G. Chittibabu, K. Nath, D. Das, Feasibility studies on the fermentative hydrogen production by recombinant *Escherichia coli* BL-21, Process Biochem. 41 (3) (2006) 682–688.

[95] K. Nath, A. Kumar, D. Das, Effect of some environmental parameters on fermentative hydrogen production by *Enterobacter cloacae* DM11, Can. J. Microbiol. 52 (6) (2006) 525–532.

[96] N. Kumar, D. Das, Continuous hydrogen production by immobilized *Enterobacter cloacae* IIT-BT 08 using lignocellulosic materials as solid matrices, Enzyme Microb. Technol. 29 (4–5) (2001) 280–287.

[97] J.J. Podestá, A.M. Gutiérrez-Navarro, C.N. Estrella, M.A. Esteso, Electrochemical measurement of trace concentrations of biological hydrogen produced by *Enterobacteriaceae*, Res. Microbiol. 148 (1) (1997) 87–93.

[98] L. Minnan, H. Jinli, W. Xiaobin, X. Huijuan, C. Jinzao, L. Chuannan, Z. Fengzhang, X. Liangshu, Isolation and characterization of a high $H_2$-producing strain *Klebsiella oxytoca* HP1 from a hot spring, Res. Microbiol. 156 (1) (2005) 76–81.

[99] B.E. Logan, S.E. Oh, I.S. Kim, S. Van Ginkel, Biological hydrogen production measured in batch anaerobic respirometers, Environ. Sci. Technol. 36 (11) (2002) 2530–2535.

[100] I. Hussy, F.R. Hawkes, R. Dinsdale, D.L. Hawkes, Continuous fermentative hydrogen production from a wheat starch Co-Product by mixed microflora, Biotechnol. Bioeng. 84 (6) (2003) 619–626.

[101] Y.K. Oh, S.H. Kim, M.S. Kim, S. Park, Thermophilic biohydrogen production from glucose with trickling biofilter, Biotechnol. Bioeng. 88 (6) (2004) 690–698.

[102] W. Giles, A. Bisits, M. Knox, G. Madsen, R. Smith, Kinetics of hydrogen production with continuous anaerobic cultures utilizing sucrose as the limiting substrate, Appl. Microbiol. Biotechnol. 57 (1−2) (2001) 56−64.

[103] K.S. Lee, Y.S. Lo, Y.C. Lo, P.J. Lin, J.S. Chang, Operation strategies for biohydrogen production with a high-rate anaerobic granular sludge bed bioreactor, Enzyme Microb. Technol. 35 (6−7) (2004) 605−612.

[104] F. Chang, Biohydrogen production using an up-flow anaerobic sludge blanket reactor, Int. J. Hydrogen Energy 29 (1) (2004) 33−39.

[105] C.C. Chen, C.Y. Lin, Using sucrose as a substrate in an anaerobic hydrogen-producing reactor, Adv. Environ. Res. 7 (3) (2003) 695−699.

[106] J.J. Lay, Modeling and optimization of anaerobic digested sludge converting starch to hydrogen, Biotechnol. Bioeng. 68 (3) (2000) 269−278.

[107] C.Y. Lin, C.C. Chang, C.H. Hung, Fermentative hydrogen production from starch using natural mixed cultures, Int. J. Hydrogen Energy 33 (10) (2008) 2445−2453.

[108] M. Morimoto, M. Atsuko, A.A.Y. Atif, M.A. Ngan, A. Fakhru'l-Razi, S.E. Iyuke, A.M. Bakir, Biological production of hydrogen from glucose by natural anaerobic microflora, Int. J. Hydrogen Energy 29 (7) (2004) 709−713.

[109] S. O-Thong, P. Prasertsan, D. Karakashev, I. Angelidaki, Thermophilic fermentative hydrogen production by the newly isolated *Thermoanaerobacterium thermosaccharolyticum* PSU-2, Int. J. Hydrogen Energy 33 (4) (2008) 1204−1214.

[110] T. Zhang, H. Liu, H.H.P. Fang, Biohydrogen production from starch in wastewater under thermophilic condition, J. Environ. Manage. 69 (2) (2003) 149−156.

[111] A.P. Batista, L. Gouveia, P.A.S.S. Marques, Fermentative hydrogen production from microalgal biomass by a single strain of bacterium *Enterobacter aerogenes* − effect of operational conditions and fermentation kinetics, Renew. Energy 119 (2018) 203−209.

[112] X.Y. Shi, D.W. Jin, Q.Y. Sun, W.W. Li, Optimization of conditions for hydrogen production from brewery wastewater by anaerobic sludge using desirability function approach, Renew. Energy 35 (7) (2010) 1493−1498.

[113] G.-L. Tang, J. Huang, Z.-J. Sun, Q.-Q. Tang, C.-H. Yan, G.-Q. Liu, Biohydrogen production from cattle wastewater by enriched anaerobic mixed consortia: influence of fermentation temperature and pH, J. Biosci. Bioeng. 106 (1) (2008) 80−87.

[114] Y. Yang, K. Tsukahara, S. Sawayama, Biodegradation and methane production from glycerol-containing synthetic wastes with fixed-bed bioreactor under mesophilic and thermophilic anaerobic conditions, Process Biochem. 43 (4) (2008) 362−367.

[115] M. Ferchichi, E. Crabbe, G.H. Gil, W. Hintz, A. Almadidy, Influence of initial pH on hydrogen production from cheese whey, J. Biotechnol. 120 (4) (2005) 402−409.

[116] N. Azbar, F.T. Çetinkaya Dokgöz, T. Keskin, K.S. Korkmaz, H.M. Syed, Continuous fermentative hydrogen production from cheese whey wastewater under thermophilic anaerobic conditions, Int. J. Hydrogen Energy 34 (17) (2009) 7441−7447.

[117] H. Yang, P. Shao, T. Lu, J. Shen, D. Wang, Z. Xu, X. Yuan, Continuous bio-hydrogen production from citric acid wastewater via facultative anaerobic bacteria, Int. J. Hydrogen Energy 31 (10) (2006) 1306−1313.

[118] K.W. Jung, D.H. Kim, H.S. Shin, Continuous fermentative hydrogen production from coffee drink manufacturing wastewater by applying UASB reactor, Int. J. Hydrogen Energy 35 (24) (2010) 13370−13378.

[119] C.H. Lay, J.H. Wu, C.L. Hsiao, J.J. Chang, C.C. Chen, C.Y. Lin, Biohydrogen production from soluble condensed molasses fermentation using anaerobic fermentation, Int. J. Hydrogen Energy 35 (24) (2010) 13445−13451.

[120] S. Van Ginkel, B.E. Logan, Inhibition of biohydrogen production by undissociated acetic and butyric acids, Environ. Sci. Technol. 39 (23) (2005) 9351–9356.

[121] A. Noblecourt, G. Christophe, C. Larroche, P. Fontanille, Hydrogen production by dark fermentation from pre-fermented depackaging food wastes, Bioresour. Technol. 247 (2018) 864–870.

[122] T.M. Vatsala, S.M. Raj, A. Manimaran, A pilot-scale study of biohydrogen production from distillery effluent using defined bacterial co-culture, Int. J. Hydrogen Energy 33 (20) (2008) 5404–5415.

[123] S. Venkata Mohan, G. Mohanakrishna, S.V. Ramanaiah, P.N. Sarma, Simultaneous bio-hydrogen production and wastewater treatment in biofilm configured anaerobic periodic discontinuous batch reactor using distillery wastewater, Int. J. Hydrogen Energy 33 (2) (2008) 550–558.

[124] B.S. Fernandes, G. Peixoto, F.R. Albrecht, N.K. Saavedra del Aguila, M. Zaiat, Potential to produce biohydrogen from various wastewaters, Energy Sustain. Dev. 14 (2) (2010) 143–148.

[125] N. Ren, J. Li, B. Li, Y. Wang, S. Liu, Biohydrogen production from molasses by anaerobic fermentation with a pilot-scale bioreactor system, Int. J. Hydrogen Energy 31 (15) (2006) 2147–2157.

[126] R. Saidi, P.P. Liebgott, H. Gannoun, L. Ben Gaida, B. Miladi, M. Hamdi, H. Bouallagui, R. Auria, Biohydrogen production from hyperthermophilic anaerobic digestion of fruit and vegetable wastes in seawater: simplification of the culture medium of *Thermotoga maritima*, Waste Manag. 71 (2018) 474–484.

[127] S.-E. Oh, S. Van Ginkel, B.E. Logan, The relative effectiveness of pH control and heat treatment for enhancing biohydrogen gas production, Environ. Sci. Technol. 37 (22) (2003) 5186–5190.

[128] A. Wang, Bioaugmented hydrogen production from microcrystalline cellulose using co-culture—*Clostridium acetobutylicum* X9 and *Ethanoigenens harbinense* B49, Int. J. Hydrogen Energy 33 (2) (2008) 912–917.

[129] E. Eroğlu, I. Eroğlu, U. Gündüz, L. Türker, M. Yücel, Biological hydrogen production from olive mill wastewater with two-stage processes, Int. J. Hydrogen Energy 31 (11) (2006) 1527–1535.

[130] E.C. Koutrouli, H.N. Gavala, I.V. Skiadas, G. Lyberatos, Mesophilic biohydrogen production from olive pulp, Process Saf. Environ. Prot. 84 (4) (2006) 285–289.

[131] S. O-Thong, P. Prasertsan, N. Intrasungkha, S. Dhamwichukorn, N.K. Birkeland, Improvement of biohydrogen production and treatment efficiency on palm oil mill effluent with nutrient supplementation at thermophilic condition using an anaerobic sequencing batch reactor, Enzyme Microb. Technol. 41 (5) (2007) 583–590.

[132] A.A.Y. Atif, A. Fakhru'L-Razi, M.A. Ngan, M. Morimoto, S.E. Iyuke, N.T. Veziroglu, Fed batch production of hydrogen from palm oil mill effluent using anaerobic microflora, Int. J. Hydrogen Energy 30 (13–14) (2005) 1393–1397.

[133] K. Vijayaraghavan, D. Ahmad, Biohydrogen generation from palm oil mill effluent using anaerobic contact filter, Int. J. Hydrogen Energy 31 (10) (2006) 1284–1291.

[134] D. Sivaramakrishna, D. Sreekanth, V. Himabindu, Y. Anjaneyulu, Biological hydrogen production from probiotic wastewater as substrate by selectively enriched anaerobic mixed microflora, Renew. Energy 34 (3) (2009) 937–940.

[135] G.-F. Zhu, P. Wu, Q.-S. Wei, J. Lin, Y.-L. Gao, H.-N. Liu, Biohydrogen production from purified terephthalic acid (PTA) processing wastewater by anaerobic fermentation using mixed microbial communities, Int. J. Hydrogen Energy 35 (15) (2010) 8350–8356.

[136] H.H.P. Fang, C. Li, T. Zhang, Acidophilic biohydrogen production from rice slurry, Int. J. Hydrogen Energy 31 (6) (2006) 683–692.

[137] S. Roy, S. Ghosh, D. Das, Improvement of hydrogen production with thermophilic mixed culture from rice spent wash of distillery industry, Int. J. Hydrogen Energy 37 (21) (2012) 15867–15874.

[138] H. Yu, Z. Zhu, W. Hu, H. Zhang, Hydrogen production from rice winery wastewater in an upflow anaerobic reactor by using mixed anaerobic cultures, Int. J. Hydrogen Energy 27 (11–12) (2002) 1359–1365.

[139] I. Hussy, F.R. Hawkes, R. Dinsdale, D.L. Hawkes, Continuous fermentative hydrogen production from sucrose and sugarbeet, Int. J. Hydrogen Energy 30 (5) (2005) 471–483.

[140] Y. Ueno, S. Otsuka, M. Morimoto, Hydrogen production from industrial wastewater by anaerobic microflora in chemostat culture, J. Ferment. Bioeng. 82 (2) (1996) 194–197.

[141] J. Monod, The growth of bacterial cultures, Annu. Rev. Microbiol. 3 (1) (1949) 371–394.

[142] Y. Mu, G. Wang, H.Q. Yu, Kinetic modeling of batch hydrogen production process by mixed anaerobic cultures, Bioresour. Technol. 97 (11) (2006) 1302–1307.

[143] I.K. Kapdan, F. Kargi, Bio-hydrogen production from waste materials, Enzyme Microb. Technol. 38 (5) (2006) 569–582.

[144] T.Y. Jeong, G.C. Cha, S.H. Yeom, S.S. Choi, Comparison of hydrogen production by four representative hydrogen-producing bacteria, J. Ind. Eng. Chem. 14 (3) (2008) 333–337.

[145] H. Koku, I. Erolu, U. Gündüz, M. Yücel, L. Türker, Aspects of the metabolism of hydrogen production by *Rhodobacter sphaeroides*, Int. J. Hydrogen Energy 27 (11–12) (2002) 1315–1329.

[146] Z.Y. Hitit, C.Z. Lazaro, P.C. Hallenbeck, Single stage hydrogen production from cellulose through photo-fermentation by a co-culture of c and *Rhodopseudomonas palustris*, Int. J. Hydrogen Energy 42 (10) (2017) 6556–6566.

[147] M. Abo-Hashesh, D. Ghosh, A. Tourigny, A. Taous, P.C. Hallenbeck, Single stage photo-fermentative hydrogen production from glucose: an attractive alternative to two stage photofermentation or co-culture approaches, Int. J. Hydrogen Energy 36 (21) (2011) 13889–13895.

[148] J.Z. Lee, D.M. Klaus, P.C. Maness, J.R. Spear, The effect of butyrate concentration on hydrogen production via photofermentation for use in a Martian habitat resource recovery process, Int. J. Hydrogen Energy 32 (15) (2007) 3301–3307.

[149] T. Assawamongkholsiri, A. Reungsang, Photo-fermentational hydrogen production of *Rhodobacter* sp. KKU-PS1 isolated from an UASB reactor, Electron. J. Biotechnol. 18 (3) (2015) 221–230.

[150] K. Nath, D. Das, Effect of light intensity and initial pH during hydrogen production by an integrated dark and photofermentation process, Int. J. Hydrogen Energy 34 (17) (2009) 7497–7501.

[151] E.S. Shuba, D. Kifle, Microalgae to biofuels: 'Promising' alternative and renewable energy, review, Renew. Sustain. Energy Rev. 81 (2018) 743–755.

[152] E. Eroglu, A. Melis, Photobiological hydrogen production: recent advances and state of the art, Bioresour. Technol. 102 (18) (2011) 8403–8413.

[153] J.R. Benemann, Hydrogen production by microalgae, J. Appl. Phycol. 12 (3/4) (2000) 291–300.

[154] B.K. Nayak, S. Roy, D. Das, Biohydrogen production from algal biomass (*Anabaena* sp. PCC 7120) cultivated in airlift photobioreactor, Int. J. Hydrogen Energy 39 (14) (2013) 7553–7560.

[155] S. Roy, K. Kumar, S. Ghosh, D. Das, Thermophilic biohydrogen production using pretreated algal biomass as substrate, Biomass Bioenergy 61 (2014) 157–166.

[156] J.H. Mussgnug, V. Klassen, A. Schlüter, O. Kruse, Microalgae as substrates for fermentative biogas production in a combined biorefinery concept, J. Biotechnol. 150 (1) (2010) 51–56.

[157] T.A.D. Nguyen, K.R. Kim, M.T. Nguyen, M.S. Kim, D. Kim, S.J. Sim, Enhancement of fermentative hydrogen production from green algal biomass of *Thermotoga neapolitana* by various pretreatment methods, Int. J. Hydrogen Energy 35 (23) (2010) 13035–13040.

[158] B.E. Logan, Peer reviewed: extracting hydrogen and electricity from renewable resources, Environ. Sci. Technol. 38 (9) (2004) 160A–167A.

[159] B.E. Logan, J.M. Regan, Electricity-producing bacterial communities in microbial fuel cells, Trends Microbiol. 14 (12) (2006) 512–518.

[160] S. Cheng, B.E. Logan, Sustainable and efficient biohydrogen production via electrohydrogenesis, Proc. Natl. Acad. Sci. 104 (47) (2007) 18871–18873.

[161] C.Y. Lin, S.Y. Wu, P.J. Lin, J.S. Chang, C.H. Hung, K.S. Lee, C.H. Lay, C.Y. Chu, C.H. Cheng, A.C. Chang, J.H. Wu, F.Y. Chang, L.H. Yang, C.W. Lee, Y.C. Lin, A pilot-scale high-rate biohydrogen production system with mixed microflora, Int. J. Hydrogen Energy 36 (14) (2011) 8758–8764.

[162] M. Krupp, R. Widmann, Biohydrogen production by dark fermentation: experiences of continuous operation in large lab scale, Int. J. Hydrogen Energy 34 (10) (2009) 4509–4516.

[163] Q. Zhang, C. Lu, D.-J. Lee, Y.-J. Lee, Z. Zhang, X. Zhou, J. Hu, Y. Wang, D. Jiang, C. He, T. Zhang, Photo-fermentative hydrogen production in a 4 m 3 baffled reactor: effects of hydraulic retention time, Bioresour. Technol. 239 (2017) 533–537.

[164] C. Peintner, A.A. Zeidan, W. Schnitzhofer, Bioreactor systems for thermophilic fermentative hydrogen production: evaluation and comparison of appropriate systems, J. Clean. Prod. 18 (SUPPL. 1) (2010) S15–S22.

[165] C.N. Lin, S.Y. Wu, J.S. Chang, Fermentative hydrogen production with a draft tube fluidized bed reactor containing silicone-gel-immobilized anaerobic sludge, Int. J. Hydrogen Energy 31 (15) (2006) 2200–2210.

[166] K.Y. Show, Z.P. Zhang, J.H. Tay, D. Tee Liang, D.J. Lee, W.J. Jiang, Production of hydrogen in a granular sludge-based anaerobic continuous stirred tank reactor, Int. J. Hydrogen Energy 32 (18) (2007) 4744–4753.

[167] C. Hatzis, C. Riley, G.P. Philippidis, Detailed material balance and ethanol yield calculations for the biomass-to-ethanol conversion process, Appl. Biochem. Biotechnol. 57–58 (1) (1996) 443–459.

[168] P. Kongjan, I. Angelidaki, Extreme thermophilic biohydrogen production from wheat straw hydrolysate using mixed culture fermentation: effect of reactor configuration, Bioresour. Technol. 101 (20) (2010) 7789–7796.

[169] Innovations - Eden. [Online]. Available: http://edeninnovations.com/innovations/.

[170] F. Moreno, M. Muñoz, J. Arroyo, O. Magén, C. Monné, I. Suelves, Efficiency and emissions in a vehicle spark ignition engine fueled with hydrogen and methane blends, Int. J. Hydrogen Energy 37 (15) (2012) 11495–11503.

[171] S. Kumari, D. Das, Biologically pretreated sugarcane top as a potential raw material for the enhancement of gaseous energy recovery by two stage biohythane process, Bioresour. Technol. 218 (2016) 1090–1097.

[172] T.V. Laurinavichene, B.F. Belokopytov, K.S. Laurinavichius, D.N. Tekucheva, M. Seibert, A.A. Tsygankov, Towards the integration of dark- and photo-fermentative waste treatment. 3. Potato as substrate for sequential dark fermentation and light-driven $H_2$ production, Int. J. Hydrogen Energy 35 (16) (2010) 8536−8543.

[173] K. Nath, A. Kumar, D. Das, Hydrogen production by *Rhodobacter sphaeroides* strain O.U.001 using spent media of *Enterobacter cloacae* strain $DM_{11}$, Appl. Microbiol. Biotechnol. 68 (4) (2005) 533−541.

[174] X.H. Li, D.W. Liang, Y.X. Bai, Y.T. Fan, H.W. Hou, Enhanced $H_2$ production from corn stalk by integrating dark fermentation and single chamber microbial electrolysis cells with double anode arrangement, Int. J. Hydrogen Energy 39 (17) (2014) 8977−8982.

[175] L. Lu, N. Ren, D. Xing, B.E. Logan, Hydrogen production with effluent from an ethanol-$H_2$-coproducing fermentation reactor using a single-chamber microbial electrolysis cell, Biosens. Bioelectron. 24 (10) (2009) 3055−3060.

[176] J.L. Varanasi, S. Roy, S. Pandit, D. Das, Improvement of energy recovery from cellobiose by thermophillic dark fermentative hydrogen production followed by microbial fuel cell, Int. J. Hydrogen Energy 40 (26) (2015) 8311−8321.

# Chapter 4

# Energy Storage Using Hydrogen Produced From Excess Renewable Electricity: Power to Hydrogen

Marcelo Carmo[1], Detlef Stolten[1,2]

[1]*Forschungszentrum Jülich GmbH, Jülich, Germany;* [2]*RWTH Aachen University, Aachen, Germany*

## MOTIVATION

In recent decades the threat of climate change, the potential of renewable energy in terms of capacity as well as cost reduction, has been significantly underestimated. Besides the COP 21 agreement that focuses on the reduction of $CO_2$ emissions as a major driver for developed countries, there is a second equally important driver emerging from developing countries in the form of the reduction of local emissions. The $CO_2$ issue, while certainly serious, is typically perceived by the layman as being somewhat remote, whereas the issue of local emissions is a great deal more tangible. In real terms the latter is therefore more pressing, and as such it requires long-term political focus. Nonetheless, the prospect of rising sea levels has given impetus to the drive to reduce $CO_2$ emissions, in both developed and developing nations. This is, of course, more difficult to address in developing countries because of more stringent cost constraints on mitigating efforts. The use of renewable energy serves to bring down expenditure on fuel imports, potentially also on infrastructure such as electric grids and comprehensive gas grids for developing countries. There is therefore a necessity and an opportunity for worldwide efforts to combat climate change in all nations, regardless of the economic status.

This purpose of this chapter is to elucidate the opportunities, requirements, and constraints involved in the use of renewable sources for energy storage. It will focus on chemical storage, which is suitable for *long-term* storage. Short-term storage options will be briefly outlined, including power to heat options.

Science and Engineering of Hydrogen-Based Energy Technologies. https://doi.org/10.1016/B978-0-12-814251-6.00004-6

Long-term strategies are subject only to relatively minor losses during storage—such as self-discharge or gas slip—and very low specific investment cost for the storage units because the turnover frequency of the storage capacity is low, increasing the specific cost of the energy stored.

## RENEWABLE ENERGY, VOLATILITY, AND STORAGE

The paradigm of the existing power supply chain is load-following of the demanded power production. In a fully or mostly renewable power production setup, the power input is determined by the weather situation. Hence, temporal over- and underproduction are inherent, changing the paradigm completely into harnessing most of the renewable energy attainable, regardless of the actual demand at the time.

Realizing that the electrical grid provides no storage capability—other than a gas grid does—the electrical energy harnessed needs to be used instantaneously at some point of the grid. Relevant options are setting up a strong transportation grid to equilibrate renewable energy input of different nature—e.g., wind versus solar—or different climate zones. The latter generally require transportation over very long distances like in the thousand kilometer range. The other set of options is storage of the power overproduced.

If that route is pursued, underproduction can be covered either by using fossil fuels such as natural gas or by reconverting the stored energy back again to electricity. Natural gas provides great flexibility options particularly when transforming the power production to renewables. The energy from temporal overproduction can then be fed back to other energy sectors such as transportation, households, or industry. The scheme is more often referred to today as sector coupling. Beyond improving the actual energy balance, it provides the opportunity to interconnecting the power sector already at an early stage of transformation with the other energy sectors.

### Grid Stabilization and Short-Term Storage

Currently, grids get stabilized by a rotating momentum created by the rotational energy of the turbines and the generators. In case of increasing power demand, the rotating frequency decreases, which is mirrored in the generated frequency for the grid and serves as a signal to increase the power input by gearing up the power plant until the desired standard frequency is reached. The property of rotating masses exhibiting inertia is equivalent to describing them in the amount of stored energy. For this reason the existing grades are pretty resilient to very fast load changes and need to be supported by additional input only in the range of a few minutes and not instantaneously. This is handled today, for example, by swiftly adjusting steam power plants that can store energy in the steam vessels or by pumped hydropower plants. Thereafter, further power sources can be activated in the range of approximately 15 min.

The scheme is expected to change dramatically if no rotating masses are grid-connected anymore. Experience with a long-distance DC—DC transport line in China toward Hong Kong indicates that without rotating masses—or in this case, negligible rotating masses, which would also be the case in renewable scenarios—the response time needs to be in a 10 to 100 millisecond range rather than in the minute range. This is important to note for the development of electrolyzers that might help stabilize the electric grid alongside battery units or other devices such as fly wheels. Whereas batteries can instantaneously respond to load changes they are limited in sustaining the response over a longer time owing to the limited capacity, the use of electrolyzers for grid stabilization is still to be proven. Batteries, if not fully charged, can operate in either direction to compensate a positive residual load, i.e., less energy provided in the grid than demanded, or a negative residual load, i.e., less demand that the amount of energy produced at that moment in time. Electrolyzers can basically deliver the same service when being operated below the full capacity. When operated at full capacity, electrolyzers can only shed off load and compensate that way for positive residual load. These dynamic processes need further investigation on the grid side and particularly for electrolyzer systems.

## Energy Security and Long-Term Storage

Whereas the short-term, i.e., hourly and daily, fluctuation of renewable energy is very obvious and palpable. It is less intensely discussed that renewable energy also has strong longer-termed fluctuations, leading to seasonal over- or underproduction of power as well as to zero or close to zero production for sustained periods, such as multiple days or even multiple weeks. Just to elucidate the order of magnitude of required storage for backing up a 2-month period of no power input, one can take one-sixth of the annual power consumption of a country. This would, for instance, for Germany result into a storage requirement of about 80 TWh. Assuming that 50% of the regular power requirement can be covered by other measures such as power transportation from other regions or smart reduction of the power consumption for the time, it would still amount 40 TWh. Different assumptions for storage requirement in the long run for Germany differ from 30 to 60 TWh. This does not yet include the requirements of transportation and industry. Comparing these values with the existing capacity of pumped hydro storage of 0.04 TWh, there are obviously three orders of magnitude between the requirement and the existing storage capabilities of pumped hydro, disqualifying this technology for long-term storage of energy notwithstanding its paramount role for grid stabilization. When looking at hydrogen storage, the two questions arising from these considerations are whether the chemical storage of hydrogen de-livers higher storage densities than mechanical storage does and whether there are viable concepts of storing large quantities of hydrogen. A brief example might show the enormous energy density of gas storage. Hydrogen contains 3 kWh per standard cubic meter and gets compressed for mass storage to at least 100 bars, leading to an energy content of 337 kWh per cubic meter stored and real gas behavior considered. For pumped hydro storage in comparison, the potential energy of an elevated basin can be harnessed. This leads to about 0.8 kWh for 1 m$^3$ stored and a difference in altitude of the two water basins of 300 m assumed. Hence, the gas storage provides about 400 times the energy per unit volume, nearly delivering the three orders of magnitude required. The second question to be answered is whether there are appropriate means of storage for hydrogen in very large quantities.

The basic storage options for large quantities of hydrogen are underground storage of gaseous hydrogen, liquid hydrogen storage in large containers, or chemical storage of hydrogen, such as in liquid organic hydrogen carriers (LOHCs) or as chemical components such as methanol or Dimethyl ether (DME). This order also reflects the efficiency of the storage pathways. For liquid hydrogen storage, it is to be considered that active cooling to the level of liquefaction is most efficient at very large installations. Hence, in all downstream steps in shipping and storing up to the point of usage, cooling is performed passively by using the evaporation heat originating from hydrogen boil off. In this case the efficiency is dependent on whether the hydrogen can be effectively used elsewhere in the system or needs to be discarded, i.e., burnt. A similar argument is valid for the liquid hydrogen organic carriers that need heat to release the hydrogen. The heat amounts to about one-third of the energy of the hydrogen stored. If this heat is available from other sources, such as off-heat in industrial processes or geothermal energy, it might not harm the energy efficiency much. As for the chemical components such as methanol, the energy pathway includes the production of the chemical component, yet also the hydrogen production and storage as well as the provision of the carbon reaction partner from biomass or $CO_2$ is to be considered.

Finally, the choice of a storage option also depends on the end use of the hydrogen. Table 4.1 shows a collection of typical data for road transportation.

First of all it is important to note that hydrogen production via electrolysis is listed in Table 4.1 with an efficiency of 70% [1−3]. This is due to the fact that the lower heating value of hydrogen is considered because, in transportation applications, polymer electrolyte membrane fuel cell (PEMFC) condensation of the water produced is not feasible. When comparing hydrogen applications in the stationary sector with condensing boilers, the higher heating value (HHV) can be taken. Hence, 70% efficiency considering the lower heating value and 80% considering the HHV can be achieved with the same electrolyzer. The storage and distribution chain of gaseous hydrogen can be assumed to be at an efficiency of 90%, encompassing transportation to a storage site, compression to storage pressure of about 100−200 bar, and transportation to its end-use site. The value of 90% is a good value for any first estimation, yet it might vary strongly depending on the specific conditions [4].

**TABLE 4.1 Impact of Different Storage and End-Use Options on the Total Efficiency**

| Applications | Electrolysis (%) | Methanation/ Fuel Production (%) | Storage Distribution (%) | Fuel Cell/ Electric Drive Train (%) | Internal Combustion Engine + Drive Train (%) | Total Efficiency (%) |
|---|---|---|---|---|---|---|
| 1. GH$_2$ + FC Passenger car | 70 | – | 90 | 60 | – | 38 |
| 2. GH$_2$ + FC Truck | 70 | – | 90 | 50–60 | – | 32–38 |
| 3. GH$_2$ + ICE Passenger car | 70 | – | 90 | – | 25 | 16 |
| 4. GH$_2$ + ICE Truck | 70 | – | 90 | – | 40 | 25 |
| 5. GH$_2$ + methanation + ICE | 70 | 80 | 90 | – | 25 | 13 |
| 6. GH$_2$ + methanation + ICE Truck | 70 | 80 | 90 | – | 40 | 20 |
| 7. LH$_2$ + FC Passenger car | 70 | – | 63 | 60 | – | 26 |
| 8. Liquid biofuel | | 50 | 95 | | 25 | 12 |
| 9. Liquid CO$_2$-based fuel | 70 | 55–80 | | | 25 | 10–14 |
| 10. Gasoline | | 90 | 90 | | 25 | 20 |
| 11. Battery electric cars | – | – | 80 | 85 | | 68 |

For cars and trucks, efficiencies can be estimated using standard driving cycles, such as the NEDC for Europe, JCO8 for Japan, US06 for the United States, and the most recent Worldwide Harmonized Light Duty Vehicles Test Procedure (WLTP) [5]. All of them have in common that they test the vehicles under different conditions of speed and acceleration trying to represent the average use of vehicles in cities and on county roads and highways. There are notable differences in individual use to be expected, yet these standards are the best measures for comparison available. As older standards were soft on acceleration and top speed, the WLTP is more realistic.

Average efficiencies over a driving cycle for the different propulsion technologies are given in Table 4.1 as 60% for the fuel cell drive train, 25% for the drive train of internal combustion engines for vehicles, and 40% for long-haul trucks owing to the fact that these have more efficient diesel engines and that highway driving predominates. The values given here are estimates based on literature sources that typically provide data on efficiencies for specified points of operation or fuel economy values [6−9]. In the power-to-fuel route, hydrogen and $CO_2$ are used to produce synthetic fuels. Starting from hydrogen, methanation can achieve efficiencies of approximately 80%, while liquid fuel synthesis varies between 55% and 80% depending on the process route and the selected fuel [10,11]. An optimized Fischer−Tropsch process with exceptional recycling loops can achieve a plant efficiency of approximately 80% and a total efficiency of 14%. Fischer−Tropsch biofuels produced via gasification of biomass reach similar well-to-tank efficiencies of approximately 50%. Extraordinarily efficient is the use of electricity in a battery electric drive train with 68% in total, resulting from 80% storage and distribution efficiency and 85% efficiency of the drive train [12].

Hence, the battery electric drive train is the benchmark for what is achievable with fuel based on electricity, the so-called electrofuels. Efficiency considerations are sometimes criticized as of secondary order, yet with renewable energy efficiency stands as a proxy for cost and spatial requirements for the installations because renewable energy is to be harnessed at a very low energy density of the respective energy source such as wind or sun. The low energy density translates into relatively high energy cost through the capital expenditure (CAPEX) and into relatively large areas needed for the installations to harness the energy, such as photovoltaic (PV) fields or wind farms. This leads to the requirement of higher efficiency in other parts of the energy pathway to keep the total cost at bay.

## Hydrogen Applications

This paragraph discusses the implications of the different options using renewable hydrogen. The most efficient use of hydrogen in transportation is as gaseous hydrogen for vehicles with a fuel cell (#1, Table 4.1), resulting in an efficiency of 38%. This compares with 20% efficiency in vehicles with internal combustion engines today based on gasoline, representing the lower benchmark. When using an internal combustion engine, the efficiency would drop to 16%, owing to the lower top efficiency and particularly to the lower part load efficiency (#3, Table 4.1). Because natural gas combustion engines already exist and the storage of natural gas aboard a vehicle is somewhat easier than that of hydrogen, methanation of the hydrogen to use synthetic natural gas (SNG) could be an option. With an efficiency of methanation of 80%, the efficiency of the whole energy pathway would drop to 13% and the cost would rise accordingly, provided the $CO_2$ for the synthesis is taken for granted or even more if it is to be paid for.

This comparison shows that it is sensible to stick with the simplest fuel, which is hydrogen, and electrochemical conversion via a fuel cell instead of an internal combustion engine, as long as these technologies apply for the use case. Fuel cell vehicles are being constantly introduced into the market by Hyundai and Toyota these days and are being developed by many other car manufacturers. The situation is different for trucks, however. To clarify the difference, long-haul trucks are being compared with the statement for passenger vehicles.

Long-haul trucks run on a much higher efficiency of the diesel engine of about 40%, need about 200–300 kW of power and a lifetime expectancy of more than 20,000 h compared with 5000 h for passenger cars. Long-haul trucks also run on high power for most of the part of their operation. Bearing in mind the characteristics of internal combustion engines, which provide an efficiency of 47% in the most efficient point of operation [6] and fuel cells and exhibit particularly benign part-load characteristics and a decreasing efficiency with increasing load [9], the advantage of efficiency of fuel cells diminishes for this application. That can be compensated by choosing a larger fuel cell, which in turn leads to higher investment, making hydrogen combustion engine an alternative to be considered. The latter would produce no soot as fossil fuel motors do, yet it would need $NO_x$ clean-up. Another alternative would be the route via methanation using $CH_4$ as a fuel with a combustion engine. For trucks these options would compare efficiency-wise to 32%–38% for gaseous $H_2$ with a fuel cell [#2, Table 4.1], with a combustion engine to 25% and for SNG with a combustion engine to 20%. Combining the use of SNG with a fuel cell is likely to be referred to niche markets because the reforming process with an efficiency of 80% would further reduce the total efficiency to 16%. This shows that for long-haul trucks the picture is not yet clear, further development and demonstration is needed, and barriers that have been overcome already for passenger cars are still an issue for the much higher requirements.

Whereas hybridization is not so effective at a constant high power demand because the limited battery capacity is quickly drained, it comes in strongly in city traffic, enabling regenerative braking and powerful acceleration with relatively small engines, be it fuel cell engines or motors. For this reason, fuel cell buses are viable for public transport. As for the efficiency comparison in this use case, fuel cells are ahead of internal combustion engines such as in passenger cars; however, combustion engines can also be hybridized. Buses and delivery trucks are much closer to passenger vehicles, and indeed most of the urban fuel cell buses are equipped with derivatives of fuel cell systems from passenger cars, sometimes using the same hardware. Coaches on the other hand are similar in their characteristics to long-haul trucks.

Liquid hydrogen can be considered an alternative over gaseous hydrogen for its high energy density in the use case of a passenger car. Liquefaction consumes about one-third of the energy content of the hydrogen energy, leading to a total efficiency of 26% for the passenger car with a fuel cell compared with 38% when fueled with gaseous hydrogen. For sake of easy use for the consumer, the fuel would be gasified and compressed for storage in a car anyway to get around the issue of the necessary hydrogen evaporation for cooling a liquid tank, which leads to fuel consumption when the car is not used for a longer time and bars access to public parking garages. The picture of efficiency changes, if imported hydrogen is considered. Aboard naval vessels, hydrogen will not be transported as compressed gas, but in the liquid stage or by liquid organic hydrogen carriers or as chemical substances like methanol. Taking the case of liquid hydrogen delivered to a port, the efficiency argument held against liquefaction above is no longer valid because the step had to be taken for overseas transport anyway and advantage of the easier distribution of liquid hydrogen can be exploited. Yet, what about cost and specific land use at the site of production owing to the less efficient step? That can be compensated for by choosing preferable sites for harnessing renewable energy.

As a case in point, wind energy can be harnessed in Patagonia at 5000 to 5500 full load hours, whereas in the North Sea about 4000 full load hours are available, making it nearly up for liquefaction in the case of hydrogen production in Patagonia; where the specific land use is a lot higher compared with the hydrogen production in the North Sea. Compared with 2000 full load hours of wind power production at many sites in the world, liquefaction for transportation poses no barrier, neither in terms of efficiency nor cost.

The last area for application here is what is called the power-to-fuel route, making a liquid fuel out of electricity by hydrogenating biomass or using $CO_2$ as a carbon source. The efficiency of these liquid fuels drops down to 12% and 9%, respectively. Yet, this is not a reason to rule out these fuels. Whereas they will hardly be competitive where easier alternatives exist, like the option [#1, Table 4.1] for passenger cars, $CO_2$-lean liquid fuels will be needed in aviation and some special applications. For the time being, a fuel switch in aviation is inconceivable until 2050 and not even likely in the decades thereafter, as aviation relies even more on the high energy density of liquid fuels than any other sector of transportation.

This discourse took just one property to elucidate options for transportation trying to depict the potential of hydrogen and the vectors for its application.

Designed to structure decision-making on the one hand, it revealed the complexity involved. What can be taken from this exercise is mainly the following:

- The complete energy pathway is to be considered; otherwise the results will very likely be wrong.
- Whether an energy pathway makes sense or not, strongly depends on the use case.
- For the sake of efficiency—as a proxy of cost and specific land use—the simplest fuel pathway is the best. The use case defines whether it can be applied or not.
  - Direct power use provides efficiencies around 90%−95%.
  - Power with battery storage provides efficiencies around 85%.
  - Fuel cells with hydrogen achieve efficiencies of 38%.
  - Hydrogen with methanation in a passenger car achieves of an efficiency of 13%.
- Use cases similar at the first glance can be very different, such as passenger cars and long-haul trucks, either ones being road transportation vehicles.
- Use cases looking dissimilar, such as passenger cars and urban buses, can be pretty similar.
- Hydrogen is the basis for synthetic fuels.
- Synthetic fuels attain only about half of the energy pathway efficiency of today's energy pathways.
- Nonetheless, synthetic fuel needs to be considered for specific applications such as aviation.

When assessing these fuel choices, many more properties need to be considered, yet this simple approach already provides some structure to the issue, shows the complexity, and provides insight how valuable hydrogen is for the future fuel infrastructure, be it as elementary hydrogen, syngas, or liquid fuel. The primary renewable energy source for hydrogen worldwide will be renewable power, and the primary technology the hydrogen will be produced of power is electrolysis [13]. Thus, the technological specifics of electrolysis will be outlined in the second part of this chapter.

## Water Electrolysis—A "Game Change" Technology

For long-term storage, hydrogen is an essential building block along the energy pathway. It can be generated from electricity via electrochemical water splitting, i.e., water electrolysis, or by applying thermochemical cycles. Whereas water electrolysis is a well-established process, thermochemical water splitting is still under development. Thermochemical water splitting works well together with concentrated solar power installations, which provide the required high temperature. Because neither concentrated solar power installations nor the thermochemical processes are established technologies, this route will not be further discussed in detail, yet it should be kept in mind for further development.

Although water electrolysis is a very well-established technology for producing high-grade hydrogen in the chemical and process industries, the worldwide share of hydrogen produced by electrolysis is only about 4%, whereas steam methane reforming covers about 96%. Steam methane reforming is cheaper owing to the fact that natural gas is cheaper compared with electrical power and that electrolyzers are not yet being mass-produced and optimized for low cost application, which is crucial for the generation of hydrogen as an energy carrier. Hence, the second part of this chapter will focus on electrolysis technology with regard to cost savings through design, material choices, and operating conditions.

## HYDROGEN GENERATION VIA ELECTROLYSIS

### Brief History of Water Electrolysis

Technologies to split water into hydrogen and oxygen using electricity have been present in our industry realm for more than 100 years. However, unfortunately, clean hydrogen produced by water electrolyzers in the 21st century is still very marginal compared with full hydrogen demand worldwide. The main reason is that splitting water to hydrogen with any sort of available electricity is from a cost point of view still less favorable than catalytically reforming methane or natural gas into hydrogen. To date, hydrogen is primarily produced by steam reforming of natural gas, via partial oxidation of mineral oil or coal gasification. To some extent, there is, however, nowadays an important market for the onsite generation of hydrogen using electrolyzers, pure hydrogen that is typically needed in the glass, steel, food industry, power plants for generator cooling and in the manufacture of electronic components, life support for military applications, and other small niches.

Clearly, the balance between electrolytic hydrogen and hydrocarbon-derived hydrogen (catalytic reformed) depends very much on the relative costs of the fossil-based hydrocarbon (typically very unstable) and electricity (partially stable when supplied, for instance, by a hydropower plant in a country with reliable energy infrastructure). After all, advantages for producing electrolytic hydrogen include the simplicity of production, the lack of pollution, easy scalability, and the availability of raw materials necessary for the electrochemical reaction to take place. After many decades of its demonstration and industrial usage, state-of-the-art technologies of water electrolysis are still alkaline (using a liquid alkaline caustic electrolyte), and acidic electrolysis (where a polymeric electrolyte membrane [PEM] electrolysis) is used. In addition, a strong market niche also exists using electrolysis to produce chlorine for the chemical industry, the so called chlor-alkali electrolysis, a technology, however, that is not the scope of this chapter.

## *Alkaline Water Electrolysis*

The principles of alkaline water electrolysis were introduced more than 100 years ago, and many electrolysis plants have been constructed and reliably operated to date. By the end of the last century, a few electrolyzer facilities were still in operation, such as the one in Rjukan in Norway exceeding 60,000 $Nm^3$/h and in Aswan Egypt with a capacity of up to 30,000 $Nm^3$/h. More facilities shall still be operational in Reykjavik Iceland, in Cuzco Peru, and in other countries around the globe. These plants were usually dedicated to generate hydrogen to be subsequently used for ammonia synthesis and finally for the fertilizer production, always when and where cheap electricity from a hydropower plant was locally available. Large-scale electrolyzer facilities have mainly used low-pressure electrolyzers, with the capacity of these electrolyzers ranging around 200 $Nm^3$/h of hydrogen. The Lurgi pressurized electrolyzer shown in Fig. 4.1A produced hydrogen at 740 $Nm^3$/h, which corresponds to an electrical output of approximately 3.6 MW. The stack was assembled with 560 cells having an active area diameter of 1.60 m and, depending on the number of cells, the stack length could achieve 10 m. In contrast to the Lurgi electrolyzer, the Bamag electrolyzer operated at atmospheric pressure (Fig. 4.1B) using rectangular electrodes with an active area of approximately 3 $m^2$ and usually with 100 cells with a production capacity of approximately 330 $Nm^3$/h of hydrogen.

In the 1980s and 1990s, large research projects were running as a natural response to the second oil crisis including a variety of R&D projects within Jülich and DLR; a 10-kW Electrolyzer DLR HYSOLAR with 500 $cm^2$, a 10-kW Electrolyzer FZJ 3 bar with 1000 $cm^2$, a 26-kW Electrolyzer PHOEBUS 7 bar with 2500 $cm^2$, and 5-kW Electrolyzer PHOEBUS 120 bar with 500 $cm^2$ active area. These research projects had the main goal to demonstrate innovative approaches to increase the power density and operating pressure of alkaline electrolysis. Fig. 4.2 shows pictures of these projects, demonstrating the complexity of these systems and the R&D trend toward larger electrode active

**FIGURE 4.1** Lurgi pressurized electrolyzer (A) 740 $Nm^3$/h of hydrogen and a Bamag atmospheric electrolyzer (B) with a capacity of 300 $Nm^3$/h of hydrogen.

areas and higher operating pressure. In addition, it was also the aim to increase the current density, lower the cell voltages, and increase operating temperatures to reduce capital and operational costs of alkaline water electrolyzers.

Alternatively, an incremental modification of the cell configuration to minimize the ohmic drop and the development of novel electrodes to reduce the sum of the anodic and cathodic overpotentials were consistently pursued. In cooperation with the center in Jülich, Lurgi demonstrated that, by using Raney Ni electrodes and NiO diaphragms (Fig. 4.3), it was possible to reduce the single cell voltage from 1.92 to 1.6 V at a constant current density of 0.2 A/cm² or to 1.72 V at 0.4 A/cm², where the gain was obtained by means

*10 kW Electrolyzer DLR HYSOLAR 500cm²*

*10 kW Electrolyzer FZJ 3 bar 1.000 cm²*

*26 kW Electrolyzer PHOEBUS 7 bar with 2.500 cm²*

*5 kW Electrolyzer PHOEBUS 120 bar with 500 cm²*

**FIGURE 4.2** Electrolysis activities in Jülich (electrodes, diaphragm, stack, and system development) from 1979 until 2002.

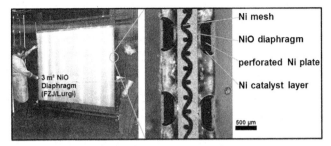

**FIGURE 4.3** Research and Development in Jülich of an NiO diaphragm for alkaline water electrolyzers.

of a zero-gap cell configuration (minimal distance between electrodes and diaphragm). As part of a government-funded project, this technology led to the construction of a 32 bar pressurized electrolyzer with an output of 1 MW.

In other German national projects (HySolar [14], SWB [15], PHOEBUS [16]), different alkaline water electrolyzers were developed, constructed, and tested. Nowadays, according to NEL, the cost of large-scale alkaline electrolyzers will be less than 700 $/kW by 2020[1]. Today, alkaline electrolyzers are supplied commercially by a few companies, and a not extensive overview of leading manufacturers is shown in Table 4.2. Most manufacturers currently specify a limited lower partial load range for the alkaline electrolyzers. The lower partial load range seems to be detrimental when coupling in particular to renewable energy sources where a nonnegligible fraction of the capacity for electrolysis cannot be used for hydrogen production. In addition, current alkaline electrolyzers are limited to still reasonably low current densities, with nominal loads around 400 mA/cm$^2$, with a few companies presenting values as low as 200 mA/cm$^2$. A low current density profile will naturally limit the ability to variate the power input when coupling alkaline electrolyzers with intermittent wind energy. The only way to circumvent this will be by increasing the installed capacity, consequently increasing the investments costs.

## Polymer Electrolyte Membrane Water Electrolysis

Around the 1960s, the development of an acidic solid polymer electrolyte by DuPont allowed the introduction of another concept of water electrolyzers. Polymer electrolyte membranes (PEMs) based on perfluorosulfonic acids that enabled general electric to realize PEM water electrolysis for the first time, a novel concept where a very thin solid membrane was coated with iridium and platinum powders forming the anode and cathode electrodes, respectively [13]. In the 1970s, other developments were demonstrated in a laboratory scale, where a single electrode surface of approximately 2300 cm$^2$ over a perfluorosulfonic acid (PFSA) membrane was fabricated. GE showed current

---

1. www.fch.europa.eu.

**TABLE 4.2** Overview of Leading Manufacturers of Alkaline Electrolyzers

| Manufacturers | Series/Operating Pressure | $H_2$ Rate ($Nm^3/h$) | Energy Consumption ($kWh/Nm^3$ $H_2$) | Partial Load Range (%) | Electrolyte/ Temperature | $H_2$ Purity (%) | Comment |
|---|---|---|---|---|---|---|---|
| Hydrogenics[a] | HYSTAT/10–25 bar | 10–60 Maximum 15 per stack | 4.9–5.4 (AC, system—all included) | 40–100 (25–100 optional) | 30% wt. KOH in $H_2O$ | 99.9; $H_2O$ saturated, $O_2 < 1.000$ ppm (before HPS) | Cell parts are all Ni coated—200 mA/ $cm^2$ at 2.2 V |
| Hydrotechnik GmbH[b] | EV Series atmospheric | 40–220 | 5.28 | 20–100 | 30% wt. KOH in $H_2O$ at 80°C | 99.9 | |
| McPhy[c] | McLyzer | 10–400 | 4.43–5.25 (DC at nominal flow rate) | — | — | — | |
| NEL[d] | A-Range | 50–485 | 3.8–4.4 (DC) | 15–100 | 25% wt. KOH in $H_2O$ at 80°C | 99.9 before HPS | |
| Nuberg PERIC[e] | ZDQ Series | 5–600 | 4.6 | | 30% wt. KOH in $H_2O$ at 95°C | 99.8 | |

| | | 1.5–5 | 5 | 0–100 | 65°C | 99.9 | Lifetime 30,000 h |
|---|---|---|---|---|---|---|---|
| Sagim S.A.[f] | M Series/7 bar monopolar | 1.5–5 | 5 | 0–100 | 65°C | 99.9 | Lifetime 30,000 h |
| Teledyne Energy Systems[g] | Titan EL-N/up to 9 bar | 100–500 | – | – | Water fed with min 1 M $\Omega$-cm | 9.999 | |
| Tianjin Mainland Hydrogen Equipment[h] | FDQ Series/up to 50 bar | 2–1000 | 4.4–4.8 (DC) | 40–100 | 30% wt. KOH in $H_2O$ at 90°C | 99.9 | Asbestos/ nonsbestos |

[a]www.hydrogenics.com.
[b]http://www.en.ht-hydrotechnik.de.
[c]www.mcphy.com.
[d]www.nelhydrogen.com.
[e]www.nubergindia.com/nuberg-hydrogen-brochure.pdf.
[f]http://www.sagim-gip.com.
[g]http://www.teledynees.com.
[h]http://www.cnthe.com.

densities close to $1.8 \, A/cm^2$ at a cell voltage of 2 V with an operating temperature of 80°C and using Nafion membranes from Dupont that were 250 μm thick. Over the following years, many other companies and research institutes dedicated themselves to further develop PEM-based electrolyzers, including but not limited to ABB in Switzerland (around 1970 and 1980) where two 100-kW systems were installed and tested. These systems were run by Stellram SA in Nyon (Switzerland) from 1987 to 1990 and by Solar-Wasserstoff-Bayern GmbH (SWB) in Neuenburg vorm Wald (Germany) from 1990 to 1996. The MEMBREL systems consisted of four modules with about 30 cells each and with an active surface area of $400 \, cm^2$ each arranged in two vertical stacks. In Japan, Fuji Electric Corporate also worked to develop a PEM electrolyzer with a single-electrode surface of $2500 \, cm^2$ as part of the World Energy Network (WE-NET) project from 1993 to 1998 funded by the Japanese Ministry of Economy, Trade and Industry (METI). With a five-cell short stack, cell voltages of 1.57 V were achieved at 80°C and a current density of $1 \, A/cm^2$ under atmospheric conditions. The objective of the WE-NET project was to develop PEM electrolyzers with a single cell area of $10,000 \, cm^2$, current densities of $1-3 \, A/cm^2$, and an efficiency of $>90\%$ relative to the HHV of hydrogen. GenHyPEM was an STREP program supported by the European Commission in the course of the sixth framework research program. GenHyPEM gathered partners from Belgium, Germany, Romania, Federation of Russia, and France. The main goal of the project was to develop low-cost and high-pressure (50 bars) PEM water electrolyzers for the production of up to $1 \, Nm^3 H_2/h$. Over the next decades after the first GE demonstrations and work from many other groups around the world, the use of PEM water electrolyzers is still limited to small niches, such as military and space program applications, and other onsite gas generation where the electricity cost is not an issue. As discussed above, "cheap and dirty" hydrogen coming from natural gas reforming is even to this date a strong competitor to supply hydrogen to the industry. The only thing that has essentially remained is the know-how that came from these R&D activities since 1960 on PEM water electrolyzers and a handful of companies still working in this field (Fig. 4.4).

## Lessons Learned From the Past

There has been an overwhelming number of scientific publications, projects, conference presentations, and much more on low-temperature fuel cells (both PEM and alkaline) since its inception. The overwhelming attention has been driven by the need of an alternative energy source, for both stationary and transport applications, a way to avoid the still dependence on pollutant fossil−based fuels. Nevertheless, over the decades we have set aside the fact that all these fuel cell hydrogen−based technologies are dependent on the same hydrogen, a fuel that presently needs to be produced from a fossil-free-based

**(A)**                                             **(B)**

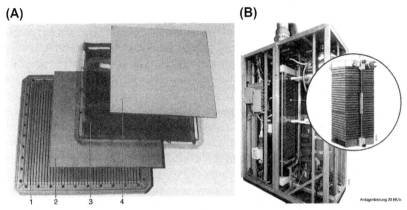

Anlagenleistung 20 N³/h

**FIGURE 4.4** Components of a 400-cm$^2$ MEMBREL cell module (A): 1, bipolar plate made of graphite/plastic; 2, cathode current collectors made of graphite/plastic; 3, membrane; and 4, anode current collector made of sintered titanium; (B) 100-kW MEMBREL pilot plant for 20 Nm$^3$ h$^{-1}$ at an operating pressure of 1−2 bar.

energy source. There has been always the hope that our so dreamed "hydrogen economy" would be still fulfilled, where even the production of hydrogen using nuclear-based energy was considered. This has led to a "knowledge" gap, where unfortunately not enough resources (both industry and governmental) have been allocated to the R&D of water electrolyzers, reflected by the marginal number of projects and consequently lack of publications and know-how dedicated to the study of both alkaline and PEM electrolyzers. Today, the prospects have finally changed, fueled by the recent "Energiewende" concept in Germany, where gigawatt seasonal and intermittent renewable electricity (wind and solar) will need to be somehow stored and where water electrolyzers will fatally play a key role in our future energy scheme. What happens now is that in the next few decades (until 2050), the technology needs to be sustainably available both in terms of efficiency and durability, and most importantly at feasible capital or installation costs. This gives us (researchers and industry) only a few more years to potentially turn both "ancient" technologies into "game changer" technologies. There is not much time left, and clear breakthroughs are necessary to, for instance, quickly increase the efficiencies to values above 70% (perhaps 80%), guarantee long-term durability of above 100,000 h, and reduce capital costs to values under 500 €/kW. Because of a said lack of consistent R&D in the past, the only thing we know is that both technologies are still based on rudimentary cell and stack designs, high load of noble metals, costly titanium-based components, and moderate to sometimes miserable performance profiles.

These conditions surely not enough to couple to renewable energy sources such as wind and PVs and be able in the future to generate a reasonable business case for large-scale (gigawatt range) water electrolyzers.

## Key Players in 2018 (Research and Industry)

The constant increase in the share of renewable but intermittent energy sources will fatally change the energy matrix, where water electrolysis starts to play a key role, storing gigawatts of energy in the form of hydrogen. This is already a shortcoming reality in countries such as Germany and Denmark where a few large-scale demonstration projects already exist. For these reasons, many small, median, and fairly large enterprises are emerging in the field of water electrolyzers to be ready for a hydrogen market to come. Some of them have kept their original names, some have changed, possibly as a marketing strategy, and a few have merged or were absorbed by other companies. Table 4.3 presents a list (in alphabetic order) of the key players involved in the research, development, and/or commercialization in the field of water electrolyzers. This is not an exhaustive list, and the main objective is to present a quick overview about the main players. For obvious reasons, specific and detailed information about the companies will have to be accessed by directly contacting their specific offices, and only public information freely provided in their websites (2018) shall be presented.

**TABLE 4.3** List of Key Players Involved in the Research, Development, Fabrication, and/or Commercialization in the Field of Water Electrolyzers

| Companies | Businesses and Products |
| --- | --- |
| 3M, USA | Membrane electrode assemblies (MEAs) |
| AGFA, Belgium | Diaphragms for alkaline electrolyzers |
| Areva H2Gen, France | PEM electrolyzers |
| Asahi Kasei, Japan | Chlor-alkali electrolysis, components |
| Baltic Fuel Cells, Germany | Testing hardware |
| Bekaert, Germany | Components |
| Borit NV | Bipolar plates |
| Chemours, France | Membranes |
| Covestro Deutschland AG, Germany | Chlor-alkali Electrolyzers |
| De Nora, Italy | Components for water electrolyzers |
| Dioxide Materials, USA | MEAs |
| Freudenberg Performance Materials SE & Co. KG, Germany | Components for water electrolyzers |
| FuelCon AG, Magdeburg-Barleben, Germany | Testing hardware and test stations (alkaline and PEM) |

**TABLE 4.3** List of Key Players Involved in the Research, Development, Fabrication, and/or Commercialization in the Field of Water Electrolyzers—cont'd

| Companies | Businesses and Products |
|---|---|
| FUMATECH BWT GmbH, Germany | Components for water electrolyzers |
| Giner ELX Inc. Newton, MA, USA | Electrolyzer stacks and systems |
| GKN, UK | Components |
| Greenerity® GmbH, Germany | MEAs |
| GreenHydrogen.dk ApS, Kolding, Denmark | Electrolyzers |
| Greenlight Innovation, Burnaby, BC, Canada | Testing hardware and test stations (alkaline and PEM) |
| Haldor Topsoe, Denmark | High-temperature electrolysis |
| Heraeus Deutschland GmbH & Co. KG, Germany | Catalysts and components |
| H-TEC SYSTEMS GmbH, Lübeck, Germany | PEM electrolysis stacks and systems |
| Hydrogenics, Belgium | Alkaline and PEM electrolyzers |
| HydrogenPro, Porsgrunn, Norway | High-pressure alkaline electrolyzer plants |
| HyPlat (Pty) Ltd, Cape Town, South Africa | MEAs, platinum-based catalyst |
| HIAT GmbH, Schwerin, Germany | MEAs, platinum-based catalyst |
| iGas energy GmbH, Stolberg, Germany | Electrolyzer plants for hydrogen production |
| ITM Power Plc, Sheffield, United Kingdom | PEM electrolyzers |
| Johnson Matthey, United Kingdom | Precious metals and catalysts |
| Kumatec GmbH, Germany | High-pressure alkaline electrolyzers |
| McPhy Energy SA, Italy | Alkaline electrolyzers |
| Mott, USA | Porous transport layers and components |
| Nel Hydrogen, Notodden, Norway | Alkaline electrolyzers |
| Oerlikon Metco Coatings GmbH, Germany | Coatings |
| Pajarito Powder, LLC, Albuquerque, NM, USA | Catalysts |
| Precors GmbH, Germany | Bipolar plates |
| Proton OnSite, USA | PEM electrolyzers |

*Continued*

**TABLE 4.3** List of Key Players Involved in the Research, Development, Fabrication, and/or Commercialization in the Field of Water Electrolyzers—cont'd

| Companies | Businesses and Products |
| --- | --- |
| Siemens, Germany | PEM electrolyzers |
| Sunfire GmbH, Germany | High-temperature electrolyzers |
| Thyssenkrupp Uhde Chlorine Engineers GmbH, Germany | Chlor-alkali electrolyzers |
| Tianjin Mainland Hydrogen Equipment, Tianjin, P.R. China | — |
| Toray, Japan | Components for electrolyzers |
| TreadStone Technologies | MEAs and bipolar plates |

## Principles of Water Electrolysis

The decomposition of water using free electrons consists of two electrochemical partial reactions separated by an ion-conducting electrolyte. The three typical processes used to split the water into hydrogen and oxygen are distinguished by the choice of electrolyte, with the electrolyte defining all other components inside the cell, stack, and balance of plant. In addition, the electrolyte choice shall also define the operating conditions. All these three technologies are shown in Fig. 4.5, with their respective partial reactions for the hydrogen evolution reaction (HER) and the oxygen evolution reaction.

Both alkaline and acidic (PEM) electrolyzers require liquid water to solvate the ions passing through the diaphragm (alkaline) or membrane (PEM). The water boiling temperature point will for both cases limit its operating temperature because water in the liquid form is required for transport of ions. For the solid oxide cell, $O^{2-}$ is transported across a dense ionic conductor consisting of $ZrO_2$ doped with $Y_2O_3$ that only occurs between 700 and 1000°C, limiting again its operating temperature. For the alkaline type, usually 25%–35% wt. KOH in water is used as an electrolyte, running usually at 80–90°C, from atmospheric to pressures as high as 200 bar. Only materials that can sustain the harsh caustic conditions are selected, usually Ni- or steel-based electrodes; asbestos-, NiO-, or $ZrO_2$-based diaphragms; and potassium hydroxide–resistant polymer materials as frames and/or gaskets. For the PEM type, the acidic condition provided by the perfluorosulfonic acid membrane and ionomer and the high potential at the anode side (oxygen evolution) will require the use of precious metals based on iridium and platinum and the use of titanium-based components.

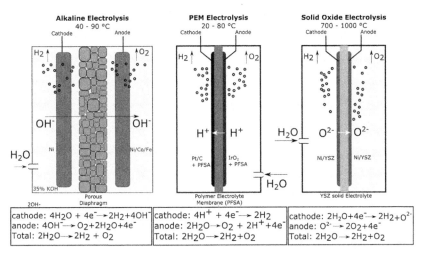

**FIGURE 4.5**  Operating principles of the different types of water electrolysis.

## Thermodynamics

$H_2O$ is a very thermodynamically stable molecule in our nature, a fluid responsible for the maintenance of life of most living organisms in our planet. It covers roughly 70% of the earth's surface, and around the same percentage is found inside our body. However, it contains a "valuable fuel," hydrogen. One way to harvest the $H_2$ molecule is by overcoming the equilibrium cell voltage, $E^0$, or the electromotive force by means of using free electrons, therefore electrolysis.

If two electrodes are immersed in a water-based electrolyte, and with the absence of current flow, the equilibrium potential is established, cell voltage is defined by the potential difference between anode and cathode, which turns into

$$E^0 = E^0_{anode} - E^0_{cathode} \qquad (4.1)$$

In addition, Eq. (4.2) relates the change in the standard Gibbs free energy ($\Delta G^0$) of the electrochemical reaction to the equilibrium cell voltage as follows:

$$\Delta G^0 = -nFE^0 \qquad (4.2)$$

where n is the number of moles of electrons transferred in the reaction and F is the Faraday constant (96,485C).

One can calculate at a fixed temperature the $\Delta_r G^0$ for any specific reaction from

$$\Delta_r G^0 = \Delta_r H^0 - T \cdot \Delta_r S^0 \qquad (4.3)$$

where $\Delta H^0$ is the standard variation of enthalpy, T the temperature, and $\Delta S^0$ the variation of entropy. Now using Table 4.4 with the $\Delta H^0$ of formation and

**TABLE 4.4** Thermodynamic Values Related to the Reaction $H_2O \rightarrow H_2 + \frac{1}{2} O_2$

|  | $\Delta H^0_f$ (kJ/mol) | $\Delta G^0_f$ (kJ/mol) | $S^0$ (J/mol K) |
|---|---|---|---|
| $H_2O$ liquid | −285.83 | −237.129 | 69.91 |
| $H_2O$ vapor | −241.818 | −228.572 | 188.825 |
| $H_2$ gas | 0 | 0 | 130.684 |
| $O_2$ gas | 0 | 0 | 205.138 |

$S^0$ for the reactants and products in the water (liquid) splitting reaction, one can calculate the final $\Delta_r G^0$ at 25°C (298K) using

$$\Delta_r G^0 = \sum_{R,P} \left( \upsilon_P \Delta_f H^0_P - \upsilon_R \Delta_f H^0_R \right) - T \cdot \sum_{R,P} \left( \upsilon_P S^0_P - \upsilon_R S^0_R \right) \quad (4.4)$$

and by filling the terms giving

$$\Delta_r G^0 = (0 \text{ kJ/mol} + 1/2 \cdot 0 \text{ kJ/mol} - (-285.83 \text{ kJ/mol})) - 298.$$
$$(+0.130684 \text{ kJ/mol} + 1/2 \cdot 0.205138 \text{ kJ/mol} - 0.06991 \text{ kJ/mol})$$
$$= \Delta_r G^0 = +237.153 \text{ kJ/mol(for } H_2O \rightarrow H_2 + 1/2O_2)$$

Finally, using Eq. (4.2), one can calculate the theoretical electromotive force (emf) for splitting liquid water at 25°C, giving

$$E^0 = -\Delta G^0/nF = -237153 \text{ J/mol}/2 \cdot 96485$$
$$= -1.223 \text{ V}(H_2O \rightarrow H_2 + 1/2O_2)$$

We can easily approximate overall water electrolysis cell potential to $E^0(25°C) = 1.23$ V and the change in Gibbs free energy as $+237.2$ kJ/mol, which is the minimum amount of electrical energy required to produce hydrogen and oxygen.

From the calculations above, we can start playing with different thermodynamic conditions to obtain the maximum efficiency to convert the electrical work into chemical energy (electrolysis process) and vice versa. The first thing we must notice is that the water splitting reaction is thermodynamically unfavorable and can only occur when sufficient electrical energy is supplied to overcome the reversible cell potential of 1.23 V, only for the electrochemical reaction to start. Second, the total reaction enthalpy must be supplied in the form of electrical energy, and therefore a thermoneutral voltage (1.48 V) has to be respected if one wants to avoid the generation of heat. Above 1.48 V, the reaction will be exothermic, which can be in some cases beneficial if extra heat can be supplied from another source to compensate the $T\Delta S$ term that kicks in

**TABLE 4.5** Different Thermodynamic Operating Conditions for the Water Splitting Reaction

| | $\Delta H^0$ (kJ/mol) | $E^0$ (V) | $\Delta G^0$ (kJ/mol) | $E^0$ (V) |
|---|---|---|---|---|
| Higher heating value using liquid water | 285.83 | −1.48 | 237.15 | −1.23 |
| | Pure electric energy supply Base case for the electrolysis process | | $T\Delta S$: heat supply $H_2O_{vapor}$: electric $H_2O_{splitting}$: electric Use of waste and/or ambient heat | |
| Lower heating value using liquid water | 241.81 | −1.25 | 228.57 | −1.19 |
| | $T\Delta S$: electric $H_2O_{vapor}$: heat supply (vapor) $H_2O_{splitting}$: electric | | $T\Delta S$: heat supply $H_2O_{vapor}$: heat supply (vapor) $H_2O_{splitting}$: electric High-temperature electrolysis | |

(use of waste or ambient heat). The same happens if, for example, water vapor can be supplied, giving another gain in terms of efficiency, definitely not negligible, in case electricity cost is a concern.

Table 4.5 summarizes this discussion, giving an overview of different possible thermodynamic conditions to operate the water electrolyzers.

In addition to Table 4.5, Fig. 4.6 shows the reversible cell potential dependence on the temperature typically found in the literature.

One can clearly notice that if only the reaction enthalpy is considered to calculate the cell voltage, a temperature increase will have a marginal effect on the final cell voltage and a significant variation is only obtained with regard to the equilibrium voltage. Nevertheless, even when the equilibrium potential is met, beyond this point the electrode reactions are inherently slow and over potentials ($\eta_n$) above the equilibrium cell voltage is needed to perform the reaction ($\eta = E_{cell} - E_0$). The overall cell overpotential (what comes above, for example, 1.23 V) can be understood as a sum of the high activation energy barrier to split the atoms (cathode + anode), ohmic resistances (electric and ionic), and mass transport limitations inside the cell/stack, giving

$$\eta_{Total} = \eta_{Anode} + \eta_{Cathode} + \eta_{Ohmic} + \eta_{Mass} \qquad (4.5)$$

where

- $\eta_{Total}$ is the total cell overpotential;
- $\eta_{Anode}$ is the overpotential related to the activation losses at the anode;
- $\eta_{Cathode}$ is the overpotential related to the activation losses at the cathode;

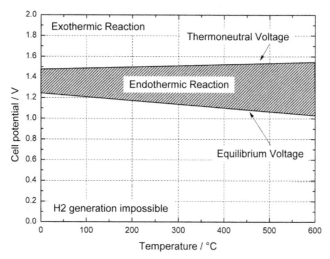

**FIGURE 4.6** Cell potential at the equilibrium versus temperature for the water electrolysis reaction.

- $\eta_{Ohmic}$ is the overpotential related to the sum of electric and ohmic resistances; and
- $\eta_{Mass}$ is the overpotential related to losses due to mass transport limitations.

## Activation Overpotential

In simple terms, the activation overpotential corresponds to the energetic barrier to be overcome to begin any oxy–redox electrochemical reaction, specifically related to the electron transfer that happens at the electrode interface. Fig. 4.7 shows a typical scheme to demonstrate this phenomenon and the key role of an electrocatalyst to lower the energy barrier for any reaction of interest.

If one considers a simple reaction where only the transference of one electron takes place in the oxy–redox reaction, we have

$$A + ze^{-} \underset{k_{anode}}{\overset{k_{cathode}}{\longleftrightarrow}} B$$

where $z$ is the number of electrons; in this case 1, A is the oxidized specie, B is the reduced specie, $k_{cathode}$ is the rate constant for the cathodic reaction, and $k_{anode}$ is the rate constant for the anodic reaction.

The goal now is to determine the reaction rates for both anode and cathode, and this only considers a reaction of the first order, giving for both cathode and anode

$$v_{cathode} = k_{cathode}[A]_{interface} \tag{4.6}$$

$$v_{anode} = k_{anode}[B]_{interface} \tag{4.7}$$

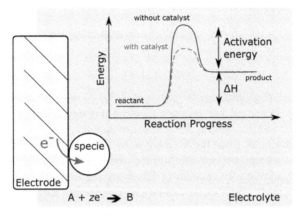

**FIGURE 4.7** Activation overpotential in electrochemical reactions. An activation energy barrier has to be overcome for the reaction to take place.

By using Arrhenius, and introducing now the activation energy terms for each reaction, we can obtain each equation for the reaction rate, remaining

$$v_{cathode} = k_1^{cathode}[A]e^{-\left(\Delta G_{cathode}^{\#}/RT\right)} \tag{4.8}$$

$$v_{anode} = k_1^{anode}[B]e^{-\left(\Delta G_{anode}^{\#}/RT\right)} \tag{4.9}$$

If we now use the Faraday law ($i = F \cdot v$) for the one electron transfer reaction above, we can obtain the necessary current dependence:

$$i_{cathode} = F \cdot v_{cathode} \tag{4.10}$$

$$i_{anode} = F \cdot v_{anode} \tag{4.11}$$

The next step is to introduce the Gibbs free energy so that the activation energies for the cathode and anode in Eqs. (4.8) and (4.9) can be deconvoluted toward voltage dependence, where we use

$$\Delta G_{cathode}^{\#} = \Delta G_{cathode*}^{\#} + F\beta(E_0 + \eta) \tag{4.12}$$

$$\Delta G_{anode}^{\#} = \Delta G_{anode*}^{\#} - F(1 - \beta)(E_0 + \eta) \tag{4.13}$$

Finally, by combining Eqs. (4.8), (4.10), and (4.12) for the cathode and Eqs. (4.9), (4.11), and (4.13) for the anode and assuming equilibrium state where $\eta = 0$, $i = 0$, and $i_{anode} = i_{cathode} = i_0$, we merge into the famous Butler–Volmer equation:

$$i = i_0\left\{e^{-(\beta F\eta/RT)} - e^{((1-\beta)F\eta/RT)}\right\} \tag{4.14}$$

## Ohmic Overpotential

The ohmic overpotential is essentially, as its name suggests, given by the pure Ohm's law, where

$$\eta_{ohmic} = RI \tag{4.15}$$

where R is the resistance (electronic and/or ionic) and I is the current. At higher current densities, the primary electron transfer rate (activation overpotential) is no longer limiting. Hence, overpotentials will arise through slow transport of reactants from the electrolyte to the electrode surface and also obviously the transport from the surface, away from the electrode to the electrolyte. In addition, limitations or losses arising from electron resistances to/from cell or stack components can be also responsible for an increase in the ohmic overpotential.

## Mass Transport Overpotential

As implied by its definition, mass transport limitation is driven by the inertia to transport species (reactants and/or products) to the electrode surface (reactants) and/or out of the electrode surface. In other words, at high current density, the electron transfer and ohmic transport will not be rate-determinant and the ability to provide new species will be the rate-determining step.

The mass transference step can be governed by three different mechanisms: (1) diffusion overpotential; (2) convection overpotential (natural and/or imposed); and (3) migration overpotential (governed by an electric field gradient).

## *The Nernst Equation*

The standard cell potentials, which were discussed above, refer to cells in which all dissolved substances are at unit activity, which essentially means an "effective concentration" of 1 mol/L. The same approximation was used for gases that take part in the electrochemical reaction, where an effective pressure (known as the fugacity) of 1 atm is usually considered. Naturally, if different values of concentration and/or pressures are taken, the cell potential for a given electrochemical reaction will be affected, which either brings the free energy change $\Delta_r G$ toward more negative values than $\Delta_r G^0$ or more positive values than $\Delta_r G^0$. This change in the relation of $\Delta_r G$ to $\Delta_r G^0$ will consequently affect the relation from cell voltage E to the standard cell potential $E^0$ as well. Now, with this in mind we can combine $\Delta G = -nFE$ and $\Delta G = -nFE^0$ (Eq. 4.2) with

$$\Delta G = \Delta G^0 + RT\ln Q \tag{4.16}$$

and obtain

$$-nFE = -nFE^0 + RT\ln Q \tag{4.17}$$

which can be rearranged into

$$E = E^0 - \frac{RT}{nF}\ln(Q) \tag{4.18}$$

This is the very celebrated Nernst equation, which importantly relates the cell potential to the standard potential and to the activities of the electroactive species. As discussed above, we must notice that the cell potential will be the same as $E^0$ only if Q is unity. The Nernst equation is more commonly written in base 10 log forms and at 25°C we obtain

$$E = E^0 - \frac{0.059}{n} \log Q \tag{4.19}$$

This relation has a very crucial significance, where a cell potential will change by 59 mV per 10-fold change in the concentration of a substance involved in a one-electron oxidation or reduction; for two electron processes, the variation will be 28 mV per decade concentration change.

## Cell Potentials Versus pH—Water Stability Diagram

Because we are strictly dealing with water oxidation here (electrolysis), naturally the electron transfer reactions will involve the transference of hydrogen ions (PEM) and hydroxide ions (alkaline). The standard potentials for these reactions therefore refer to the pH, either 0 or 14, at which the appropriate ion has unit activity. Because multiple numbers of $H^+$ or $OH^-$ ions are often involved, the potentials given by the Nernst equation can vary greatly with the pH. It is frequently useful to look at the situation in another way by considering what combinations of potential and pH allow the stable existence of particular species. This information is most usefully expressed by means of an E versus pH diagram, also known as a Pourbaix diagram.

The reduction reaction using an acidic electrolyte can be written as

$$2H^+ + 2e^- \rightarrow H_2(g)$$

or in neutral or alkaline solutions as

$$H_2O + 2e^- \rightarrow H_2(g) + 2OH^-$$

These two reactions are equivalent and follow the same Nernst equation:

$$E_{\frac{H^+}{H_2}} = E^0_{\frac{H^+}{H_2}} + \frac{RT}{nF} \ln \frac{[H^+]^2}{p_{H_2}} \tag{4.20}$$

which at 25°C and $H_2$ with unit partial pressure gives

$$E = E^0 - 0.059 \, pH = -0.059 \, pH \tag{4.21}$$

Analogously, we will have for the oxidation of water

$$H_2O \rightarrow O_2(g) + 4H^+ + 2e^-$$

the following:

$$E_{\frac{O_2}{H_2O}} = E^0_{\frac{O_2}{H_2O}} + \frac{RT}{nF} \ln p_{O_2} \cdot [H^+]^4 \tag{4.22}$$

From this information we can construct the stability diagram for water that is shown in Fig. 4.8.

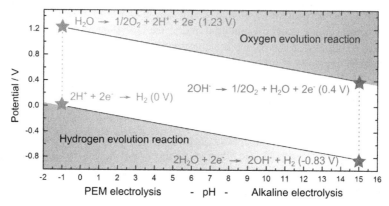

**FIGURE 4.8** Stability (Pourbaix) diagram for water.

## Faraday's Laws of Electrolysis

The known Faraday constant 96,485 C/mol denoted by the symbol F, or also called 1 F, corresponds to the amount of electricity that is carried by 1 mol of electrons. The number comes from the charge of a single electron $1.6023 \times 10^{-19}$ C times the number of electrons in 1 mol, given by the Avogadro number $6.02 \times 10^{23}$ electrons.

According to the Faraday's first law, in a given electrochemical reaction the mass (m) of substance that is deposited or released at the electrode surface is directly proportional to the amount of electricity or charge (Q) that passes through it.

$$m \, \alpha \, Q \text{ or } m = Z \cdot Q \qquad (4.23)$$

where m is the mass in grams, Q is the charge in Coulombs, and Z is the proportionality number given in g/C. The proportionality number can be translated as the electrochemical equivalent (E), which is the mass produced or consumed at the electrodes by 1 C of charge. This means that 1 C will liberate 1 g equivalent of a substance divided by 96,485, where Z can be finally expressed as Z = E/96,485. Now, if we rearrange Eq. (4.23) using Q = I·t, we obtain

$$m = \frac{E \cdot I \cdot t}{F} = \frac{M \cdot I \cdot t}{F \cdot z} \qquad (4.24)$$

where m is the mass of the substance liberated at the electrode (in grams), M is the molar mass of the substance (in g/mol), I is the current (in amperes), t is the time (in seconds), F is the Faraday constant or 96,485 C, and z is the valence number of ions of the substance or the number of electrons transferred per ion.

## Basic Principles of Alkaline Water Electrolysis

An alkaline electrolyzer is characterized by having two electrodes (cathode and anode) immersed in a liquid alkaline electrolyte consisting of a caustic potash solution at a level of 20%–35% KOH. Today, microporous diaphragms are almost exclusively used as separators/diaphragms in alkaline electrolysis (Fig. 4.5) with the function of keeping the product gases apart from one another for the sake of efficiency and safety. The diaphragm must also be permeable to the hydroxide ions and water molecules. Three major challenges are normally associated with alkaline electrolyzers, low partial load range, limited current density, and low operating pressure. First, the diaphragm does not completely prevent the product gases from permeating through it, leading to safety-related shutdowns at approximately 2 vol.% $H_2$ in $O_2$. The diffusion of oxygen into the cathode chamber reduces the efficiency of the electrolyzer because oxygen will be catalyzed back to water with the hydrogen present on the cathode side. Additionally, extensive mixing (particularly hydrogen diffusion to the oxygen evolution chamber) also occurs and must be avoided to preserve the efficiency and ensure a safe operation. This is particularly severe at lower current densities ($<100$ mA/cm$^2$) where the oxygen production rate decreases, thus drastically increasing the hydrogen concentration to unwanted and dangerous levels. The second drawback for alkaline electrolyzers is the low maximum achievable current density because of the high ohmic losses across the liquid electrolyte and diaphragm. Because the $ZrO_2$ particles and polymer mesh that gives mechanical stiffness to the diaphragm are not $OH^-$ conductive, it contributes to an additional loss (ohmic), dramatically reducing the cell voltage efficiency at a given current. The high porosity will also lead to another challenge, the inability to operate at high pressure, which makes for a bulky stack design configuration. To operate at higher pressures, the electrodes are designed so that a given distance is obtained between them, which is a distance or spacing that will minimize the extensive mixing of gases produced at each electrode. In addition to this, electrodes that are apart from the diaphragm will eventually lead to a "curtain" of gas bubble between electrode and diaphragm, also contributing to an increase in the ohmic overpotential (the so-called "bubble resistance").

Fig. 4.8 shows an illustration with typical overpotentials for the anode and cathode of an alkaline electrolysis cell, where the most widely used electrode material is nickel because of its stability and favorable activity in high concentrated KOH solutions. One can notice that, in contrast to PEM electrolysis (acidic media), the cell voltages are more or less distributed between cathode and anode. In general words, however, the overpotential of oxygen evolution is more challenging for real cells than that related to the HER (cathode). Moreover, the oxidizing regime will fatally reduce any catalyst material to its oxidized form, severely reducing the surface area because of a passivation phenomenon on the catalyst surface. On the cathode side, hydride phase formation that leads to electrode deactivation seems to be a concern, especially if pure Ni electrodes are used.

### Basic Principles of Polymer Electrolyte Membrane Water Electrolysis

The concept idealized by Grubb at GE, where a solid perfluorosulfonic membrane is used as an electrolyte, also referred to as proton exchange membrane or polymer electrolyte membrane (PEM) water electrolysis, is responsible for providing high proton conductivity, relatively low gas cross-over, compact system design, and high pressure operation. The thin polymer membrane (typically $200 \, \mu m$ thick) is in part the reason for many of the advantages of PEM electrolyzers. In contrast to alkaline electrolysis, it can operate at much higher current densities, capable of achieving values above $10 \, A/cm^2$, drastically reducing the investment cost of electrolyzers. The solid polymer membrane allows for a much thinner electrolyte than the alkaline electrolyzers.

The relatively low gas crossover rate of the polymer electrolyte membrane (yielding hydrogen with high purity) allows for the PEM electrolyzer to work under a wider range of power input. This is due to the fact that the proton transport across the membrane responds quickly to the power input, not delayed by inertia as in liquid electrolytes blocked by porous and high ohmic resistant diaphragms. As discussed before, in alkaline electrolyzers operating at low load, the rate of hydrogen and oxygen production reduces, while the hydrogen permeability through the diaphragm remains constant, yielding a larger concentration of hydrogen on the anode (oxygen) side, thus creating hazardous and less efficient conditions. In contrast with alkaline electrolysis, PEM electrolysis covers practically the full nominal power density range, eventually reaching values over 100% of nominal rated power density, where the nominal rated power density is derived from a fixed current density and its corresponding cell voltage. This is due to the lower permeability of hydrogen through Nafion compared with, for example, Zirfon diaphragms for alkaline electrolyzers. A solid electrolyte allows for a compact system design with strong/resistant structural properties, in which high operational pressures (equal or differential across the electrolyte) are achievable. The high pressure operation of an electrolyzer brings the advantage of delivering hydrogen at a high pressure (sometimes called electrochemical compression) for the end user, thus requiring less energy to further compress and store the hydrogen. It also diminishes the volume of the gaseous phase at the electrodes, thus significantly improving product gas removal, which follows Fick's law of diffusion. In a differential pressure configuration, only the cathode (hydrogen) side is under pressure, which can eliminate the hazards related to handling pressurized oxygen and the possibility of self-ignition of Ti in oxygen. The pressure increase minimizes the expansion and dehydration of the membrane, preserving the integrity of the catalytic layer. Problems related to higher operational pressures in PEM electrolysis are also present, such as cross-permeation phenomenon that increases with pressure. Pressures above 100 bar will require the use of thicker membranes (although more resistant) and internal recombination catalysts to maintain the critical concentrations (mostly $H_2$ in $O_2$) under safety threshold (2 vol.% $H_2$ in $O_2$). Lower gas permeability through the membrane (crossover) can be obtained by incorporating miscellaneous fillers inside the membrane material, but this normally leads to less conducting materials. The corrosive acidic regime provided by the proton exchange membrane requires the use of distinct materials. These materials must not only resist the harsh corrosive low pH condition but also sustain the high applied voltage, especially at high current densities. Corrosion resistance applies not only for the catalysts used but also for the porous transport layers (PTLs) and bipolar plates. Only a few materials can be selected that would perform in this environment. This will demand the use of scarce, expensive materials and components such as noble catalysts (platinum group metals [PGMs], e.g., Pt, Ir, and Ru), titanium-based current collectors, and bipolar plates. Iridium has one particular limitation because it is one of the rarest elements in the earth's crust, having an average mass fraction of 0.001 ppm in crustal rock. Conversely, gold and platinum are 40 times and 10 times more abundant, respectively.

## Design and Operation of Cells, Stacks, and Systems

When running electrochemical reactions in real single cells and stacks, an extensive variety of aspects and parameters when designing cell and stack hardware ought to be considered. Properly defining the operating conditions is also crucial, so that maximum efficiency, both voltage and faradaic, can be achieved, and acceptable durability is demonstrated. As discussed above, undesirable overpotentials shall arise when electrochemical systems are out of equilibrium. If one manages to accurately design cells and stacks, and intelligently choose the operating conditions, we will surely demonstrate cost-effective electrolysis devices that present low ohmic losses and mass transport limitations.

As discussed above, what essentially differentiates the two technologies is the use of caustic KOH solution for the alkaline type and the high anodic overpotential and acidic condition for the PEM. These requirements will impose the use of specific materials and components and consequently limit its design and choice of materials. As we expand the active area of electrodes and components, complexity in terms of manufacturing and assembling starts to play a crucial role. As these cells become "too large" with areas beyond 1 $m^2$, the costs involved will eventually turn scaling-up approaches impracticable.

Much more challenging if, for instance, high pressure operation is required. Cells and stacks have been historically manufactured using a round design, so that the pressure distribution inside the stack is adequate to provide distributed electrical contact and reactants flow, avoid the leakage of gas and fluids, prevent failures, and provide long-term durability under high pressure operation. As an advantage, there is a consequent waste of materials when cutting, a cost that increases almost linearly with scaling-up. That is the reason why multikilowatt/multimegawatt stacks are preferably rectangular. However, tradition in manufacturing alkaline stacks still maintains its round shape because dealing with liquid KOH in a rectangular shape and under pressure is not trivial.

When putting components together, either at the cell or stack level, it is crucial to obtain very low contact resistances between the components. That is the reason why PEM water electrolyzers use a catalyst-coated membrane concept, where the electrode is directly coated on the membrane, reducing the ohmic drop between them. Next is to obtain a good electrical contact between the electrode and titanium PTL and finally between the PTL and bipolar plate. When assembling the cell or stack, too less torque (or cell pressure) will lead to unacceptable low electrical conductivities across the layers and potential leakage issues. On the other hand, too much torque will transfer stress to the least mechanical resistant components such as catalyst layer, membrane, and carbon PTLs. This will cause its destruction, collapse of the necessary porosity avoiding the entrance of reactants, membrane creping, and a rapid increase in the mass transport losses inside the stack. Besides the critical loss in the cell efficiency, too much stress shall also limit the lifetime of the components because an optimal humid condition is necessary to guarantee the lifetime of electrodes and membranes. The same reasoning applies for the alkaline type, where a compromise between the mechanical integrity of the components and contact pressure has to be determined.

As an advantage for the alkaline case, liquid KOH will guarantee good $OH^-$ transport across all components, as long as KOH can fully achieve the reaction sites. In this case, the only thing to avoid is the lack of electrical contact between electrodes and PTLs (Ni mesh), especially when using a bipolar plate configuration.

To enhance the contact between the PTL and catalyst layer, and between the PTL and bipolar plate, it is sometimes required to coat the PTL with PGM-based materials. This also to increase the lifetime of PEM stacks, by reducing the overtime passivation and dissolution of Ti-based PTLs. In addition, PGM coating should also reduce embrittlement of the materials on the cathode side. PGM coating on both PTL and bipolar plates is also a must, if cell voltages beyond 2 V are obtained, increasing the chances of irreversible degradation inside the components at the anode side of PEM electrolyzers. In an alkaline regime, one approach to reduce the cost is to substitute the use of raw Ni to fabricate the bipolar plates, by using stainless steel plates coated with Ni. Important here is to obtain a homogeneous coating free of irregularities, which is also resistant to 30% KOH at 80−90°C.

In a bipolar plate configuration, each separator plate is responsible to separate the gases and the flow field responsible to provide an even supply of water (or reactants) over the PTL. The flow field should also release the produced gases (counterflow). Flow fields are not necessarily important when low active areas are used, as long as water is sufficiently provided to all reaction sites. However, as large electrode areas are needed for multikilowatt or multimegawatt systems, flow fields become essential so that a minimum pressure drop to circulate the reactant or cooling water is obtained. Fig. 4.9 shows the different designs or patterns that can be used in flow fields; pin-type, parallel, serpentine, parallel serpentine, eventually 3D-wire mesh, expanded metal sheets or perforated plates (Fig. 4.10).

The most common for PEM electrolysis is the serpentine type, whereas expanded mesh or perforated plates are commonly used in alkaline electrolyzers. The major drawback to the use of flow fields in separator plates is the machining cost related to its manufacturing. Machining each individual part requires large processing times and incurs rather large material waste as well as increased material thicknesses, ultimately increasing ohmic losses. A PGM coating step is also tremendously expensive and requires the use of expensive precious metals, complex equipment, and rare expertise. The use of thinner materials reducing manufacturing time can cause significant reductions in the overall stack cost. Owing to the need for flow fields to facilitate even water distribution in cells with larger active areas, a flow field must be incorporated. To include the flow field in the separator plate, the plate thickness must be at least 2−3 mm while the stamped and hydroformed components can easily

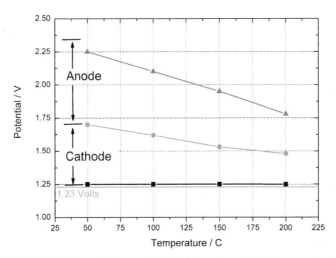

**FIGURE 4.9** An illustration of the contributions of anode and cathode overpotentials of an alkaline electrolysis cell.

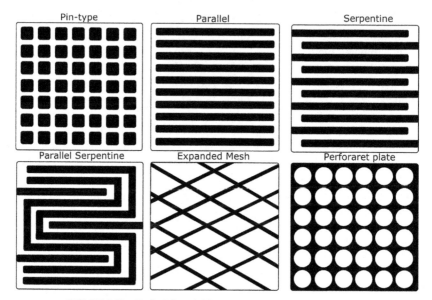

**FIGURE 4.10** Typical flow-field patterns used in water electrolyzers.

reach thicknesses as small as 0.1 mm. Avoiding the incorporation of the flow field in the separator plate as the case for alkaline electrolyzers can, however, drastically reduce the manufacturing costs, thus removing the need for expensive machining or stamping processes.

## ACKNOWLEDGMENTS

We would like to thank Prof. Ralf Peters and Dr. Thomas Grube, Forschungszentrum Jülich GmbH, Jülich, Germany, for the innumerous inspiring discussions and assistance in the preparation of this chapter.

## REFERENCES

[1] M. Robinius, et al., Linking the power and transport Sectors-Part 2: modelling a sector coupling scenario for Germany, Energies 10 (7) (2017).

[2] M. Robinius, Strom- und Gasmarktdesign zur Versorgung des deutschen Straßenverkehrs mit Wasserstoff, RWTH Aachen University, Jülich, 2015.

[3] L. Bertuccioli, A. Chan, D. Hart, F. Lehner, B. Madden, E. Standen, Development of Water Electrolysis in the European Union - Final Report, E4tech Sàrl/element energy, Lausanne/Cambridge, 2014.

[4] M. Reuss, et al., Seasonal storage and alternative carriers: a flexible hydrogen supply chain model, Appl. Energy 200 (2017) 290−302.

[5] T. Grube, Passenger car drive cycles, in: Fuel Cells: Data, Facts and Figures, Wiley-VCH Verlag GmbH & Co. KGaA, 2016, pp. 12−21.

[6] A. Boretti, Advances in hydrogen compression ignition internal combustion engines, Int. J. Hydrogen Energy 36 (19) (2011) 12601−12606.

[7] U. Eberle, B. Muller, R. von Helmolt, Fuel cell electric vehicles and hydrogen infrastructure: status 2012, Energy Environ. Sci. 5 (10) (2012) 8780−8798.

[8] T. Grube, Potentiale des Strommanagements zur Reduzierung des spezifischen Energiebedarfs von Pkw, Technische Universität Berlin, Jülich, 2014.

[9] Collaboration, J.-J.R.C.-E.-C., Well-to-Wheels Analysis of Future Automotive Fuels and Powertrains in the European Context - Tank-to-wheels Report, European Commission, Joint Research Centre, 2011.

[10] M. Schemme, P. Peters, R.C. Samsun, T. Grube, D. Stolten, Alternative transport fuels and their production using surplus electricity, water and $CO_2$, in: AIChE Annual Meeting, 2016. San Francisco.

[11] W.L. Becker, et al., Production of Fischer-Tropsch liquid fuels from high temperature solid oxide co-electrolysis units, Energy 47 (1) (2012) 99−115.

[12] N. Merdes, R. Pätzold, N. Ramsperger, H.-G. Lehmann, Die neuen R4-Ottomotoren M270 mit Turboaufladung, in: ATZ - Automobiltechnische Zeitschrift, 2012, pp. 58−63.

[13] M. Carmo, et al., A comprehensive review on PEM water electrolysis, Int. J. Hydrogen Energy 38 (12) (2013) 4901−4934.

[14] W. Hug, et al., Highly efficient advanced alkaline electrolyzer for solar operation, Int. J. Hydrogen Energy 17 (9) (1992) 699−705.

[15] A. Szyszka, Schritte zu einer (Solar-) Wasserstoff-Energiewirtschaft. 13 erfolgreiche Jahre Solar-Wasserstoff-Demonstrationsprojekt der SWB in Neunburg vorm Wald, 1999. Oberpfalz.

[16] H. Barthels, et al., Phoebus-Julich: an autonomous energy supply system comprising photovoltaics, electrolytic hydrogen, fuel cell, Int. J. Hydrogen Energy 23 (4) (1998) 295−301.

# Chapter 5

# Hydrogen Energy Engineering Applications and Products

Hirohisa Uchida[1], Makoto R. Harada[2]
[1]*Professor, School of Engineering, Tokai University/President & CEO, KSP Inc., Japan;*
[2]*Research Adviser, National Institute of Advanced Industrial Science and Technology (AIST), Research Institute for Chemical Process Technology*

## INTRODUCTION

This chapter describes the engineering application of hydrogen energy. Applications of hydrogen energy technologies fall into four categories: hydrogen production, storage, transportation, and stationary use. Each field has undergone independent development.

Hydrogen production has a long history. The hydrogen production from fossil fuels exceeds a 100-year history. From the 1900s, research and development for mass production, advanced industrial hydrogen production, and progress has been continued to the present. This technology is currently making a great contribution to hydrogen energy technology and has future prospects of additional progress, including hydrogen production from biomass.

Hydrogen storage is an indispensable technology for fuel cell mobile applications. Various techniques for storing hydrogen, such as hydrogen storage alloys, high-pressure hydrogen gas storage, liquid hydrogen storage, and the like, have been developed and are currently being used at hydrogen refueling stations. These technologies have enriched the prospects for fuel cell mobile applications, including current fuel cell vehicles. Hydrogen station deployment is actively being carried out in East Asia, Japan, South Korea, and China, as well as in the United States and Europe.

Transportation of hydrogen has a very important significance for the supply of feedstock. Currently, research on how to transport hydrogen is under way. Currently, two major projects are progressing in Japan. The first one is a development plan to transport liquid hydrogen by ship, and the other one is hydrogen production using organic hydrides.

Stationary fuel cell applications have initially unveiled hydrogen utilization by the society. But at present, mobile/transportation fuel cells, typified by fuel cell vehicles, forklift, buses, and cars, are taking the lead. Also, due to the

Science and Engineering of Hydrogen-Based Energy Technologies. https://doi.org/10.1016/B978-0-12-814251-6.00009-5

massive production of hydrogen, hydrogen combustion in turbines for power generation will soon start, inducing an expansion of hydrogen energy applications. Concerning vehicles, Toyota first commercialized Mirai in 2014, then Honda began selling Clarity in 2016. Subsequently, Toyota began selling buses in March 2018. In 2019 fuel cell trucks are planned to be sold, and in the near future demand for transportation fuel cells will grow steadily. Sales of forklifts have also started in 2016.

As described above, development of hydrogen utilization technology is progressing continuously with the goal of being significantly implemented by 2030, and this trend will continue. Even from a global view, many countries are working on applied technologies of hydrogen energy, fuel economy, testing cogeneration power generation with stationary fuel cells for several world metropolitan cities. Technology development and deployment road maps focus on long-term schedules, up to 2050.

Application technology of hydrogen energy has dramatically improved in the past decade, and it is speculated that this trend will continue in the future.

In this chapter, the main features concerning applied technology of hydrogen energy are described.

Chapter 5.1

# Hydrogen Production Technology From Fossil Energy

## INTRODUCTION

Hydrogen ($H_2$) production from fossil fuels has a long history and has been industrialized since the beginning of the 20th century. Various reaction types or plants have been proposed, but here we will summarize the technically established methods.

Hydrogen is presently artificially manufactured in large scale. For this reason, various technological developments have been carried out so far. It has mainly been produced from the electrolysis of water and from fossil fuels [1−3], as will be described here.

Hydrogen as an energy carrier is usually found in the state of a compound bonded to other elements. Therefore, for hydrogen to become an energy source, the hydrogen compound must be efficiently decomposed with the use of some external energy.

Hydrogen is a well-known and versatile energy carrier that has been produced in large scale, with global amounts exceeding 55 Mt, mainly from fossil fuels [4]. Hydrogen production methods such as steam reforming, partial oxidation and auto-thermal reforming, ATR, are used, however, the steam reforming of natural gas is mostly employed, making it less expensive than other procedures [5]. Hydrogen atoms of natural gas are separated into hydrogen molecules in the course of the reaction, and at the same time, carbon dioxide is produced as a by-product. For this reason, the development of $CO_2$ -free hydrogen by renewable energy has been advanced [6].

Here we will explain the hydrogen production method from fossil fuel by explaining the hydrogen production reaction, the reaction process, the catalyst, and so on.

## CHARACTERISTICS OF HYDROGEN PRODUCTION PROCESSES

First, the various hydrogen production methods will be explained, including steam reforming using steam reforming reaction (partial reforming), partial oxidation using oxygen as an oxidizing agent, and autothermal utilization of reaction heat generated when hydrogen is produced. A combined reforming method combines a steam reforming method and an ATR method [7].

As shown in Fig. 5.1, energy, including hydrogen, can be manufactured industrially from various fossil fuels through various hydrogen production methods. These processes are already established industrially.

In this chapter, the technology of production of hydrogen from fossil fuels is discussed, mainly the steam reforming reaction, which is a hydrogen production process also used for fuel cell systems.

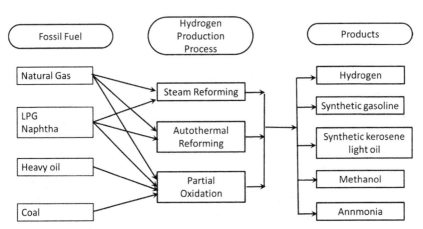

**FIGURE 5.1** Relationship between types of fossil fuels and various hydrogen production processes.

## Hydrogen Production Reaction

### Catalytic Steam Reforming

The steam reforming process has been carried out for more than 100 years and is the most historically old and industrially constructed process. Therefore, since this hydrogen production method has been described in many places so far, we will first describe the reaction, and then explain items deeply related to hydrogen production such as catalyst and poisoning.

It is thought that the steam reforming reaction proceeded simultaneously with the following two reactions.

$$CH_4 + H_2O \rightarrow CO + 3H_2 - 206 \text{ kJ/mol} \tag{5.1}$$

$$CO + H_2O \rightarrow CO_2 + H_2 + 41 \text{ kJ/mol} \tag{5.2}$$

Eq. (5.2) is called a shift reaction, which is a CO concentration reduction reaction that generates $CO_2$ from CO.

ALthough $CO_2$ has become a global warming problem, there is also a reaction called carbon dioxide reforming reaction:

$$CH_4 + CO_2 \rightarrow 2CO + 2H_2 - 248 \text{ kJ/mol} \tag{5.3}$$

This reaction is used in research to convert $H_2$ and $CO_2$ into useful compounds, which has been drawing attention in recent years. Since this reaction is an endothermic reaction larger than the steam reforming reaction and since carbon deposition is liable to occur in the catalyst layer, a catalyst capable of suppressing carbon deposition currently has to be used. Fig. 5.2 shows a

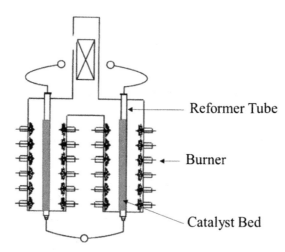

Radiant Wall Type

First Reformer

**FIGURE 5.2** Radiant wall-type reformer. *Based on Miyasugi et al., J. Jpn. Petrol. Int. 25 (1982) 260.*

schematic diagram of this reactor, but it is an energy-consuming reactor that heats a catalyst bed reactor with several burners from the outside.

## Partial Oxidation

In this method, oxygen ($O_2$) is used as an oxidizing agent, and hydrogen is produced by utilizing a partial oxidation reaction of a fossil fuel as a feedstock. The amount of $O_2$ to be introduced is an amount corresponding to 30%–50% of the theoretical oxygen amount required for complete combustion. When a flame occurs, $CO_2$ is introduced and the flame is controlled. Most types of fuels can be used. This method is characteristic of a noncatalytic reaction. The partial oxidation reaction formula is

$$C_mH_n + \frac{m}{2}O_2 \rightarrow mCO + \frac{n}{2}H_2 \tag{5.4}$$

In case the fuel is methane, it becomes

$$CH_4 + \frac{1}{2}O_2 \rightarrow CO + 2H_2 + 35.7 \text{ kJ/mol} \tag{5.5}$$

Fossil fuel and $O_2$, which is the oxidizer, as well as steam, are injected from the burner to perform partial oxidation reaction. In the partial oxidation reaction, slight carbon deposition occurs. Normally, about 3 wt% of carbon is precipitated with respect to the fuel. Carbon in this product gas is removed as a carbon slurry by water quenching or water washing. Fig. 5.3 shows a

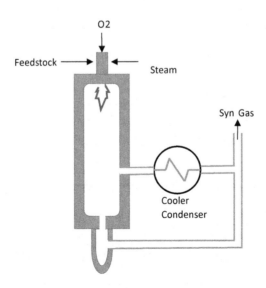

**FIGURE 5.3** Partial oxidation hydrogen production secondary reformer. *Based on Dry et al., Industrials chemicals via C1 processes, ACS Symp. Ser. 328 (1987) 18.*

schematic diagram of this reactor. Although it sometimes fills ceramic balls with very high heat resistance, it is fundamentally a very simple structure [7].

## Autothermal Reforming

ATR is a hydrogen production method combining a partial oxidation method and a steam reforming method. At the inlet of the catalyst layer, $O_2$ (which is an oxidizing agent), fuel, and steam are introduced; part of the fuel is burned; and the steam reforming reaction is carried out by utilizing the reaction heat. Therefore, the partial oxidation reaction mainly occurs on the upstream side of the reactor and the steam reforming reaction mainly occurs at the downstream side of the reactor. The overall reaction is called ATR reaction.

In the ATR method, the following reaction occurs.

Partial oxidation reaction: $CH_4 + 3/2O_2 \rightarrow CO + 2H_2O + 519$ kJ/mol  (5.6)

Steam reforming reaction: $CH_4 + H_2O \rightarrow CO + 3H_2 - 206$ kJ/mol  (5.7)

Shift reaction: $CO + H_2O \rightarrow CO_2 + H_2 + 41$ kJ/mol  (5.8)

$O_2$ is used as the oxidizing agent as in the partial oxidation reactor described earlier, as the temperature at the inlet of the reactor reaches 1000°C or higher. For this reason, a metallic reactor cannot be used. The inner wall of the reactor of the partial oxidation reaction and ATR reaction is coated with refractory brick or refractory cement and a reactor with high heat resistance is used. Fig. 5.4 shows a schematic diagram of this reactor.

## Combined Reforming

This process is used in the synthesis of ammonia and methanol. Part of fossil fuel as a primary reformer is a method of decomposing fuel using the steam reforming reaction method and then reforming it by the ATR method as a secondary reforming. Fig. 5.5 shows a schematic diagram of this reactor, but basically it is a combination of the steam reforming and ATR reactors.

In the secondary reforming, when synthesizing ammonia, air is used as an oxidant, and the gas composition at the ATR reactor outlet is adjusted such that the molar ratio of $H_2$ and $N_2$ required for ammonia synthesis is 1:3. Since the oxidizing agent of the ATR method is air, $O_2$ from the air is utilized, and an expensive oxygen separating apparatus is unnecessary. For this reason, a gas for ammonia synthesis can be produced economically.

On the other hand, when methanol synthesis is carried out, $O_2$ is used as an oxidizing agent. The gas composition at the outlet of the ATR reactor has a molar ratio of $H_2$ and $CO$ required for methanol synthesis of 2:1, and it is possible to obtain a gas easily meeting methanol synthesis reaction conditions. This chemical reaction formula is a partial oxidation reaction formula.

$$CH_4 + 1/2O_2 \rightarrow CO + 2H_2 + 35.7 \text{ kJ/mol} \qquad (5.9)$$

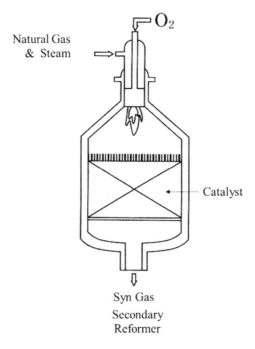

**FIGURE 5.4**  Autothermal hydrogen production secondary reformer. *Based on B. Glöckler, A. Gritsch, A. Morillo, G. Kolios, G. Eigenberger, Autothermal reactor concepts for endothermic fixed-bed reactions, Chem. Eng. Res. Des. 82 (2) (February 2004) 148–159.*

Features of the four hydrogen production methods are summarized in Table 5.1.

Table 5.2 also summarizes the industrial conditions for producing each product including hydrogen [8].

## Industrial Hydrogen Production Process

The history of hydrogen production catalysts used for the steam reforming method, partial oxidation method, and others is very old, and even in the experiment note of Michael Faraday, hydrogen, CO, $CO_2$, and the like are generated when steam is brought into contact with heated coal. In these notes, there is a description that white smoke is adsorbed on a metal such as Ni, and the phenomenon of a predictable catalyst is noted.

In order to produce hydrogen from fossil fuel, it is essential that the fossil fuel gas component and water vapor are adsorbed to the catalyst; the fossil fuel, which is a hydrocarbon, decomposes and reacts comprehensively with the oxygen molecules in the water vapor. Also, considering chemical reactions, production of CO is essential. Therefore, in the hydrogen production process,

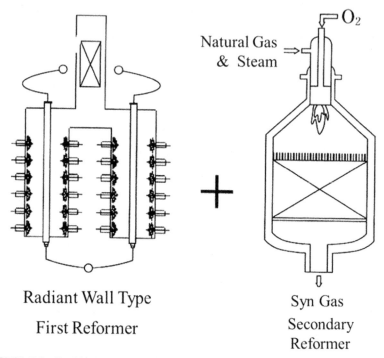

Radiant Wall Type

First Reformer

Syn Gas

Secondary Reformer

**FIGURE 5.5** Combined reforming reactor. *Based on H. Ebrahimi, A. Behroozsaranda, A. Zamaniyan, Arrangement of primary and secondary reformers for synthesis gas production. Chem. Eng. Res. Des. 88 (10) (October 2010) 1342—1350.*

a shift reactor is always arranged downstream of the reforming process. Fig. 5.6 shows the industrial hydrogen production process by the steam reforming process. Temperature, pressure, and gas concentration shown in the process are typical examples of the plant actually being operated. Other hydrogen production methods such as the partial oxidation method only replace the steam reforming reactor with other reactors, and there is no change in the process [9]. The hydrogen production process is divided into four steps: a desulfurization step for removing sulfur compounds, a reforming step for producing hydrogen, a shift step for reducing the concentration of carbon monoxide, and a purification step for removing impurities. These four steps will be described.

## Desulfurization Step

$H_2S$ is contained in natural gas, and sulfide typified by methyl mercaptan is contained in a gas state in other hydrocarbon fuels. This sulfur compound adsorbs to the reforming catalyst on the downstream side and the catalyst for low temperature carbon monoxide shifts to lower the activity. In order to

**TABLE 5.1** Hydrogen Production Methods from Fossil Fuels

| Process | Steam Reforming System | Partial Oxidation System | Autothermal Reforming System | Two-Stop Reforming System |
|---|---|---|---|---|
| Available fossil fuel | Natural gas, LPG, naphtha up to C6 saturated hydrocarbons. Unsaturated hydrocarbons are difficult to use. | Hydrocarbons, heavy oil, Town gas | Natural gas, LPG, naphtha | Natural gas, LPG, naphtha, saturated fossil fuel, and fossil fuel with few unsaturated hydrocarbons |
| Reactor type | External heat exchange–type fixed bed reactor | Adiabatic fixed bed reactor | Adiabatic fixed bed reactor | (Type- I) Two-stage external heat exchanger–type reactor or (type-II) isothermal steam reforming reactor and autothermal reforming–type reactor |
| Main chemical reaction | Catalytic steam reforming reaction $CH_4 + H_2O \rightarrow CO + H_2 - 206$ kJ/mol shift reaction $CO + H_2O \rightarrow H_2 + CO_2 + 41$ kJ/mol | Catalytic partial oxidation reaction $CH_4 + 1/2O_2 \rightarrow CO + H_2 + 38$ kJ/mol | Partial oxidation reaction and catalytic steam reforming reaction | Catalytic steam reforming reaction |
| Oxygen plant | Not required | Required | Required | Not required for type-I, but required for type-II |

Continued

**TABLE 5.1** Hydrogen Production Methods from Fossil Fuels—cont'd

| Process | Steam Reforming System | Partial Oxidation System | Autothermal Reforming System | Two-Stop Reforming System |
|---|---|---|---|---|
| Catalyst | $Ni/Al_2O_3$, $Ru/Al_2O_3$ | $Ni/Al_2O_3$, $Rh/Al_2O_3$ | $Ni/Al_2O_3$, $Ni/MgAl_2O_4$, $Rh/Al_2O_3$, | $Ni/Al_2O_3$ |
| Operating conditions | Maximum bed temperature 750—950°C | Maximum bed temperature 900—1150°C | Maximum bed temperature 900—1150°C | First steam reformer bed temperature 500°C (type-I) |
| Temperature and pressure | Reactor pressure 1.5—3 MPa | Reactor pressure 3—7 MPa | Reactor pressure 2—6 MPa | Second steam reformer bed temperature 750—950°C Reactor pressure 1.5—3 MPa (type-II) Second reactor max. bed temperature 900—1150°C Reactor pressure 2—6 MPa |
| Hydrogen/CO ratio | 2.8—4.8 (mol/mol) | 1.7—2.0 (mol/mol) | 1.8—3.8 (mol/mol) | 2.2—4.4 (mol/mol) |
| Hydrogen production scale per series | 800—80,000 $Nm^3/h$ | 7000—70,000 $Nm^3/h$ | 10000—7,50,000 $Nm^3/h$ | 50,000—3,00,000 $Nm^3/h$ |

**TABLE 5.2** Examples of Industrial Conditions for the Production of Pure Hydrogen and of Hydrogen for the Production of Other Chemical Components

| Condition | $H_2$ | $H_2$ for Ammonia Production | $H_2$ for Methanol Production[a] | $H_2$ for Syngas Production[b] |
|---|---|---|---|---|
| Feedstock | Natural gas | Natural gas | Natural gas | Natural gas |
| Pressure | 40 | 34 | 16.5 | 27 |
| $H_2O/C$ (mol/atom) | 6.5 | 3.5 | 2.5 | 0.6 |
| $CO_2/C$ (mol/atom) | 0 | 0 | 0.3 | 0.1 |
| Space velocity (volCH$_4$/volcat) | 700 | 1170 | 1425 | — |
| Maximum temperature of Reference | 850 | 795 | 950 | 1050 |
| $H_2/CO$ (product) | 4.4 | 4.5 | 2.1 | 2 |

[a]W.D. Deckwer, et al., Kinetics studies of Fischer-Tropsch synthesis on suspended iron/potassium catalyst-rate inhibition by carbon dioxide and water, Ind. Eng. Chem. Process Des. Dev. 25(3) (1986) 643–649.
[b]I. Dybkjaer, T.S. Christensen, Syngas for large scale conversion of natural gas to liquid fuels, Stud Surf. Sci. Catal. 136 (2001) 435.

avoid sulfur poisoning, desulfurization is carried out. The sulfur compound is converted to $H_2S$ by the hydrodesulfurization catalyst and absorbed by ZnO. Through this catalytic reaction, it is possible to reduce sulfur compounds to about 20–50 ppb.

## Reforming Reaction Process

Since it has already been explained in the previous section, the reaction is different between the primary reformer and the secondary reformer, although omitted here. Particularly in the primary reformer when the molecular weight of the hydrocarbon is increased, carbon deposition tends to occur easily.

## Shift Reaction Process

The composition of the reformed gas at the reforming process outlet is 72% hydrogen, 3% methane, 13% Carbon monoxide, and 12% Carbon dioxide. Since Carbon monoxide is toxic and we wish to reduce the concentration and increase $H_2$ concentration, the shift reaction shown in Eq. (5.2) is generated. The high temperature Carbon monoxide transformer operates at 300°C–350°C

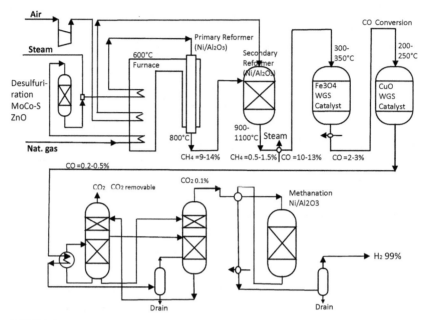

**FIGURE 5.6** Hydrogen production process diagram, operating conditions: $S/C = P(H_2O)/P(CH_4) = 2.5-4.0$, T(exit) $= 900-1100°C$, P (exit) $= 2-3$ MPa. *From C.H. Bartholomew, R.J. Farrauto, Hydrogen production and synthesis gas reactions, in: Fundamentals of Industrial Catalytic Processes, second ed., John & Wiley Sons, Inc., 2006, pp. 339–371 (chapter 6).*

and reduces the outlet Carbon monoxide concentration to about 2%−3%. Then, we operate the low temperature Carbon monoxide transformer packed with Cu-based catalyst at about 200°C and reduce the outlet Carbon monoxide concentration from 0.2% to 0.4%. The Carbon monoxide shift reaction is advantageous at lower temperatures, but the reaction rate is slower.

### Purification Step

Conventionally, this was a system using decarboxylation with an amine type absorbent and a methanation reaction with a $Ni/\gamma-Al_2O_3$ catalyst, but recently it was cooled down to room temperature and passed through a pressure swing adsorption device, whereby a purity of 99.9% of hydrogen is obtained.

The results are shown in Table 5.3, of representative catalysts used for desulfurization, reforming, CO shift, and methanation reaction.

## THERMODYNAMICS

The steam reforming reaction, which is a hydrogen production reaction, and the cooccurring shift reaction are chemical reactions dominated by equilibrium. The equilibrium constant of each reaction is shown in Table 5.4 at

**TABLE 5.3** Industrial Hydrogen Production Reaction and Typical Catalyst

| Process | Reaction | Industrial Typical Catalyst |
|---|---|---|
| Hydrode sulfurization | $R-H + 2H_2 \rightarrow H_2S + H-R-H$ | $CoMo/Al_2O_3$ |
| Adsorption | $ZnO + H_2S \rightarrow ZnS + H_2O$ | $ZnO$ |
| Primary reforming | $HC + H_2O \rightarrow H_2 + CO + CO_2 + CH_4$ | $Ni/MgO$ |
| Secondary reforming | $2CH_4 + 3H_2O \rightarrow 7H_2 + CO + CO_2$ | $Ni/MgAL_2O_4,$ $Ni/CaAl_2O_4$ |
| High-temperature CO shift | $CO + H_2O \rightarrow H_2 + CO_2$ | $Fe_3O_4/Cr_2O_3$ |
| Low-temperature CO shift | $CO + H_2O \rightarrow H_2 + CO_2$ | $Cu/ZnO/Al_2O_3$ |
| Methanation | $CO + 3H_2 \rightarrow CH_4 + H_2O$ | $Ni/Al_2O_3$ |

**TABLE 5.4** Equilibrium Constants

| °C | $K_R$[a] | $K_S$[b] |
|---|---|---|
| 0 | $0.066 \times 10^{-29}$ | $4.555 \times 10^5$ |
| 100 | $0269 \times 10^{-19}$ | $3.587 \times 10^3$ |
| 200 | $0.487 \times 10^{-13}$ | $2299 \times 10^2$ |
| 300 | $0.682 \times 10^{-9}$ | $3.973 \times 10$ |
| 400 | $0.618 \times 10^{-6}$ | $1.192 \times 10$ |
| 500 | $1.021 \times 10^{-4}$ | $4.999$ |
| 600 | $0.873 \times 10^{-3}$ | $3.53$ |
| 700 | $1.305 \times 10^{-1}$ | $1.541$ |
| 800 | $1.759$ | $1.059$ |
| 900 | $1.533 \times 10$ | $7.692 \times 10^{-1}$ |
| 1000 | $0.955 \times 10^2$ | $5.923 \times 10^{-1}$ |

[a]Steam reaction equilibrium constant.
[b]Shift reaction equilibrium constant.

temperatures up to 1,000°C. In the steam reforming reaction, the equilibrium constant rises as the temperature rises. This is because the steam reforming reaction is an endothermic reaction [10].

The equilibrium constant is very small up to about 400°C, indicating that the reaction hardly proceeds in terms of equilibrium. The inlet temperature of the industrial reformer is set to 450°C or higher, but this equilibrium constant is determined. The fact that the equilibrium constant is small means that the target amount of hydrogen is small, so that it is industrially meaningless to react at 450°C or lower.

Since the equilibrium constant of the steam reforming reaction exceeds 1 at 700°C or higher, it is understood that the objective hydrogen production can be performed smoothly. Therefore, the maximum temperature of the catalyst layer in any reformer is 700°C or more. In addition, since the steam reforming reactor is reacted at a high pressure of about 2−3.5 MPa, it is disadvantageous for hydrogen production in equilibrium, so the maximum temperature of the catalyst layer is set as high as 800°C −1000°C.

On the other hand, the shift reaction decreases as the temperature rises. Therefore, the shift reaction proceeds at a low temperature, and the reverse shift reaction proceeds at a high temperature of 800°C or higher. Therefore, in the reformer outlet gas, CO and $CO_2$ are detected. CO is processed by a shift reactor installed on the downstream side as shown in Fig. 5.7. The hydrogen production reaction is an equilibrium reaction according to thermodynamics.

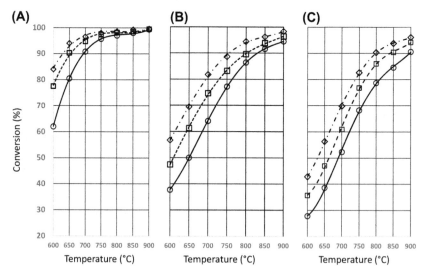

**FIGURE 5.7** Simulation results of conversion of steam reforming reaction of methane. Operating conditions (A) 0.1 MPa, (B) 0.5 MPa, (C) 1 MPa, ○ S/C = 2, □ S/C = 3, ◇ S/C = 4, S/C = steam to carbon ratio.

The higher the pressure and the higher the steam to carbon ration, S/C, the higher the conversion. Using this property, the stationary fuel cell system operates at a low pressure, improves the methane conversion rate, and improves the operation efficiency of the plant.

## INDUSTRIAL CATALYST DESIGN

An overview of industrial hydrogen production catalyst is outlined. Hydrogen production catalysts are used industrially for the production of methanol and ammonia. Technologies are also established. Especially with old steam reforming catalysts, in addition to Ni, precious metals such as Rh and Ru are sometimes used and are commercialized. Industrially, since the Ni-type catalyst is used most frequently, the components and compositions of representative Ni-type catalysts used industrially are shown in Table 5.5. Although some alkaline earth metals, which are basic oxides, have been introduced as additives (as promoters), there was a catalyst that had previously been added with alkali metals. However, at present, no alkali metal is used as a conversion agent since the alkali metal gradually leaves the catalyst and causes corrosion of the metal piping on the downstream side of the reactor, which causes a failure of the plant.

On the other hand, alumina is mostly used as a catalyst carrier, but there are catalysts in which $MgAl_2O_4$ or $MgO$ having a spinel crystal structure is used. Hydrogen production is made at 800°C or more, in some cases 1000°C or more, so high temperature heat resistance is required [11]. It is also required to have resistance against carbon deposition described later or resistance to sulfur poisoning. After past decades of development, the present industrial catalyst has been established. Therefore, it seems that the current catalyst will continue to be used in the future.

## DEACTIVATION

### Carbon Formation

Carbon deposition is considered to be an inevitable phenomenon in hydrogen production reactions. Carbon deposition occurs in the steam reforming reaction depending on operating conditions. Precipitated carbon is classified into three types. Table 5.6 describes these carbons [12]. The most typical carbon deposition is called whisker carbon and occurs on the surface of the Ni catalyst. Carbon deposition amount increases due to diffusion of C in the fuel. In addition, there are carbon deposits in which carbon is generated as a core of the feedstock radical called encapsulating polymers. This phenomenon is seen as the activity gradually decreases. It is a phenomenon observed at a low temperature of 500°C or less. The third carbon deposition is called pyrolytic carbon, which is generated by thermal decomposition of hydrocarbons as a

**TABLE 5.5** A Representative Industrial Ni Catalyst of Steam Reforming

| Feedstock | Catalyst Composition wt% | | | | | | |
|---|---|---|---|---|---|---|---|
| | NiO | Al$_2$O$_3$ | MgO | MgAl$_2$O$_4$ | CaO | SiO$_2$ | CeO$_2$ |
| Natural gas | 12 | 88 | | | | | |
| Natural gas | 20 | 80 | | | | | |
| Natural gas | 20 | | 80 | | | | |
| Natural gas | 20 | 40 | | | | | 40 |
| Natural gas | 20 | | | 80 | | | |
| Natural gas | 30 | 20 | 20 | | | | 30 |
| LPG | 30 | 30 | | | | | 40 |
| LPG | 35 | 30 | | | | | 35 |
| Naphtha | 80 | 20 | | | | | |
| Naphtha | 25 | 25 | 10 | | 15 | 25 | |
| Naphtha | 40 | 30 | 30 | | | | |
| Light hydrocarbon (C8) | 60 | | 40 | | | | |
| Light hydrocarbon (C10) | 75 | | 25 | | | | |

**TABLE 5.6** Different Routes for Three Types of Carbon Formation in Hydrogen Production Reactions [15]

|  | Whisker Carbon | Encapsulating Polymers | Pyrolytic Carbon |
|---|---|---|---|
| Formation | Diffusion of C through Ni-crystal: nucleation and whisker growth with Ni crystal at top | Slow polymerization of hydrocarbon radicals on Ni-surface into encapsulating film | Thermal cracking of hydrocarbon: deposition of C-precursors on catalysts |
| Temperature | >300°C | <500°C | >600°C |
| Phenomenon | Deactivation breakdown of catalyst and increasing ΔP | Gradual deactivation | Encapsulation of catalyst particle: deactivation and increasing ΔP |
| Critical parameters | Low steam/carbon Low activity Aromatic feed in feedstock Abrupt temperature change | Low temperature Low steam/carbon Low H₂/carbon in feedstock Aromatic reed in feedstock | High temperature High void fraction Low steam/carbon high-pressure deactivation |

feedstock. For this reason, this phenomenon occurs in a temperature range of 600°C or more.

Usually carbon deposition in hydrogen production utilizing steam reforming reaction generates whisker carbon. It depends on contamination of unsaturated hydrocarbons in the feedstocks. Fig. 5.8 is a graph showing the time course of the carbon precipitation amount when using a Ni catalyst as the feedstock of unsaturated hydrocarbon. It is understood that carbon precipitates in a very short time.

Fig. 5.9 is the scanning electron microscope (SEM) image of the surface of the Ni catalyst. (A) is a new catalyst. It is evident that the Ni particles are highly dispersed on the alumina carrier. In this way, the Ni particles are initially in a highly dispersed state on the surface of the carrier.

Normally, since the Ni catalyst is used at a high temperature of 450°C or higher, sintering occurs and the particle diameter becomes coarse. That is, the Ni particles move on the surface of the carrier. Therefore, the activity decreases as compared with the initial stage. Fig. 5.9(B) is a SEM image of this sintered surface. Ni particles present a liquid-like behavior at high temperatures, 450°C or higher. Ni particles will coarsen by sintering because their

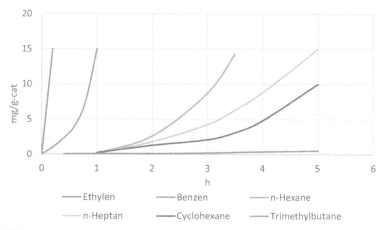

**FIGURE 5.8** Carbon formation from different feedstock. Thermogravimetric studies. 0.7 g Ni/MgO/Al$_2$O$_3$ catalyst. S/C = 2, 0.1 MPa, 500°C. *Based on J.R. Rostrup-Nielsen, New aspects of syngas production and use, Catal. Today. 63 (2−4) (2000)159.*

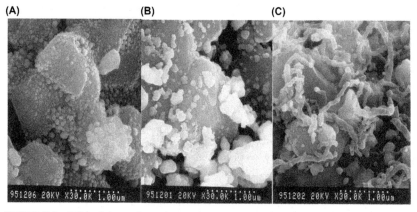

**FIGURE 5.9** SEM of the surface of the Ni catalyst. (A) New catalyst surface. (B) Sintered surface of the catalyst. Ni particles become large according to reaction heat. (C) Surface of the carbon deposit. It is usually caused by a decrease in activity. Plant steam reforming catalyst. S/C = 2.5, P = 0.1 MPa, Ni/Al$_2$O$_3$ catalyst. Feedstock gas is propane.

surface energy is decreased. For this reason, there is a phenomenon in which the activity decreases.

When the activity declines, carbon precipitates on the surface of the catalyst as shown in Fig. 5.9(C). The carbon deposition shown in Fig. 5.9(C) is a typical whisker carbon, and filamentary nanotubes are observed. Also, Ni particles acting as a catalyst exist at the tip of the filament. Carbon deposition is remarkable, especially when subjected to sulfur poisoning described later. Usually, in a reformer of a fuel cell that is a hydrogen energy device, carbon deposition often occurs on the inlet side of the catalyst layer, and most of it is whisker carbon. The cause is considered to be due to low S/C (molar ratio of steam to carbon) at the start of the reformer, and when carbon is generated, it cannot be removed and thus the reformer is replaced [13].

## Poisoning

Sulfur is a severe poison for steam reforming catalyst. Sulfur compounds are strongly chemisorbed on the active metal surface so that this activity is for deactivation [14].

Sulfur is the element that takes the most activity in steam reforming catalyst. Sulfur compounds adsorb to the active sites on the active metal surface and inhibit the reaction to block adsorption of reactants. That is, it is the greatest cause of deteriorating the catalytic function. Sulfur poisoning is an inevitable reason to decrease steam reforming catalyst activity since sulfur compounds are contained in small amounts in fuels [15].

For this reason, usually sulfur compounds are treated with sulfur in a desulfurizer located upstream of the plant. However, even if high-performance desulfurization is carried out, $H_2S$ passes through 5—30 vol. ppb sulfur poisoning, in which sulfur poisoning is adsorbed on the reforming catalyst packed on the downstream side of the desulfurizer and decreases the activity of the reforming catalyst in addition to causing precipitation of carbon. Generally, as the concentration of $H_2S$ increases, the sulfur poisoning of the reforming catalyst becomes more intense. $H_2S$ decomposes on the Ni surface and chemical sulfur chemisorbs.

$$H_2S + Ni \rightarrow S - Ni + H_2 \qquad (5.10)$$

This reaction is usually carried out at 300°C, but since the equilibrium constant at this temperature is $K_p = 5.9 \times 10^6$ the reaction proceeds sufficiently. The concentration of sulfur in the fuel at the inlet of the reformer is about 10—50 vol ppm. Fig. 5.10 shows the result of simulating the situation of sulfur poison of the loaded catalyst in the steam reforming reactor. The horizontal axis represents the length of the catalyst layer, and the vertical axis represents the coverage. This coverage is a ratio of how much active sites of Ni surface are adsorbed with sulfur. The operating conditions are as follows: inlet temperature 500°C, outlet temperature 800°C, pressure 3.4 MPa, S/C = 3.3.

Even at any concentration of sulfur compounds, the curve is convex downward from the reactor inlet side toward the outlet side. This indicates that when the sulfur compound enters the catalyst bed, sulfur is adsorbed to nickel so that the concentration of the sulfur compound is reduced. As described

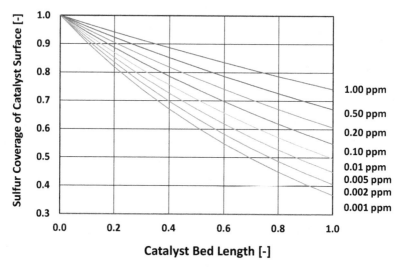

**FIGURE 5.10** Sulfur coverage of catalyst surface in tubular reformer [15].

earlier, sulfur poisoning is strongly poisoned at the inlet of the catalyst layer, and the influence of poisoning toward the outlet tends to weaken. This specific example will be shown in many papers. Here, it is pointed out from the result of the actually operated reformer that the activity decreases as the sulfur poisoning amount increases [16].

## CONCLUSION

Hydrogen production reaction and process were reviewed. The hydrogen production reaction is a reaction controlled by chemical equilibrium, and the process utilizing chemical equilibrium is systemized. Although research in this field has a long history, many themes remain to be studied, which will continue in the future.

Chapter 5.2

# Hydrogen Storage and Transport Technologies

Chapter 5.2.1

# High Pressure H₂ Storage and LH₂ Storage for Transport Technology

## INTRODUCTION

A hydrogen energy system consists of hydrogen production, transportation/ storage, utilization, and so on. Technology for efficiently transporting and storing hydrogen with a low energy density is especially important. At present, high-pressure hydrogen gas, storage alloy, liquid hydrogen ($LH_2$), complex type, carbon type, and the like are being researched and developed as a transportation storage medium of hydrogen, but it is clear that each method is lower than the energy transport density of fossil fuel. Among them, $LH_2$, having a density about 800 times that of hydrogen gas at normal temperature and atmospheric pressure, is a promising method of transporting hydrogen.

Here, the storage of hydrogen will be described for the cases of gaseous and liquid hydrogen, $LH_2$. Recent technical developments regarding the development of pressure vessels for gaseous and liquid hydrogen storage and the physical properties and characteristics of $LH_2$ are discussed.

## DEVELOPMENT OF TECHNOLOGY FOR HIGH PRESSURE GAS HYDROGEN CONTAINERS

In the development of high-pressure technology for compressed hydrogen container system, the following five items of research have been investigated and put to practical use.

Science and Engineering of Hydrogen-Based Energy Technologies. https://doi.org/10.1016/B978-0-12-814251-6.00010-1
**221**

## Selection of Liner Material

Research was conducted on hydrogen embrittlement behavior under a super high-pressure hydrogen environment and hydrogen embrittlement resistant materials were selected.

## Selection of Metallic Materials for Parts

Embrittlement tests of metallic materials used for parts, such as valves of 70 MPa class for compressed hydrogen storage system, were carried out and materials presenting no problem in practical use were selected.

## Development of Sealing Material

Sealing performance was evaluated for rubber O-rings. As a result, no-leakage sealing material was developed.

## Development of 70 MPa Class Hydrogen Container

The manufacturing technology for 70 MPa class hydrogen containers was specified. This technology was implemented with the intention of reducing costs. As a result, tanks with resistance to practical problems have been developed.

## Development of Temperature Prediction Model for Gas and Container During Filling

When rapidly filling high-pressure hydrogen containers with hydrogen, the hydrogen temperature in the container rapidly increases, and there is a concern about the bad influence it might have on the container integrity. Therefore, in order to improve the prediction accuracy, large capacity data on thermal property values is introduced, and a model was constructed.

Based on this, high-pressure resistant hydrogen tank, up to 82 MPa, shown in Fig. 5.11, has been developed and is currently in practical use. Fig. 5.12 shows the developed simulation model.

## STORAGE EFFICIENCY OF LH$_2$

LH$_2$ is a liquefied combustible gas somewhat similar to the commonly used fuels liquefied natural gas (LNG) or liquefied petroleum gas (LPG). A comparison between the physical properties of liquefied gases is shown in Table 5.7 [18,19].

Lighter and smaller volumes have advantages over transportation equipment fuel.

**FIGURE 5.11** Ultra-high-pressure hydrogen gas tank. *Based on http://www.jfecon.jp/product/hpc/fcv.html; http://www.hydrogencarsnow.com/index.php/hydrogen-fuel-tanks/.*

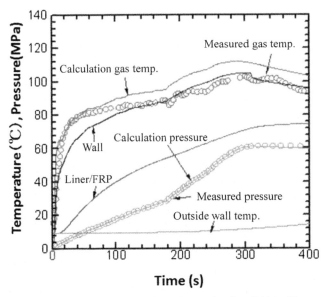

**FIGURE 5.12** Developed simulation model for hydrogen charging. *T. Nejat, Transportation fuel hydrogen, Based on Energy Technol. Environ. 4 (1995) P2712—P2730.*

**TABLE 5.7** Physical Properties of Liquefied Gases [18]

|  | LH$_2$ | LNG | LPG |
|---|---|---|---|
| Chemical formula | H$_2$ | CH$_4$ | C$_3$H$_8$ |
| Molecular weight (g) | 2 | 16 | 44 |
| Boiling point (°C) | −253 | −161 | −42.1 |
| Saturated liquid density v (kg/m$^3$) | 70.8 | 442.5 | 552 |
| Latent heat (kJ/kg) | 31.4 | 225.9 | 426 |
| Low calorific value (MJ/L) | 8.5 | 22.1 | 25.6 |
| Carnot work (W) | 13.8 | 1.69 | − |
| Minimal liquefaction work (kJ/g) | 12.0 | 1.09 | − |
| Transport proper efficiency (−) | 1 | 0.61 | 0.62 |

LH$_2$, liquid hydrogen; LNG, liquefied natural gas; LPG, liquefied petroleum gas.

As a measure of its superiority, a transportation coefficient [18] is defined to compare LH$_2$ to other liquid fuels. As seen, the transportation coefficients for liquefied natural gas and for liquefied petroleum gas are, respectively, 0.61 and 0.62, while it reaches 1.0 for liquid hydrogen, demonstrating that it presents high transportation suitability.

However, since the physical properties of LH$_2$ are low boiling point and low latent heat, it is easy to evaporate. Furthermore, since the amount of heat generated per unit volume of LH$_2$ is about one-third that of other liquefied gases, advanced transportation storage technology is required.

Weight efficiency (wt%) and volumetric efficiency (kg/m$^3$) of hydrogen storage alloy, chemical carrier, hydride, compressed hydrogen gas, LH$_2$, and other storage mediums are targeted for in-vehicle hydrogen storage (storage amount 5−7 kg). Fig. 5.13 also shows targets for the U.S. Department of Energy (DOE) 2010 and 2015 [19].

System efficiency drops below the efficiency shown in Table 5.7 because hydrogen storage/desorption devices, reformers, and such are added. As a result, the system efficiency of any medium is lower than the DOE target value (e.g., 6 wt% in 2010, 45 kg/m$^3$). However the LH$_2$ tank shows a numerical value close to the DOE target value by volume/weight efficiency, and further improvement will exceed the target value.

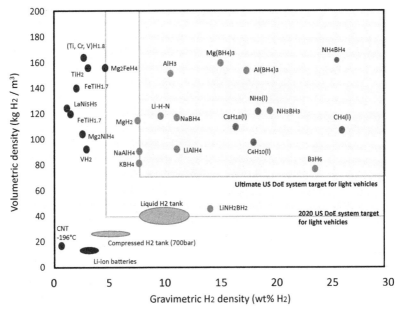

**FIGURE 5.13** Overview of selected materials and their volumetric and gravimetric hydrogen density. The U.S. Department of Energy targets for the hydrogen storage system are also shown for comparison [20].

## CURRENT LH$_2$ SYSTEM

LH$_2$ has been used as a rocket launch fuel because of its high thrust. The system flow of the LH$_2$ storage and supply facility and the overall view of the facility are shown in Fig. 5.14. The LH$_2$ storage and supply system is divided into LH$_2$ system and high pressure hydrogen gas system. The storage tank usually has a filling capacity of 500 m$^3$ and the average evaporation rate is 0.1%/day (in the case of pearlite vacuum insulation). High-pressure hydrogen gas is produced by boosting LH$_2$ up to 25 MPa with a liquid boost pump. For liquid transfer to the rocket tank, the storage tank is pressurized and pumped with gas evaporated from the LH$_2$ service tank.

Each type of hydrogen station is being built in each country as the scale of public road driving tests of hydrogen cars expands. Among them, the off-site type LH$_2$ station is the same as that of the launch facility LH$_2$ system.

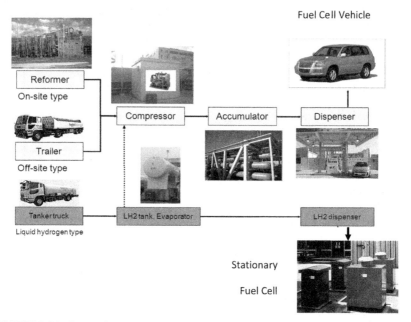

**FIGURE 5.14** System flow of the LH$_2$ storage and supply facility and the overall view of the facility [21].

**FIGURE 5.15** German hydrogen station. It is used in conjunction with a gas station. *Ref. https://sgforum.impress.co.jp/news/4196.*

Fig. 5.15 [22] shows flow diagrams and pictures of the LH$_2$ station at Munich Airport, which has been in operation since 1999. This station is a hybrid system having both on-site and off-site methods, except the LH$_2$ storage tank (storage capacity 12 m$^3$) of the station and the liquid boosting pump boosting pressure of 35 MPa and capacity 120 L/h (25 MPa, 0.5 $m^3$/h × 2 units) of the liquid boosting pump. Future civilian $LH_2$ stations are required to improve functions such as rapid filling, liquid loss reduction, operability, and economy, but it is predicted that $LH_2$ technology of the launching facility will be utilized.

## FUTURE $LH_2$ SYSTEM

Large-scale introduction of hydrogen in the future makes the $LH_2$ system advantageous in terms of supply cost, $CO_2$ emissions, and the like. Efficient systems coupling with existing LNG systems are important if hydrogen introduction gradually progresses in sync with the conversion policy to low-carbon fuels such as LNG.

The actual LNG system from the fuel source to the utilization system and the hydrogen system are very similar processes. In the process leading to fossil fuel or a hydrogen utilization system, energy loss caused by liquefaction power, evaporation, transfer, and precooling is considered to be a problem. The magnitude of energy loss varies depending on the system scale, transportation means, and so on. As an example Table 5.8 shows the energy balance of a small-scale $LH_2$ system [23].

In this case, each hydrogen source (high calorific value) with hydrogen source as coke by-product gas (hydrogen purity: about 57%), liquefaction scale 1.2 ton/day, transportation equipment lorries (13 $m^3$), transport distance 20 km, indicates a breakdown of process consumption power and calorie consumption. Reduction of liquefaction loss is a major technical subject, but there is also an estimate that the liquefaction power will be about one-third of the conventional one if LNG's cold heat is used. Improved techniques of conventional liquefaction methods and magnetic refrigeration techniques for high efficiency are currently being studied.

**TABLE 5.8** Energy Balance of a Small Scale $LH_2$ System [23]

|  | Unit (%) |
| --- | --- |
| Hydrogen purification | 12.9 |
| Liquefaction | 29.6 |
| Storage and withdrawal | 7.8 |
| Residual calorie | 54.7 |
| Total | 100 |

## DEVELOPMENT OF LH$_2$ TRANSPORTATION AND STORAGE TECHNOLOGY

Heat flow characteristics, high heat insulation, and the light weight of the LH$_2$ container tank (important equipment for LH$_2$ transport) are important technologies.

Actually carrying out a transportation test on a public road running test of a distance of about 600 km showed that the pressure drop is kept below 0.1 MPa and LH$_2$ can be transported safely. The background of the success of this test is the development of high insulation technology. LH$_2$ is stored in a double container (outer tank, inner bath) with a vacuum heat insulating layer to suppress evaporation. In the case of mobile tanks (car-mounted containers, etc.), the heat input rate from the support structure part increases because the rigidity of the support structure is increased. The majority of heat input is radiant heat. Laminated vacuum insulation consisting of a spacer (e.g., polyester net) and radiation shield (e.g., aluminum vapor deposited film) is used to prevent radiant heat input. The heat insulation performance depends on the number of layers, but heat input is about 1 to 0.5 W/m$^2$, which greatly differs depending on the type of laminated vacuum insulation material and construction method. This insulation performance level, where the temperature decreased about 1°C, corresponds to the heat insulation performance when the hot water boiled at 100°C is left for 1 month.

Weight reduction of the container weight means to reduce the weight of the tank while maintaining the tank strength. This means that the weight efficiency increases and the heat capacity becomes smaller, which makes liquid hydrogen transport more effective.

## CONCLUSION

The current research situation of the LH$_2$ system was explained. The LH$_2$ system is similar to LNG with a small environmental burden and features such as being able to efficiently utilize the existing infrastructure network. In order to introduce the LH$_2$ system to the hydrogen society, a system that is excellent in energy efficiency and economy by developing high performance of each component needs to be constructed.

Chapter 5.2.2

# Hydrogen Storage and Transport by Organic Hydrides and Application of Ammonia

## INTRODUCTION

Global warming is receiving worldwide attention as society is shifting toward an energy scenario that positively utilizes renewable energy. However, since renewable energy has accompanying fluctuation, unless energy is stored and stabilized, it is difficult to use, and a carrier is required to carry this energy. Hydrogen energy is expected as a carrier of renewable energy, but what hinders this development is the cost of transporting hydrogen with low volumetric energy density. Reduction in hydrogen supply cost is recognized as a major problem to realize a hydrogen energy society. To overcome this problem, organic hydrides, which are liquid hydrides, and ammonia are attracting attention.

## ORGANIC CHEMICAL HYDRIDE METHOD

The organic chemical hydride method (OCH) is a method of converting an aromatic group such as toluene to a hydrogenation reaction, converting it to a saturated cyclic compound such as methyl cyclohexane, and using this as a hydrogen storage medium at room temperature and atmospheric pressure. It is a method of storing and transporting in the liquid state and generating hydrogen by dehydrogenation reaction at the place of use.

Since OCH can be stored and transported in the liquid state at normal temperature and normal pressure, there is little latent risk involved, and transportation is performed similarly to conventional liquid chemicals in chemical tankers and chemical lorries. In such a situation there is reduced risk and more efficiency as compared to the transportation of compressed or liquid hydrogen. It is easier to compare and it is considered to be particularly suitable for storing and transporting hydrogen in large quantities. Representative hydrogenation/dehydrogenation reaction pairs include methylcyclohexane (MCH) type, cyclohexane type, and decalin type as shown in Eqs. (5.10) to (5.12) in Fig. 5.16.

These hydrogenation/dehydrogenation reaction pairs have comparatively high values of weight storage density and volume storage density. Fig. 5.17

Science and Engineering of Hydrogen-Based Energy Technologies. https://doi.org/10.1016/B978-0-12-814251-6.00011-3
**229**

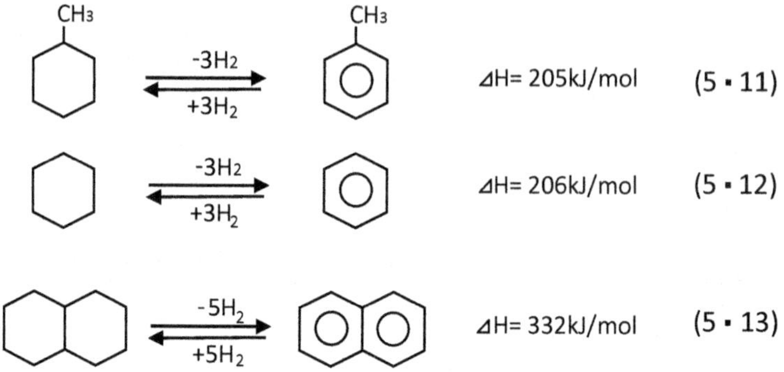

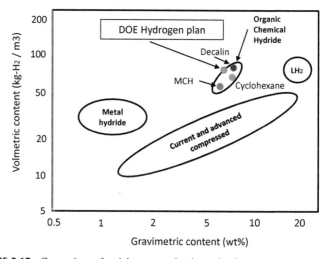

**FIGURE 5.16** Representative hydrogenation/dehydrogenation reaction pairs include methylcyclohexane type, cyclohexane type, and decalin type.

**FIGURE 5.17** Comparison of weight storage density and volume storage density [25,26].

shows a comparison of weight storage density and volume storage density of various hydrogen storage and transport methods [24]. The OCH method has a high value of 7.1 wt%, 55.5 kg-$H_2$/m$^3$ in cyclohexane system, 6.1 wt% in MCH system, 47.0 kg-$H_2$/m$^3$. In addition, the hydrogen storage density of the decalin system is as high as 7.2 wt% and 64.9 kg-$H_2$/m$^3$, compared with 6.5 wt%, which is the development target value of the U.S. Department of Energy, 62.0 kg-$H_2$/m$^3$.

The MCH system of the OCH method has been compared and studied together with the liquid ammonia method and the liquefied hydrogen method in the Euro-Québec hydrogen project from the 1980s [25,26]. However, since a dehydrogenation catalyst that operates stably has not been developed, it is a method that is not technically established.

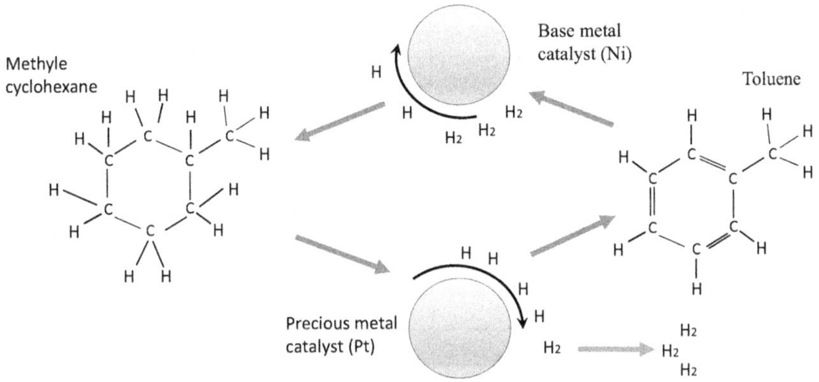

**FIGURE 5.18** Schematic diagram of dehydrogenation reaction and hydrogen addition reaction.

Schematic diagrams of dehydrogenation reaction and hydrogen addition reaction are shown in Fig. 5.18 [25,26].

In the dehydrogenation reaction, a 10 wt% Pt/C activated carbon cloth catalyst is used, but the C—H bond of MCH is broken, hydrogen atoms are dissociated, and toluene is produced. And $H_2$ is generated at the same time. In the hydrogen addition reaction, it is common to use Ni or Pd catalyst [27,28].

## DEHYDROGENATION DEVICE AND HYDROGEN REFINERY

### Performance of Dehydrogenation Catalyst

The life test results of the dehydrogenation catalyst in the MCH dehydrogenation reaction are as follows. Reaction conditions are 320°C, atmospheric pressure, Liquid Hourly Space Velocity (LHSV) $= 2.0 \, h^{-1}$, supplying a refining reagent of MCH as a raw material, and as a conventional means of suppressing carbon deposition, 5%−20% of hydrogen based on the raw material MCH cosupplied. The catalyst stably maintained performance of MCH conversion rate of 95%, toluene selectivity of 99.9% or more, and hydrogen evolution rate of $1000 \, Nm^3/m^3$−cat/h or more for more than 6000 h. Also, when reducing the amount of cosupplied hydrogen from 20% to 5% in stages, the performance was almost stably maintained even at 5%.

The by-products of this catalyst are methane and benzene, and the production amount of methane is about 100−200 ppm at the initial stage of the reaction. Also, at this time, the number of moles of methane and benzene formed is approximately equal.

It is presumed that methane and benzene will be formed by hydrogenating the side chain of MCH being cleaved. It is considered that separating by-produced methane and hydrogen from the reactor distillate is not preferable

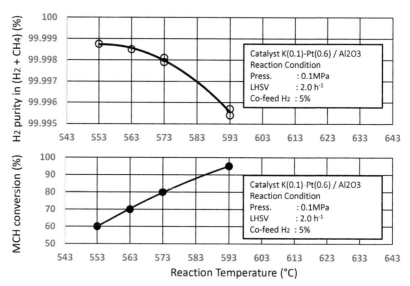

**FIGURE 5.19** Reaction temperature dependence of hydrogen purity and methylcyclohexane conversion rate depending on the amount of methane and hydrogen produced. *R.W. Coughlin, K. Kawakami, A. Hasan, Activity, yield patterns, and coking behavior of Pt and PtRe catalysts during dehydrogenation of methylcyclohexane: I. In the absence of sulfur, J. Catal. 88 (1984) 150; W.J. Doolittle, N.D. Skoularikis, R.W. Coughlin, Reactions of methylcyclohexane and n-heptane over supported Pt and PtRe catalysts, J. Catal. 107 (2) (1987) 490.*

from the viewpoint of economic efficiency. It is also unlikely that a trace amount of methane will disturb when using hydrogen. Fig. 5.19 shows the reaction temperature dependence of hydrogen purity and MCH conversion calculated from the amounts of hydrogen and methane to be purified. The hydrogen purity at 320°C (595 K) is 99.995 or more, and it is considered that hydrogen for fuel cells is sufficiently pure [29,30].

## Energy Efficiency of Hydrogen Supply Facility

The energy efficiency of the hydrogen supply facility was calculated. The energy flow of this facility is shown in Fig. 5.20, but ① to ⑧ are defined as follows.

ⓘ Input MCH combustion heat quantity
② Input heat quantity: Power consumption of catalytic heater
③ Auxiliary machine electric energy: Power consumption of pump, control panel, cooling water chiller, etc.
④ Cooling water flow rate calorific value
⑤ Hydrogen combustion heat generation

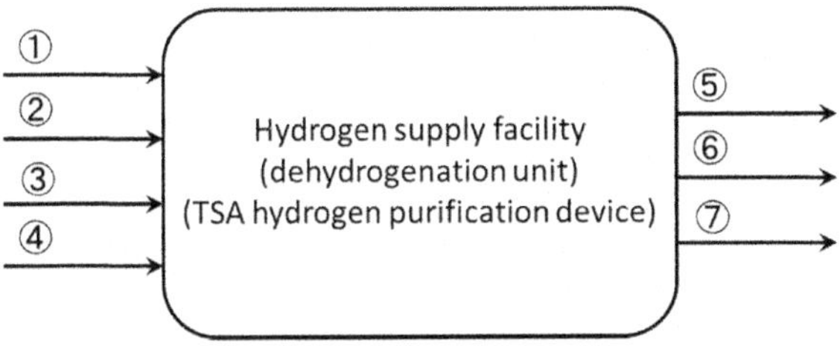

**FIGURE 5.20**   Energy flow diagram of hydrogen supply facility. TSA: Temperature Swing Adsorption.

⑥ Generation Toluene (TOL) combustion heat quantity

⑦ Cooling water outflow calorie

⑧ Heat recovered: The amount of heat recovered by cooling water

Here, ⑦ − ④ = ⑧

$$\text{(Energy efficiency)} \ = \ \frac{⑤+⑥+⑧}{①+②+③} \qquad (5.14)$$

$$\text{(Energy efficiency)} \ = \ \frac{⑤+⑥+⑧}{⑤+⑥+②+③} \ = \ \frac{⑤+⑧}{⑤+②+③} \quad (5.15)$$

The energy efficiency in the hydrogen supply facility can be obtained by Eq. (5.14). In addition, ⑤ and ⑥ are equivalent to ①. If ⑥ = 0, Eq. (5.14) becomes Eq. (5.15).

Calculation of energy efficiency was calculated using Eqs. (5.16) and (5.17) for cases where recovery heat is not used and cases where it is used. Further, the calorific value in the unit volume was calculated using Eq. (5.18).

(Recovered heat is not taken into account)

Energy efficiency =

$$\left( \frac{\text{Generated hydrogen combustion heat quantity}}{\text{Generated hydrogen combustion heat amount} \ + \ \text{input heat amount} \ + \ \text{auxiliary power}} \right)$$

$$(5.16)$$

(Recovered heat is taken into account)

Energy efficiency 2 =

$$\frac{\text{Generated hydrogen combustion heat amount} \ + \ \text{recovered heat amount}}{\text{Generated hydrogen combustion heat amount} \ + \ \text{input heat amount} \ + \ \text{auxiliary machine power}} \right)$$

$$(5.17)$$

**TABLE 5.9** Efficiency of Hydrogen Supply Facility

| Temperature °C | | 250 | 275 | 300 | 325 |
|---|---|---|---|---|---|
| Hydrogen combustion | Heat (HHV) [MJ] | 620 | 1390 | 1955 | 2130 |
| Input heat quantity | Heater [MJ] | 900 | 1190 | 1430 | 1530 |
| Auxiliary machine electric energy [MJ] | Motive power | 94 | 118 | 172 | 180 |
| | Controller | 21 | 21 | 21 | 21 |
| | TSA | 577 | 590 | 632 | 650 |
| Recovered heat [MJ] | Cool water | 132 | 156 | 175 | 190 |
| Energy efficiency [%] | Efficiency 1 | 28 | 42 | 46 | 47 |
| | Efficiency 2 | 34 | 47 | 50 | 51 |
| Hydrogen production unit | Volume heat [MJ] | 32 | 18 | 15 | 14 |

Heat input per 1 Nm$^3$ of hydrogen =

$$\left( \frac{\text{Input heat quantity} + \text{Auxiliary equipment power quantity}}{\text{generated hydrogen quantity}} \right) \quad (5.18)$$

The energy efficiency of this system is shown in Table 5.9, where TSA stands for Temperature Swing Adsorption. The power consumption of the control panel did not change greatly with temperature, but the amount of power required for power, chiller, and TSA increased with temperature rise [31].

## AMMONIA AS ENERGY CARRIER

Research on synthesis and decomposition of ammonia is very energetically performed. Technical description presents ammonia as normally produced by the catalytic reaction of nitrogen and hydrogen. Although process technology has improved over the years, the basic chemistry is identical to the process developed by Haber and Bosch in the early 20th century [32].

$$N_2(g) + 3H_2(g) \rightarrow 2NH_3(g) \quad \Delta H = -92 \text{ kJ/mol} \ (-46 \text{ kJ/mol}$$
$$\text{for 1 mol of } NH_3) \quad (5.19)$$

The reaction is typically carried out over iron catalysts at temperatures around 400°C −600°C and pressures ranging from 200 to 400 atm.

In practice, ammonia synthesis is usually coupled with hydrogen production to increase efficiency. The hydrogen is typically produced from natural gas, but it can also be produced from other fuels, such as petroleum coke or biomass, or from water through electrolysis. In case of hydrocarbon feedstocks, they are generally gasified to form synthesis gas (CO and $H_2$), which

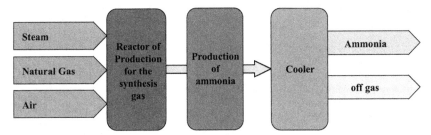

**FIGURE 5.21**  An example of a block diagram of ammonia synthesis.

can then be reacted with water and nitrogen to produce ammonia. The Mineral Commodity Summaries 2018 of the United States Geological Survey [33] indicates that the world production of ammonia in 2017 was equal to 150 Mtons of contained nitrogen.

Numerous papers have been published about the ammonia synthesis process [34−37]. There is a site called Ammonia Industry [38] that contains useful information, part of which was used to draw Fig. 5.21 in a block diagram format. It shows a rather complex process that has the advantage of being energy efficient.

## AMMONIA DECOMPOSITION

Research to decompose ammonia is very old, and it has been done since the beginning of synthesis research. For this reason, various catalysts have been proposed and numerous papers have been compiled.

Ammonia decomposition (cracking) is simply the reverse of the synthesis reaction.

$$NH_3(g) \rightarrow 1/2N_2(g) + 3/2H_2(g) \ \Delta H = +46 \, kJ/mol \qquad (5.20)$$

Note that the reaction is endothermic. The temperature required for efficient cracking depends on the catalyst. There are a wide variety of materials that have been found to be effective, but some (e.g., supported Ni catalysts) require temperatures above 1000°C [29]. Others have high conversion efficiency at temperatures in the range of 650°C −700°C. As these temperatures are well above polymer electrolyte fuel cell (PEMFC) operating temperatures, some of the fuel, or perhaps the fuel cell purge gas, would need to be burned to maintain an efficient reaction [39].

The theoretical adiabatic efficiency for the thermocatalytic reaction is about 85% relative to the energy (LHV) of the released hydrogen. If no other energy source were available, at least 15% of the available hydrogen energy content would have to be burned to supply the heat of reaction. Of course, additional energy would be required to overcome thermal losses in the cracking reactor. Since the reaction occurs at high temperature, this heat would likely come from the combustion of ammonia and/or hydrogen in onboard storage applications.

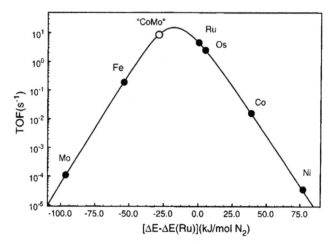

**FIGURE 5.22**   Relationship between the turnover frequency of different metals for the $NH_3$ synthesis reaction at 400°C with respect to their nitrogen adsorption energy [41].

A requirement for bimetallic catalytic systems is a high stability under reaction conditions versus segregation into monometallic particles that would alter the surface properties and thus catalytic activity [40]. Additionally, in some supported bimetallic systems, the conditions of thermal treatment or high reaction temperatures can promote the formation of less reducible oxides such as Fe—Al, Co—Al, Ni—Al, and Co—Si when supported on alumina or silica, resulting in considerably lower activity. Fig. 5.22 shows the volcano-type relationship between the catalytic activate (measured as turnover frequency) of different monometallic systems for the synthesis of ammonia and their respective nitrogen binding energy. Of the systems investigated, ruthenium-based catalysts present an optimum value as previously discussed [41].

As can be seen from Fig. 5.22, it is understood that a catalyst for synthesizing ammonia such as Mo, Fe, Ru, Os, Co, and Ni is effective for decomposition of ammonia. In addition, it is understood that the catalyst composed of two or more kinds of metals shown above has very high activity, and CoMo is a promising catalyst. Research is expected to progress toward such bimetric catalyst in the future.

## CONCLUSION

Industrial technical challenges of hydrogen storage and transportation require advanced technology, and the development of this field will continue in the future. In particular, the development of new technologies that have never been done, such as mass transport of hydrogen, has great promise in solving energy problems.

Chapter 5.3

# Utilization of Hydrogen Energy

Chapter 5.3.1

# Hydrogen Refueling Stations and Fuel Cell Vehicles

## INTRODUCTION

A fuel cell vehicle (FCV) is an electric car powered by a fuel cell (FC). Because energy conversion efficiency of FC itself is high, high energy efficiency can be expected as a whole. It is expected to be a next-generation car because it will help prevent global warming and protect the air environment by not generating greenhouse gases and air pollutants during driving.

However, because the cost of FCs and hydrogen tanks is high, the vehicle itself becomes expensive. In the early 2000s, lease sales were mainly used, but in December 2014 general sales of passenger car—type FCV (Toyota MIRAI) began for the first time. Today, in addition to Toyota, Honda is also selling FCVs in Japan. In addition to ordinary cars and buses, scooters running with FCs as prototype have also been announced.

Based on this trend, FCVs using fuel cells as the driving force are expected to develop in the future.

## FUEL CELL VEHICLE

In Japan, the hydrogen and fuel cell demonstration project (Japan Hydrogen and Fuel Cell Demonstration Project; JHFC) was implemented over an 8-year period from 2002. It is a demonstration test research project on fuel cell systems implemented by the government, and it was a project consisting of FCV and hydrogen infrastructure demonstration research.

This project involves 10 automotive companies, including the method of producing hydrogen from various feedstocks, the performance of FCVs under the actual conditions of use, the environmental characteristics, and the basic

Science and Engineering of Hydrogen-Based Energy Technologies. https://doi.org/10.1016/B978-0-12-814251-6.00012-5

Honda's prototype vehicle

**FIGURE 5.23** Fuel cell vehicle produced by Honda. It is a four-seater ordinary passenger car. Performance list is shown [42].

**FIGURE 5.24** Full view of the fuel cell bus and performance list. It does not look like a regular bus, and there is no need for special driving operation. In Tokyo it is being adopted as a public transport vehicle. In the future, this type of bus will increase [43].

energy efficiency and safety collecting data; sharing that data; and promoting the practical application of fuel cell vehicles. At the time of this project, similar projects were done in the United States and Europe, and the technical framework of the present FCV was completed. Figs. 5.23 and 5.24 show an FCV used in this JHFC project.

Ten kinds of cars were prototyped in the project. Here, Honda's prototype vehicle as a car and Toyota—Hino's FC bus as a large vehicle were shown.

A prototype car of such a sedan-type passenger car was produced in the 2000s. In Japan many automobile companies have started to develop FCVs and are acquiring technical skill. As a result, the current FCV system is completed.

## FUEL CELL VEHICLE TECHNOLOGY

### High-Pressure Hydrogen Tank

As explained earlier, FCV has a high-pressure hydrogen tank for storing gaseous hydrogen. Hydrogen tanks that can withstand higher pressures have been developed because the more hydrogen that can be stored, the mileage they can run with one filling will be extended.

On the other hand, hydrogen gas is liable to leak from the tank because its molecular size is small. Hydrogen atoms also penetrate between metal atoms, causing hydrogen embrittlement, which reduces mechanical strength. Furthermore, there is a possibility of causing explosive combustion with oxygen in the air.

Safety has always been the most important technical issue in FCVs that handle hydrogen with such properties. In addition to simply storing safely, it is necessary to have a structure that does not lead to leakage or explosion even in the event of an accident. At present, hydrogen gas filling is allowed at 70 MPa (about 700 atm). This is a standard.

For example, the hydrogen tank of FCV of Toyota has a three-layer structure of a plastic liner having gas barrier function, high-strength carbon fiber reinforced plastic, and glass fiber reinforced plastic to protect the surface.

Improvement in performance of this tank itself can prevent leakage and instantly detect leakage. In addition, structural measures have been taken so that the leaking gas immediately diffuses out of the vehicle.

## Fuel Cell Stack

The fuel cell is frequently named fuel cell stack because it is made by stacking a plurality of cells of the smallest unit responsible for electrochemical reaction. Inside, hydrogen and oxygen, separated by the polymer membrane of the electrolyte, exchange ions and electrons and an electrochemical reaction occurs. Important technologies are developed, such as sealing methods for hydrogen as fuel and flow path shapes in which reaction proceeds smoothly.

## Power Control Unit

Since the FCV is an electrically powered vehicle, it has battery or capacitor that accumulates electricity generated onboard or obtained by regeneration of kinetic energy.[1]

Important devices include the power control unit that controls charging and discharging of the battery and the power supply to the driving motor, the inverter that generates the AC current to be supplied to the driving motor, and the boosting converter that increases the voltage. This determines the performance of the entire FCV system and the life of the equipment.

## CHARACTERISTICS OF FUEL CELL VEHICLES

Features of FCVs are shown as follows. Its characteristics are quite different from internal combustion engine vehicles.

---

1. Regeneration: Converting the kinetic energy of the vehicle to electric power when decelerating. At this time, the drive motor plays the role of a generator.

## Clean Exhaust Gas

Basically only water (water vapor) is discharged during functioning of the FCV. There are no emissions of carbon dioxide ($CO_2$) and other greenhouse gases, such nitrogen oxides ($NO_x$) or methane, nor of air pollutants, such as hydrocarbons other than methane (HC), carbon monoxide (CO), and suspended particulate matter (PM).

## High Energy Efficiency

Compared to the energy efficiency of a car running on a gasoline engine (about 15%), the FCV achieves a high energy efficiency of more than twice (about 30%). Even in the low-speed region where efficiency drops greatly in gasoline vehicles, high efficiency can be maintained in FCVs.

## Various Hydrogen Sources

Hydrogen can be obtained by reforming hydrocarbons such as natural gas and methanol. Also, various fuels other than petroleum can be used. Recently, electrolysis of water using renewable energy such as photovoltaic power generation and wind power generation, or hydrogen production from methane generated from biomass and sewage sludge has been put to practical use.

## Low Noise

Since it is driven by an electric motor, the noise is very low.

## No Charge Required

The time required for filling hydrogen is about 5 min. It is an overwhelmingly short time compared to charging time for electric cars. It's almost the same as refueling gasoline cars. The mileage from one filling is about 500 km, which is longer than electric cars and almost the same as gasoline cars.

## The FCV as an Emergency Power Supplier

For vehicles equipped with the ability to supply electric power to an external consumer, the vehicle itself can be used as an emergency power supply device in several situations, including the event of a disaster. Contrarily to that, it has been reported that diesel exhaust gas is a health hazard.

FCV, which run on electricity generated on board by electrochemically combining hydrogen with oxygen from the air, could reduce global dependence on petroleum while emitting just water from their tailpipes.

The idea of powering a car with fuel cell has been around for decades. Toyota has overcome the technological difficulties and started with general

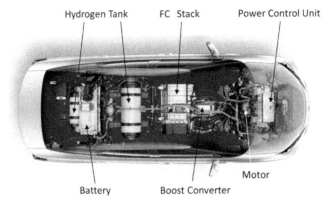

**FIGURE 5.25**  Fuel cell vehicle viewed from the top. A fuel cell that is responsible for power generation and a converter that controls electricity are combined [44].

sales in 2014. Honda started in 2016, and was followed by other automobile manufacturers that also began selling FCVs.

Fig. 5.25 is a diagram showing the structure from the top of an FCV. Every FCV has almost the same structure. Normally, the power control unit is equipped to bring out high driving performance.

Fuel cell automobiles and buses have been commercialized, and they have started real utilization on public roads in North America, Europe, Asia and other locations. Table 5.10 presents specifications of Japanese fuel cells cars and buses. The latter present more orders because require simpler operation infrastructure and strongly contribute to decreasing greenhouse gases

**TABLE 5.10** Specifications of Japanese Fuel Cell Cars and Buses

| Photo | | | |
| --- | --- | --- | --- |
| Name | Toyota FC Bus | Toyota MIRAI | Honda CLARITY |
| Size (mm) | 10,525 × 2490 × 3340 | 4890 × 1815 × 1535 | 4915 × 1875 × 1480 |
| Battery | Ni-H$_2$ | Ni-H$_2$ | Li-ion |
| Mileage (km) | 200 | 650 | 750 |
| Tank Vol. (L) | 600 | 122.4 | 141 |

emissions. Tokyo plans to have more than 100 fuel cell buses by 2020, and to increase that number sharply from there on [45].

Among others, the United States and Japan are deploying fuel cell buses.

Table 5.11 depicts demonstrations of fuel cell buses that are in operation in the United States. These will continue to focus on identifying improvements to optimize reliability and durability at the same time that new deployments are planned. In Addition to the 21 fuel cell electric buses (FCEB) presented in Table 5.11, there are plans to introduce other 44 FCEBs in subsequent years [46].

The data presented in this section represent recent results from different FCEB demonstrations. To simplify the presentation of the data, each FCEB is assigned an identifier that includes a site abbreviation followed by a manufacturer or project designation. The FCEBs presented in this section have hybrid systems that are FC dominant. Table 5.12 provides specifications for each FCEB by an unique identifier. The bus at UCI has the same configuration as the SunLine buses. Some FCEBs are pictured in Fig. 5.26.

**TABLE 5.11** Fuel Cell Transit Buses in Active Service in the United States

| | Bus Operator | Location | Active Buses | Technology Description |
|---|---|---|---|---|
| 1 | AC Transit, ZEBA | San Francisco Bay Area, CA | 13 | |
| 2 | SunLine Transit Agency (AFCB prototype) | Thousand Palms, CA | 1 | Van Hool bus and hybrid system integration, U.S. hybrid support for fuel cell |
| 3 | SunLine Transit Agency | Thousand Palms, CA | 3 | ENC/BAE Systems/Ballard next-generation advanced design to meet "Buy America" requirements |
| 4 | University of California at Irvine | Irvine, CA | 1 | ENC/BAE Systems/Ballard updated AFCB design (AFCB) |
| 5 | Massachusetts Bay Transportation Authority (MBTA) | Boston, MA | 1 | AFCB |
| 6 | Orange County Transportation Authority (OCTA) | Santa Ana, CA | 1 | AFCB |
| 7 | Flint MTA | Flint, MI | 1 | AFCB |
| | | Total | 21 | |

*ZEBA*, Zero Emission Bay Area; *AFCB*, American Fuel Cell Bus; *ENC/BAE*, El Dorado National-California/BAE Systems; *Flint MTA*, Mass Transportation Authority in Flint, MI, USA.

**TABLE 5.12** FCEB Identifiers and Selected Specifications

|  | ACT ZEBA | SL AFCB | UCI AFCB |
|---|---|---|---|
| Transit agency | AC Transit | SunLine | Anteater Express |
| Number of buses | 13 | 4 | 1 |
| Bus OEM | Van Hool | ENC | ENC |
| Model/year | A300L/2010 | Axcess/2011 and 2014 | Axcess/2016 |
| Bus length | 40 ft | 40 ft | 40 ft |
| Gross vehicle weight | 39,350 lb | 43,420 lb | 43,420 lb |
| Fuel cell OEM | UTC Power | Ballard | Ballard |
| Fuel cell model | Puremotion 120 | FC velocity HD6 | FC velocity HD6 |
| Fuel cell power (kW) | 120 | 150 | 150 |
| Hybrid system integrator | Van Hool | BAE Systems | BAE Systems |
| Design strategy | Fuel cell dominant | Fuel cell dominant | Fuel cell dominant |
| Energy storage OEM | EnerDel | A123 | A123 |
| Energy storage type | Li-ion | Li-ion | Li-ion |
| Energy storage capacity | 21 kWh | 11 kWh | 11 kWh |
| Hydrogen storage pressure (psi) | 5000 | 5000 | 5000 |
| Hydrogen cylinders | 8 | 8 | 8 |
| Hydrogen capacity (kg) | 40 | 50 | 50 |
| Technology Readiness Level (TRL) | 7 | 7 | 7 |

*ACT ZEBA*, AC Transit Zero Emission Bay; *SL AFCB*, SunLine American Fuel Cell Bus; *UCI AFCB*, University of California at Irvine American Fuel Cell Bus; *ENC*, El Dorado National-California; *OEM*, original equipment manufacturer; *Bus OEM*, bus manufacturer; *Fuel cell OEM*, fuel cell manufacturer.

## HYDROGEN REFUELING STATION

A hydrogen refueling station supplies high pressure hydrogen gas to FCV. The increase on the deployment of the network of hydrogen refueling stations has become the key factor to spread the utilization of FCV, and the public and private sectors are integrating themselves and developing them.

A hydrogen refueling station consists of a tank that stores hydrogen, a compressor that increases the pressure of hydrogen gas, a precooler that

**FIGURE 5.26** AC Transit ZEBA fuel cell electric buses (FCEB) (top), SunLine AFCB (middle), and UCI American Fuel Cells Bus (AFCB) (bottom). The AFCBs at SunLine have the higher average speed at 13.7 mph. The ZEBA buses in service at AC Transit have a lower average speed of 8.5 mph. The average speed of the UCI AFCB has not yet been determined because the hour data on the fuel cell power plant are not complete. The average monthly mileage for the group is 2637 miles per month. This continues the upward trend over time (2189 miles per month in 2014, 2464 in 2015). The average monthly miles for the ZEBA buses is 6.5% higher than last year.

preliminarily cools the gas to avoid temperature rise accompanying compression, a dispenser (injection machine), and so on. It can be classified into three types: on-site, where hydrogen is produced on site; "off-site," carried from other places; and "mobile," where equipment is loaded on a trailer.

## Safety Measures for Hydrogen Refueling Stations

Various safety measures are applied to hydrogen refueling stations.

The basic policy of safety measures is to prevent leakage of hydrogen and to detect it at an early stage, to prevent stagnation in the event of leakage, to prevent ignition, and to reduce the effects of fire.

Furthermore, installation of sprinkler facilities is mandatory by law in Japan, and seismographs are also installed. It is a hydrogen station that takes safety into consideration even on a global scale (Table 5.13 and Fig. 5.27).

In Japan, FCV manufacturers such as Toyota are seriously investing on deploying hydrogen stations [42]. The IPHE, International Partnership for Hydrogen and Fuel Cells in the Economy, announces that Japan has presently 110 HRS and might reach as many as 900 in 2030 for the utilization of 800,000 FCVs.

In addition, a service to give information on the operation status of the HRS in real time was created [47−50] and is being used by FCVs to facilitate planning for having hydrogen supply.

## HYDROGEN PURIFICATION

The purity of hydrogen is extremely important. It is known that when the purity of hydrogen is low, it directly affects the performance of the fuel cell. Particularly since $H_2S$ and CO poison Pt catalyst, which is a fuel cell electrocatalyst, performance degradation over time of fuel cells is known. This refers to the reduction of impurities in hydrogen gas. Fig. 5.28 is a graph showing the voltage drop when $H_2S$ flowed through the fuel cell anode. It is shown that the voltage decreases with time and that the voltage decreases as the concentration of $H_2S$ becomes high.

Table 5.14 presents the International Standardization for Hydrogen Fuel for Fuel Cell Vehicles. According to previous studies, if the substances shown here cannot be kept within the specified limits, the fuel is not suitable for FCVs.

In recent years, due to the development of the Internet of Things, it has been reported that it is important to measure hydrogen gas in real time during operation. If online analysis is installed, it is believed that the number of fuel-cell failures will be drastically reduced.

Table 5.15 presents the analytical methods and limits for ISO FDIS 14687-2. In fact, it also unveils for each type of contaminant in hydrogen to be used for FCVs the detection and determination limits, as well as the standard describing the test method.

Data on the amount of impurities mixed in hydrogen in actual HRS is summarized in Table 5.16. As shown in this table, data on concentration and size of particulate matter sampled at selected hydrogen fueling stations are shown. It represents an operation index of stations that are currently

**TABLE 5.13** Five Basic Components of Hydrogen Refueling Stations [47]

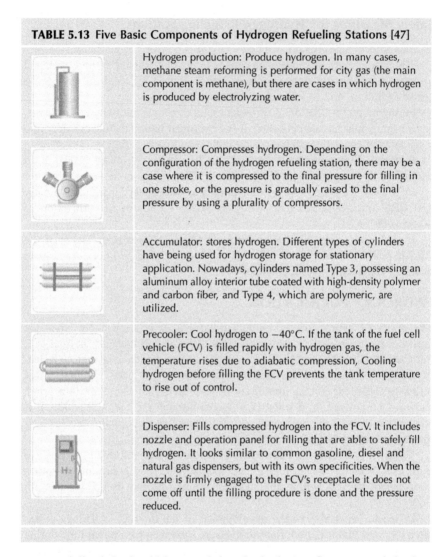

| | |
|---|---|
| | Hydrogen production: Produce hydrogen. In many cases, methane steam reforming is performed for city gas (the main component is methane), but there are cases in which hydrogen is produced by electrolyzing water. |
| | Compressor: Compresses hydrogen. Depending on the configuration of the hydrogen refueling station, there may be a case where it is compressed to the final pressure for filling in one stroke, or the pressure is gradually raised to the final pressure by using a plurality of compressors. |
| | Accumulator: stores hydrogen. Different types of cylinders have being used for hydrogen storage for stationary application. Nowadays, cylinders named Type 3, possessing an aluminum alloy interior tube coated with high-density polymer and carbon fiber, and Type 4, which are polymeric, are utilized. |
| | Precooler: Cool hydrogen to $-40°C$. If the tank of the fuel cell vehicle (FCV) is filled rapidly with hydrogen gas, the temperature rises due to adiabatic compression, Cooling hydrogen before filling the FCV prevents the tank temperature to rise out of control. |
| | Dispenser: Fills compressed hydrogen into the FCV. It includes nozzle and operation panel for filling that are able to safely fill hydrogen. It looks similar to common gasoline, diesel and natural gas dispensers, but with its own specificities. When the nozzle is firmly engaged to the FCV's receptacle it does not come off until the filling procedure is done and the pressure reduced. |

commercialized. It should be noted that the hydrogen flow rate and the impurities concentration in different HRS vary considerably. Table 5.17 presents fuel quality data from a particular HRS, the Senju Fueling Station, between 2004 and 2008.

In July 2009, major energy suppliers and industrial gas and other companies in Japan launched the Research Association of Hydrogen Supply/ Utilization Technology (HySUT) to attain a low-carbon society by establishing a hydrogen infrastructure and improving the business environment for hydrogen. Under HySUT, Japan will continue to advance deployment of hydrogen and fuel cell technologies, including hydrogen fueling stations and

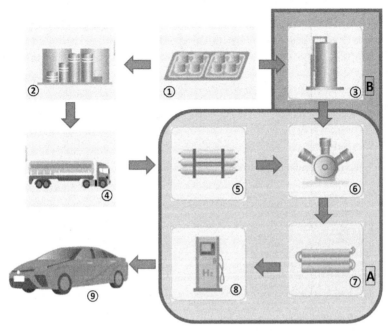

**FIGURE 5.27** Conceptual diagram of supply to hydrogen fuel cell vehicles. ①LNG base, ②Hydrogen production purification, ③Hydrogen production and purification, ④Hydrogen transport, ⑤Accumulator, ⑥Compressor, ⑦Precooler, ⑧Hydrogen dispenser, ⑨Fuel cell vehicle. (A) On-site station system, (B) Off-site station system.

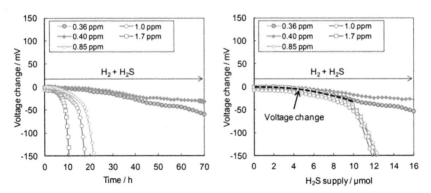

**FIGURE 5.28** Fuel cell potential variation as influenced by the concentration of $H_2S$ in the fuel supplied to the anode. Cell temperature $80°C$; anode Pt loading of $0.4$ mg $cm^{-2}$ [51].

polymer electrolyte membrane fuel cell (PEMFC) vehicles. Data from the field will be acquired so that regulations can be revised as needed. The Department of Energy (DOE) will continue to work with the appropriate agencies and organizations in Japan to facilitate international harmonization of regulations, codes, and standards, including fuel quality specifications.

**TABLE 5.14** Directory of Limiting Characteristics for Fuel Cell Vehicles From ISO FDIS 14687-2 (International Standardization for Hydrogen Fuel for Fuel Cell Vehicles)

| | Type I. Type II |
|---|---|
| Characteristics (assay) | Grade D |
| Hydrogen fuel index (minimum mole fraction)[a] | 99.97% |
| Total nonhydrogen gases | 300 μmol/mol |
| **Maximum concentration of individual contaminants** | |
| Water ($H_2O$) | 5 μmol/mol |
| Total hydrocarbons[b] (*Methane basis*) | 2 μmol/mol |
| Oxygen ($O_2$) | 5 μmol/mol |
| Helium (He) | 300 μmol/mol |
| Total nitrogen ($N_2$) and argon (Ar)[b] | 100 μmol/mol |
| Carbon dioxide ($CO_2$) | 2 μmol/mol |
| Carbon monoxide (CO) | 0.2 μmol/mol |
| Total sulfur compounds[c] ($H_2S$ *basis*) | 0.004 μmol/mol |
| Formaldehyde (HCHO) | 0.01 μmol/mol |
| Formic acid (HCOOH) | 0.2 μmol/mol |
| Ammonia ($NH_3$) | 0.1 μmol/mol |
| Total halogenated compounds[d] (*Halogenate ion basis*) | 0.05 μmol/mol |
| Maximum particulates concentration | 1 mg/kg-$H_2$ |

*NOTE:* For the constituents that are additive such as total hydrocarbons and total sulfur compounds the sum of the constituents are to be less than or equal to the acceptable limit. The tolerances in the applicable gas testing method are to be the tolerance of the acceptable limit.
[a]*The hydrogen fuel index is determined by subtracting the "total nonhydrogen gases" in this table expressed in mol% from 100 mol%.*
[b]*Total hydrocarbons include oxygenated organic species. Total hydrocarbons are measured on a carbon basis (μmol C/mol). Total hydrocarbons may exceed 2 μmol/mol due only to the presence of methane, in which case the summation of methane nitrogen and argon is not to exceed 100 ppm.*
[c]*As a minimum includes $H_2S.COS.CS2$ and mercaptans, which are typically found in natural gas.*
[d]*Includes, for example, hydrogen bromide (HBr), hydrogen chloride (HCl), chlorine ($Cl_2$), and organic halides (R-X).*

# DEVELOPMENT PLAN OF HYDROGEN REFUELING STATIONS IN THE WORLD

Hydrogen refueling stations are being developed and deployed throughout the world and more intensively in Europe, in Japan and in the EUA. There are different locations where high purity hydrogen can be supplied by the

**TABLE 5.15** Analytical Methods and Limits for ISO FDIS14687-2

| Impurities [ISO Limit] | Analytical Methods | Detection Limit (µmol/mol) | Determination Limit (µmol/mol) | Test Methods |
|---|---|---|---|---|
| Water ($H_2O$) [5 µmol/mol] | Dew point analyzer | 0.5 | 1.7 | JIS[a] K0225 |
| | GC/MS with jet pulse injection | 1 | 3 | ASTM[b] D7649-10 JIS K0123 |
| | GC/MS with direct injection | 0.8 | 2.4 | NPL[c] Report AS 64 |
| | Vibrating quartz analyzer | 0.02 | 0.07 | JIS K0225 |
| | Electrostatic capacity type moisture meter | 0.04 | 0.1 | JIS K0225 |
| | FTIR | 0.12 1 | 0.4 3 | ASTM D7653-10 JIS K0117 |
| | Cavity ring-down spectroscopy | 0.01 | 0.03 | NPL Report AS 64 |
| Total hydrocarbons (Cl basis) [2 µmol/mol] | FID | 0.1 | 0.3 | ASTM D7675-11 |
| | GC/FID | 0.01–0.1 | 0.03–1.0 | JIS K0114 |
| | FTIR | 0.01 | 0.03 | JIS K0117 |
| Oxygen ($O_2$) [5 µmol/mol] | Galvanic cell $O_2$ meter | 0.01 | 0.03 | JIS K0225 |
| | GC/MS with jet pulse injection | 1 | 3 | ASTM D7649-10 |
| | GC/PDHID | 0.006 | 0.018 | NPL Report AS 64 |
| | GC/TCD | 3 | 9 | NPL Report AS 64 |
| | Electrochemical sensor | 0.1 | 0.3 | ASTM D7607-11 |

*Continued*

**TABLE 5.15** Analytical Methods and Limits for ISO FDIS14687-2—cont'd

| Impurities [ISO Limit] | Analytical Methods | Detection Limit ($\mu$mol/mol) | Determination Limit ($\mu$mol/mol) | Test Methods |
|---|---|---|---|---|
| Helium (He) [300 $\mu$mol/mol] | GC/TCD | 3–5 | 10–15 | ASTM D1945-03 JIS K0114 |
| | GC/MS | 10 | 30 | JIS K0123 |
| Nitrogen ($N_2$). Argon (Ar) [100 $\mu$mol/mol] | GC/MS with jet pulse injection | 5 ($N_2$), 1 (Ar) 0.03 | 15 ($N_2$), 3 (Ar) 0.1 | ASTM D7649-10 JIS K0123 |
| | GC/TCD | 1–3 | 3–10 | JIS K0114 |
| | GC/PDHID | 0.001 | 0.01 | JIS K0114 |
| Carbon dioxide ($CO_2$) [2 $\mu$mol/mol] | GC/MS with jet pulse injection | 0.5 0.01 | 1.5 0.03 | ASTM D7649-10 JIS K0123 |
| | GC/FID – methanizer | 0.01 | 0.03 | JIS K0114 |
| | GC/PDHID | 0.001 | 0.01 | JIS K0114 |
| | FTIR | 0.01 0.02 | 0.03 0.06 | ASTM D7653-10 JIS K0117 |
| Carbon monoxide (CO) [0.2 $\mu$mol/mol] | GC/FID with methanizer | 0.01 | 0.03 | JIS K0114 |
| | FTIR | 0.01 0.1 | 0.03 0.3 | ASTM D7653-10 JIS K0117 |
| | GC/PDHID | 0.001 | 0.01 | JIS K0114 |

| Component | Method | | | Standards |
|---|---|---|---|---|
| Total sulfur compounds [0.004 µmol/mol] | IC with concentrator | 0.0001–0.001 | 0.0003–0.004 | JIS K0127 |
| | GC/SCD with concentrator | 0.00002 0.001 | 0.00006 0.003 | ASTM D7652-11 JIS K0114 |
| | GC/SCD without preconcentration | 0.001 | 0.003 | NPL Report AS 64 |
| Formaldehyde (HCHO) [0.01 µmol/mol] | DNPH/HPLC | 0.002–0.01 | 0.006–0.03 | JIS K0124 |
| | GC/PDHID | 0.01 | 0.03 | JIS K0114 |
| | FTIR | 0.02 0.01 | 0.06 0.03 | ASTM D7653-10 JIS K0117 |
| Formic acid (HCOOH) [0.2 µmol/mol] | IC | 0.001–1 0.002 –0.01 | 0.003–3 0.006–0.03 | ASTM D7550-09 JIS K0127 |
| | FTIR | 0.02 0.01 | 0.06 0.03 | ASTM D7653- |
| Ammonia (NH$_3$) [0.1 µmol/mol] | IC with concentrator | 0.001–0.01 | 0.003–0.03 | JIS K0127 |
| | FTIR | 0.02 0.01 | 0.06 0.03 | ASTM D7653-10 JIS K0117 |
| Total halogenated compounds [0.05 mol/mol] | IC with concentrator | 0.05 | 0.17 | JIS K0101. K0127 |
| Maximum particulate concentration [1 mg/kg-H$_2$] | Gravimetric | 0.005 mg/kg | 0.015 mg/kg | ASTM D7651-10 JIS Z881 |

*Notes:* The analysis is carried out according to the following rules [52].
[a]*JIS, Japanese Industrial Standards.*
[b]*ASTM, American Society for Testing Materials.*
[c]*NPL, National Physical Laboratory.*

**TABLE 5.16** Concentration and Size of Particulate Matter Sampled at Selected Hydrogen Fueling Stations [53]

| Flow Rate (g/s)[a] | More than 1 cm (10.000 μm) | Between 1 mm (1000 μm) and 1 cm (10.000 μm) | Between 100 μm and 1000 μm | Between 10 μm and 100 μm | Concentration (μg/L) |
|---|---|---|---|---|---|
| 29.5 | 1 | 3 | 5 | 0 | 0.019 |
| 11.9 | 0 | 2 | 6 | 4 | 0.0062 |
| 11.2 | 0 | 2 | 18 | 22 | 0.0055 |
| 9.1 | 0 | 2 | 6 | 21 | 0.0058 |
| 8.1 | 0 | 0 | 5 | 2 | 0.007 |
| 7.1 | 0 | 0 | 5 | 4 | 0.005 |
| 6.9 | 0 | 0 | 4 | 0 | 0.0044 |
| 6.68[a] | 0 | 1 | 7 | 0 | 0.0054 |
| 4.54[a] | 0 | 10 | 4 | 0 | 0.012 |
| 4.16[a] | 0 | 1 | 2 | 0 | 0.0049 |
| 4.16[a] | 0 | 2 | 1 | 0 | 0.0025 |
| 3.92[a] | 0 | 1 | 4 | 0 | 0.0025 |
| 3.3 | 0 | 0 | 5 | 0 | 0.0042 |
| 2.5[a] | 0 | 0 | 3 | 3 | 0.00041 |

All others established using a station test apparatus developed by Smart Chemistry.
[a]Flow rates determined by dispenser read-out and approximate fill time.

**TABLE 5.17** Fuel Quality Data From the Senju Fueling Station (volppm)

| Chemical | FY2004 | FY2005 | FY2006 | FY2007 | FY2008 |
|---|---|---|---|---|---|
| CO | 0.02 | 0.02 | ND | ND | 0.01 |
| $CO_2$ | ND | ND | ND | ND | ND |
| $CH_4$ | 0.08 | ND | 0.08 | ND | ND |
| $C_6H_6$ | ND | ND | ND | ND | ND |
| MeOH | ND | ND | ND | ND | ND |
| HCHO | ND | ND | ND | ND | ND |
| $CH_3CHO$ | ND | ND | ND | ND | ND |
| HCOOH | ND | ND | ND | ND | ND |
| $CH_3COCH_3$ | ND | ND | ND | ND | ND |
| $NH_3$ | ND | ND | ND | ND | ND |
| Halogen | – | – | – | ND | ND |
| $O_2$ | ND | ND | ND | ND | ND |
| $H_2O$ | 24 | 0.9 | ND | ND | ND |
| Ar | 4.95 | 0.11 | 0.73 | 1.5 | 1.34 |
| $N_2$ | 3.03 | 0.12 | 3.59 | 10.4 | 6.91 |
| He | ND | ND | ND | ND | ND |

*ND*, not detected.

hydrogen supply system technologies shown so far, and there is strong driving force for implementation in various world locations in the near future. Here, focus is given to activities in this are in the United States, in the European Union and in Japan, that concentrate most of the HRS deployed to date.

## United States

In 2002, the U.S. government concluded the "Freedom Car Partnership" to promote the use of hydrogen fuel cells between the Ministry of Energy and the three largest automobile manufacturers in the United States and accelerated the development toward commercialization of FCVs.

Since 2013, in order to expand the spread of FCVs and installation of hydrogen stations throughout the United States, the Ministry of Energy and the automobile manufacturer and the Hydrogen Energy Association of Fuel Cells established a new partnership, H2USA.

In addition to American companies, Japanese and Korean manufacturers such as Toyota, Honda, Nissan, and Hyundai also participate in H2USA. At the moment H2USA is planning to build stations throughout the United States by 2020. In addition, H2USA is also involved in the review of hydrogen station development in the states of New York, Massachusetts, Connecticut, and Rhode Island.

California is the state that is most enthusiastically working on dissemination of FCVs and hydrogen station maintenance in 50 U.S. states. California has been implementing the Zero Emission Vehicle (ZEV) regulation, which includes carbon dioxide for automobiles, since 2009. For this reason, it is also active in infrastructure development such as hydrogen station construction [53].

As of January 2016, 17 stations in California are open, 51 stations are planned by 2017, and 100 places are scheduled to be developed by around 2023. In addition, the hydrogen station set up with the subsidy of California province is supposed to derive hydrogen supply equivalent to 33% from renewable energy (wind power, sunlight, biomass, etc.).

In addition, actions were taken to introduce a total of 3300 ZEVs, including FCVs, in seven states, mostly in California and the East Coast districts, which are the provinces with foremost fuel cell activities in the United States. The spread of fuel cell automobiles is being promoted.

In September 2013, in order to expand the California hydrogen refueling station movement nationwide, a public-private partnership, H2USA, was set up with the Department of Energy and a host of official institutions and private companies. The main objective is to promote the commercial introduction and widespread adoption of hydrogen fueled fuel cell electric vehicles across the country with significant fleet through 2020 (Fig. 5.29).

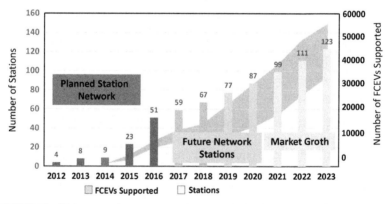

**FIGURE 5.29** Hydrogen stations are expected to spread in California. About 90 locations will be maintained by around 2020 and it is expected that it will be able to support the spread of FCVs of about 2000–30,000 [54].

## Europe

Among the countries in Europe, Germany is proceeding with plans for hydrogen energy use as soon as possible. In 2004, the Clean Energy Project (CEP), a demonstration project of FCVs and hydrogen stations, has started and the government-funded Hydrogen and Fuel Cell Technology Innovation Program (NIP) began in 2007. In the CEP framework, 50 hydrogen refueling stations were planned to be installed nationwide.

Furthermore, in 2009, H2 Mobility was established to consider infrastructure development aimed at nationwide dissemination of FCV and hydrogen stations. H2 Mobility is a unified public-private project including members from government, automakers, and energy companies. H2 Mobility announced that it is going to install 400 hydrogen refueling stations by 2023.

In addition, following the model of H2 Mobility promoted by Germany, other European countries such as the United Kingdom, France, Denmark, and others, are underway on the infrastructure development for hydrogen refueling stations. This was effective for introducing more than 100 HRS in Europe. The government and official institutions fostering Germany's initiatives, included the participation of the more active companies from this sector (Fig. 5.30).

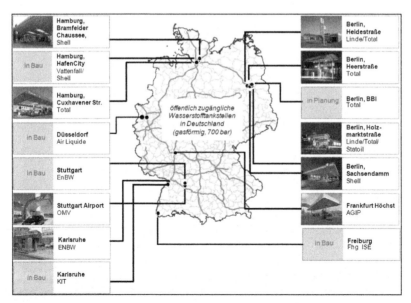

**FIGURE 5.30**  Map of hydrogen refueling stations in Germany. Hydrogen stations are deployed in major cities, and a few cities in a region are connected. As a result, the spread of hydrogen energy is made effective by generating interconnectivity for FCVs [55].

## Japan

Japan will ease up on regulations on hydrogen stations for FCVs, making the refueling points less costly to set up and operate, in hopes of helping the environmentally friendly cars catch on.

In 2018, the Ministry of Economy, Trade and Industry has revised about 20 points in rules governing the facilities. While the government is hurrying up to support electric vehicles, it also aims to lay the groundwork for the long-term growth of electric powered FCVs.

The spread of FCV has begun. In order to supply hydrogen to them, the goal was established to deploy about 100 commercial hydrogen stations within the four major metropolitan areas (Tokyo area, Chukyo area, Kansai area and northern Kyushu area), and to promote public-private cooperation to advance the project. After that, about 160 places by 2020 and 320 places by 2025 were set as goals.

In the midst of the 2020s, the number of FCVs (hydrogen demand) and the hydrogen station (hydrogen supply) will be balanced in the latter half of the 2020s, although diffusion will proceed in the form of making hydrogen stations ahead of the number of FCVs to promote FCV for the foreseeable future. Autonomous spread and development is expected. As of 2030, about 900 plants are planned to be constructed in terms of standard supply capacity hydrogen station ($300 \, \mathrm{Nm^3/h}$).

Thus in Japan, the government is showing a positive attitude toward realizing a hydrogen society. Hydrogen stations as shown in Fig. 5.31 are being constructed at a rapid pace. As a result, hydrogen stations have been built and operated in almost all areas of the country [56] (Table 5.18).

**FIGURE 5.31** Typical hydrogen station in Japan. Such a station is being built in many locations in Japan.

**TABLE 5.18** Planned Number of Hydrogen Refueling Stations (HRS) and Fuel Cell Vehicles (FCVs) in Japan

|                | FY2015 | FY2020 | FY2025  | FY2030  |
|----------------|--------|--------|---------|---------|
| Number of HRS  | 80     | 160    | 320     | 900     |
| Number of FCV  | 700    | 40,000 | 200,000 | 800,000 |

## CONCLUSIONS

In recent years, a road map has been shown from each country toward a hydrogen society or toward a renewable energy society [57–60].

In Japan, the Basic Energy Plan was announced in April 2014. This plan shows a roadmap for the realization of a hydrogen society.

Japan Fuel Cell Practice Promotion Council announced a scenario for the spread of FCVs and hydrogen stations since 2010. In this scenario, a goal was set for FCVs and hydrogen stations to be established as business in 2025 (Fig. 5.32).

In December 2014, Toyota started general sales of FCVs. They were sold as mass-produced cars for the first time in the world. Furthermore, in January 2015, Toyota announced the decision to offer the FCV license-related patents rights owned by the company free of charge.

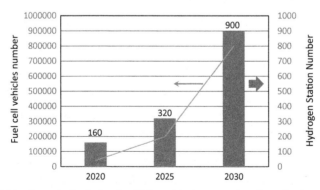

**FIGURE 5.32** Scenario for fuel cell vehicle and hydrogen station dissemination. *http://www. meti.go.jp/committee/kenkyukai/energy/nenryodenchi_fukyu/pdf/002_01_00.pdf.*

In March 2016, based on the progress made until then and the situation therein established, a scenario aiming for full-scale dissemination to contribute for future greenhouse gas emission reduction was announced, which was effectively reached.

The future price targets of domestic fuel cells mentioned previously are as follows: PEMFC type ENE-FARM will be 800,000 yen by 2019, and solid oxide fuel cell type ENE-FARM by 2021 will be 1 million yen. Also, the diffusion target of FCVs was set to about 40,000 units by 2020, about 200,000 units by 2025, and about 800,000 units by 2030. In addition, the target for deployment of hydrogen refueling stations was set to about 160 by 2020, and 320 installations by 2025.

As described earlier, in Japan, in the United States, and in Europe mainly, but elsewhere also, governments are making strategic efforts to disseminate hydrogen energy and the utilization introduction of hydrogen technologies.

On the other hand, there are many problems to be solved when disseminating hydrogen energy, and the following three points should be considered.

1. Economic efficiency. Compared with other competitive technologies, it is considered that the high cost of fuel cell and hydrogen production is a huge hurdle for large-scale dissemination, so it is desirable to reduce costs by high efficiency and technological innovation.
2. Environment. Regarding the use of hydrogen, $CO_2$ is not discharged, but when hydrogen is produced using fossil fuel as a raw material, $CO_2$ is discharged. In the future, establishment of $CO_2$ free hydrogen supply system using CCS and renewable energy is required.
3. Efforts aimed for massive imports. In the event that hydrogen demand increases in the future, it is necessary to construct a system for importing large amounts of hydrogen from overseas. Development of transportation means (liquid hydrogen ship, etc.) for that, development of international standards, and of ports infrastructure are main subjects.

Hydrogen energy and fuel cell technology are presently being introduced in the society, and are considered for utilization during the 2020 Tokyo Olympic Games in Japan. However, in order to reach broad utilization, spread in large scale, technology development and optimization must continue as a long-term effort.

Chapter 5.3.2

# Application of Hydrogen Combustion for Electrical and Motive Power Generation

## INTRODUCTION

Hydrogen energy engineering is expected to be applied to rockets, aircraft, automobiles, and power generation as a technology to convert hydrogen from chemical energy to thermal energy and then to use it.

Since the fuel cell is a power generation device that changes directly chemical energy into electrical energy, it has high power generation efficiency even on a small scale. Because it possesses no rotating parts, it makes little noise and vibration, so it is suitable for practical use as an urban energy system [61].

On the other hand, since the hydrogen turbine is easily enlarged, expectation for large-scale power generation is rising. In a hydrogen turbine, a system similar to that of a conventional Rankine cycle generator has been devised, in which hydrogen is directly introduced into a combustor, burned, and introduced to a turbine. For this reason, since the temperature in the combustor is ideally as high as about 2000°C, development of materials is necessary, but further difficulties are technologies for stably burning hydrogen [62–64].

Since high-efficiency power generation is realized when the temperature rises as the size increases, future development of concentrated power generation that reduces $CO_2$ generation is expected.

## CHARACTERISTICS OF POWER GENERATION SYSTEM

A hydrogen combustion turbine is a power generation system that drives a gas turbine or a steam turbine by utilizing combustion heat of hydrogen. Therefore, the equipment configuration is equivalent to thermal power generation. Normal thermal power generation uses natural gas as a raw material, but by replacing this natural gas with hydrogen, a hydrogen-air turbine is formed. With this combination of raw materials, NOx is generated by nitrogen and oxygen from the air, so it is necessary to place a denitration catalyst on the downstream side. It is necessary to reduce the NOx concentration in the exhaust gas. Since the combustion speed of hydrogen is faster than that of

Science and Engineering of Hydrogen-Based Energy Technologies. https://doi.org/10.1016/B978-0-12-814251-6.00005-8

hydrocarbons typified by methane, there is a possibility that the combustor can be made compact [65,66].

When oxygen is used as an oxidizing agent and introduced at a suitable ratio, exhaust gas discharges only steam. In this case, since it is possible to form a closed system, it can be the ultimate environmental power generation system. If hydrogen and oxygen can be burned completely in the combustor, it becomes possible to heat by utilizing the chemical reaction heat of the raw material, and it becomes an ideal power generation system.

The features of this hydrogen turbine using oxygen as an oxidant are as follows.

1. Since oxygen and hydrogen react with each other in the combustor to generate water vapor, the heater such as a boiler becomes compact or unnecessary.
2. Since only steam is generated, closed cycle is easy to be implemented; it eliminates the influence of the atmospheric environment.
3. In the hydrogen-oxygen combustion combined cycle type gas turbine, it is possible to directly introduce the superheated steam generated in the combustor of the gas turbine into the steam turbine, so a simple design as well as the possibility of high efficiency can be expected.

As mentioned, the hydrogen-oxygen combustion gas turbine system can construct a closed cycle system without discharging air pollutants and therefore can realize highly efficient power generation.

## DEVELOPMENT TREND

### Review of Closed Cycle

Normally, steam turbines use fossil fuels and are therefore regarded as power-generating equipment that causes environmental problems for the atmosphere. For this reason, a waste heat recovery boiler is installed and NOx reduction operation is performed to prevent air pollution. On the other hand, the exhaust gas of the hydrogen-oxygen turbine is characterized only by water vapor, and it is regarded as a power-generating device with high environmental performance, and its development is desired.

Research and development of hydrogen-fueled steam turbines and gas turbines started in the United States in the 1950s. In the 1970s, the development of combustors for small gas turbines was conducted, and then the theme shifted to the study of closed-cycle systems.

Fig. 5.33 is a conceptual diagram of a closed system. A high pressure water split electrochemical apparatus generates high pressure hydrogen and high pressure oxygen gas, and drives the turbine through the combustor. The water is recovered with a condenser, and this water is reused to make oxygen and hydrogen raw materials. In addition, it is an energy saving system that uses water generated as much as possible by utilizing it as cooling water of the

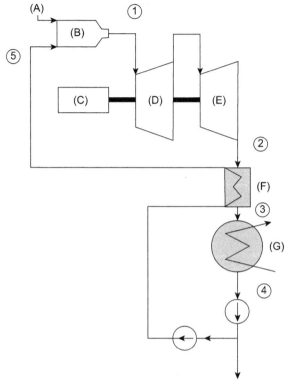

**FIGURE 5.33** Conceptual diagram of a closed system. (A) Hydrogen and oxygen, (B) combustor, (C) generator, (D) high pressure turbine, (E) low pressure turbine, (F) water supply unit, (G) cooler. ① to ⑤ at steam flow and heat are shown in Table 5.19 [67–70].

burner. What is noteworthy of this system is that the high pressure water decomposition apparatus is constituted by electrolysis. The concept of driving turbines and generating electricity using hydrogen generated from renewable energy as a raw material was invented from that time, and the possibility that turbine power generation can be applied to hydrogen energy has been studied for a long time.

## Study on Hydrogen-Oxygen Combustion Turbine System

### Simple Rankine Cycle System
The efficiency of the most basic hydrogen-oxygen combustion turbine is considered using simulation. It is a steam turbine system using steam generated by combustion of oxygen and hydrogen, and it is a simple Rankine cycle system. It is the prototype of the hydrogen-oxygen combustion turbine system and is the simplest system.

**TABLE 5.19** Design Values of Simple Rankine System [67,69]

|  | Temperature (°C) | Pressure (MPa) | Steam Flow[a] (kg/s) | Enthalpy(kJ) |
|---|---|---|---|---|
| ① | 1500 | 5 | 150 | 16,066 |
| ② | 335 | 0.005 | 150 | 8508 |
| ③ | 70 | 0.005 | 150 | 7101 |
| ④ | 33 | 0.005 | 150 | 373.5 |
| ⑤ | 215 | 5.3 | 100[a] | 1655.5 |
| ⑥ | 20 | 6 | 50[b] | 14,412.2 |

[a]The steam flow rate is the material balance of this simple Rankine system. According to ⑤, the amount of circulating water is 100 kg.
[b]⑥ indicates that hydrogen and oxygen were converted into water vapor of 50 kg. This means that 2.8 kmol/s of hydrogen and 1.4 kmol/s of oxygen flow into the combustor, respectively.

Consider the simple Rankine cycle system shown in Fig. 5.33. In addition, the simulation was carried out by setting the numerical values shown in Table 5.19 for the design conditions. Assuming that the inlet temperature of the high pressure turbine is equivalent to the combustor outlet temperature, it was 1500°C. This is because hydrogen and oxygen, which are raw material gases, become steam in the combustor. The temperature at this exit is about 1500°C, which is a temperature close to the real gas turbine inlet temperature; that is, a temperature equal to the actual design condition. Conditions at the outlet of the low pressure turbine were set to 336°C, which is the turbine outlet condition adopted for the actual machine design, and the pressure was set to a value close to the normal pressure.

It is a system that can pass through the reheater, which is a heat exchanger, and circulate the water after turning steam completely into water in the condenser. The turbine was a cascade-type high-pressure and low-pressure two-stage turbine, which is a standard specification. The electrical energy generated is roughly estimated, but it is assumed to range between 35 to 500,000 kW.

This system circulates 100 kg/s of steam from hydrogen and oxygen generated in the combustor, so a drain line and a recirculation line are provided on the downstream side of the condenser. In addition, a water flow pump is arranged in the recirculation line.

Hydrogen and oxygen, which are raw material gases, were set at 50 kg/s in terms of the amount of water vapor on the downstream side of the combustor, which means that 2.8 kmol/s of hydrogen and 1.4 kmol/s of oxygen flow into the combustor.

Next, the T-s diagram of this system is shown in Fig. 5.33. Cycle from ① to ② → ③ → ④ → ⑤ → ⑥ in order. Table 5.19 summarizes the design values at each point.

Find this simple Rankine efficiency. The power generation efficiency $\eta$ is the product of the simple Rankine efficiency $\eta RA$, the mechanical efficiency $\eta ME$, and the generator efficiency $\eta GE$. Here, it is assumed that the mechanical efficiency $\eta ME$ is 99% and the generator efficiency $\eta GE$ is 98%. Calculation was carried out with reference to the T-s diagram.

$$\eta = \eta RA \times \eta ME \times \eta GE \; \frac{① - ②}{⑥} \; 0.99 \times 0.98 \times 100$$

$$= 0.52 \times 97.02$$

$$= 50.5\%$$

That is, the power generation efficiency is 50.5%, which is higher than that of the conventional steam turbine.

This system is the simplest Rankine cycle—the T-s diagram is a triangle, which is a shape that is not advantageous for improving efficiency. Moreover, no further improvement in efficiency can be expected.

### Reheat Rankine System

This system introduces another pair of combustors at the inlet of the low pressure turbine of the simple Rankine system of steam, introduces hydrogen and oxygen as raw materials, and reburns them.

The enclosing area of the T-s diagram expands and the thermal efficiency is advantageous. However, the configuration of the equipment becomes complicated, and when the turbine is combined with three stages and four stages, the thermal efficiency can be increased, but this method is not necessarily realistic.

Fig. 5.34 is the design value of the reheat Rankine system, but temperature and pressure were set according to the simple Rankine system. In the case of ⑧ where the amount of steam is also high, as in the simple Rankine cycle, 50 kg of steam is introduced by oxyhydrogen combustion, and in ⑨, which is low pressure, 20 kg of steam is generated by oxyhydrogen combustion.

The design values of this system are shown in Table 5.20.

According to this assumption, the mechanical efficiency $\eta ME$ is assumed to be 99%, the generator efficiency $\eta GE$ is assumed to be 98%, and the efficiency is calculated with reference to the T-s diagram.

$$\eta = \eta RA \times \eta ME \times \eta GE \; \frac{(① - ②) + (③ - ④)}{⑧ + ⑨} \times 0.99 \times 0.98 \times 100$$

$$= 0.54 \times 97.02$$

$$= 52.4\%$$

It rose about two points compared to the simple Rankine cycle system. This result indicates that the performance improves when introducing the reheat system, suggesting that the development of the reheat system is the key to improving the efficiency.

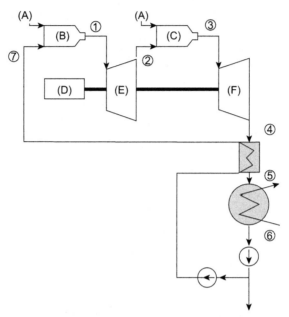

**FIGURE 5.34** Design value of the reheat Rankine system. (A) Hydrogen and oxygen, (B) and (C) combustor, (D) generator, (E) high pressure turbine, (F) low pressure turbine. ① to ⑦ at steam flow and heat are shown in Table 5.20 [67–70].

**TABLE 5.20** Design Value of Reheat Rankine System [69]

|  | Temperature (°C) | Pressure (MPa) | Steam Flow (kg/s) | Enthalpy (kJ) |
|---|---|---|---|---|
| ① | 1500 | 5 | 150 | 57,614 |
| ② | 900 | 0.4 | 150 | 11,660 |
| ③ | 1500 | 0.35 | 170 | 18,262 |
| ④ | 600 | 0.005 | 170 | 11,350 |
| ⑤ | 300 | 0.005 | 170 | 9471 |
| ⑥ | 85 | 0.005 | 170 | 424 |
| ⑦ | 270 | 5.3 | 100 | 2126 |
| ⑧ | 20 | 6 | 50[a] | 13,650 |
| ⑨ | 20 | 0.5 | 20[b] | 6595 |

[a]Hydrogen and oxygen, which are raw material gases, were designed assuming that the amount of steam on the downstream side of the combustor was generated at 50 kg/s, but it was confirmed that hydrogen and air entered the combustor at 2.8 kmol/s and 1.4 kmol/s, respectively. And it is supposed to burn completely.
[b]The amount of steam was set at 20 kg/s, which was set to fully combust the hydrogen gas flowing into the combustor at 1.12 kmol/s and oxygen gas at 0.56 kmol/s.

## Toward Future Development of Hydrogen Turbine

### Technical Tasks for Hydrogen Combustion

Hydrogen turbines can be dealt with by improving gas turbine combustors currently in practical use. However, the combustion characteristics of hydrogen gas have three problems as shown in Table 5.21, and it is necessary to correct this. Hydrogen has the disadvantage that it has combustion characteristics such as (1) low calorific value, (2) high burning rate, (3) high adiabatic flame temperature, compared to existing fuels such as natural gas. For this reason, various problems such as local hot spots due to high flame temperature and occurrence of NOx when air is used as an oxidant must be solved.

### Hydrogen Combustor in Gas Turbine

Gas turbines are roughly divided into three main types:

1. Diffusion method, which is a method of separately injecting fuel and air
2. Premixing method, which is a method of premixed injection of fuel and air
3. Dispersion mixing method, in which fuel and air are mixed rapidly and combined with floating flame technology

Since the diffusion system is a system in which fuel and air are injected separately, high temperature spots tend to occur locally, and generation of NOx becomes a problem. For this reason, NOx is reduced by water/steam injection or the like. Although this has the disadvantage of lowering the efficiency, it is possible to deal with diverse fuel types, and hydrogen

**TABLE 5.21** Comparison of Combustion Characteristics of Methane and Hydrogen Energy and Improvement of Its Combustor [71]

| Comparison of Combustion Characteristics Between Hydrogen and Natural Gas | Improvement of Hydrogen Gas Combustor |
|---|---|
| Hydrogen has a low combustion heat of about 1/3 to natural gas. | Design to increase gas flow rate. This requires ingenuity in design, such as increasing the diameter of the gas pipes and increasing the number of combustors. |
| The burning rate of hydrogen is about 7 times faster than that of natural gas. | In the combustor of the premix type, the inversion tends to occur. For this reason, damage to the combustor is remarkable. |
| Hydrogen has adiabatic flame temperature about 10% higher than natural gas. | Local hot spot generation. When air is used as an oxidizing agent, NOx is generated. |

combustion is also possible. This method has many achievements, and overseas has succeeded in the demonstration power test with hydrogen coking. For this reason, since the hydrogen concentration is variable and there is no limitation on the concentration, it is a practical and easy-to-use combustion method, and it is a technology at the practical stage. Although the efficiency decreases due to water and steam injection in the diffusion method, it is possible to deal with diverse fuel types, and there are also many achievements of hydrogen combustion: specifically, when hydrogen produced by industrial processes of ironworks and refineries is utilized for private power generation, or when coal gas is used for power generation.

The distributed mixing method is currently under development but is a method that compensates for the drawbacks of the earlier two methods. This method avoids reverse combustion and at the same time performs lean combustion of NOx.

The outline structure of the burner developed in Fig. 5.35 and the outline of technology are shown. This technology uses the rapid mixing by a multi-coaxial jet burner (cluster burner) as the core of technology. Furthermore, in order to cope with coal gasification gas, we developed a technology to adjust flame direction by combining a number of coaxial jet burners, and stably maintain the flame in a space away from the burner structure.

The cluster burner is composed of an air hole plate provided with many small diameter air holes and a large number of fuel nozzles arranged coaxially with the air hole. In the cluster burner, the fuel is dispersed and the fuel and air are mixed rapidly by the turbulent flow in the air hole so that a lean air-fuel mixture is formed with a short mixing time, and low NOx combustion becomes possible. By combining the rapid mixing technique and the flame levitation technique, it is possible to form a flame at a position determined by

**(A)**          **(B)**

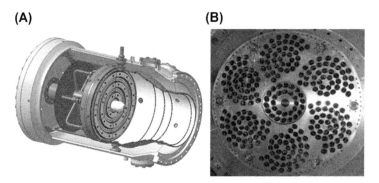

FIGURE 5.35   Image of several MW class hydrogen dedicated fire gas turbine combustor (A) and photo of hydrogen burner seen from the front (B). This burner is called cluster because there are seven holes clustered together. Reducing NOx is a dry method, aiming at combustion using only hydrogen as a fuel. Since NOx is generated by air combustion, a denitrification facility is required on the downstream side [72−74].

the flow pattern in a state in which the fuel and air are sufficiently mixed after the coaxial jet ejected from the burner structure. The reliability of the structure can be secured and the flame can be held at substantially the same position even if the combustion speed changes. Burning characteristics of energy are different between methane and hydrogen. The main points of correspondence improvement of the combustor are summarized in Table 5.21, though it cannot be unconditionally compared.

Also as shown in Fig. 5.35, by dividing the coaxial jet burner group of the cluster burner into the inner circumference system and the outer circumference system and controlling the ratio of the fuel to be allocated to each system, the hydrogen concentration changes. Then, a technology was established to stably obtain low NOx combustion even if the combustion velocity varied.

Currently, for the development of a hydrogen-denitrification dry low NOx combustor that can be applied to large gas turbines for power-generation businesses of several hundred MW class, technologies such as large temperature increase of cluster burner and evaluation of combustor characteristics were developed.

In the research of this combustor, the goal is to optimize the design and manufacture of the burner by the reduced model. In addition, it realizes stabilization of flame shape, realization of stable combustion, and reduction of NOx.

## DEVELOPMENT STATUS

Efforts are also being made to combine high efficiency and low carbon coal-fired power generation by combining integrated coal gasification combined cycle (IGCC) and Carbon capture and storage (CCS); development of hydrogen combustion gas turbines is being promoted as a part of this effort.

### Enel (Italy)

Enel of Italy started a plant demonstration project of hydrogen power generation in 2009 with a view to combining IGCC, Integrated Gasification Combined Cycle, and CCS, carbon capture and storage, in the future. After that, IGCC was put into practical use research in Japan, and development is still continuing now. And CCS was adapted experimentally to U.S. thermal power generation.

Specifically, about 12 MW is generated by hydrogen power generation using by-product hydrogen generated in the surrounding petrochemical plant as a fuel. Steam generated at that time is sent to a coal-fired power plant to generate 4 MW. It was a demonstration project of GTCC, Gas Turbine Combined Cycle, totaling 16 MW. Currently the operation has been stopped due to the end of government support.

In continuous operation for 2300 h until May 2010, we demonstrated power generation with 80% hydrogen concentration by the diffusion combustion method for the purpose of reducing nitrogen oxide emissions and so on. In addition, despite a short time and low load, a dedicated burning power generation with a hydrogen concentration of 100% was performed. There was no conspicuous damage to the burner and the wing, but loss of the thermal barrier coating in the combustor was confirmed [75].

## GE (United States)

Severe environmental regulations are being studied in the United States; for example, when new coal-fired power plants are established, carbon dioxide emissions standards that are difficult to achieve unless CCS is carried out are mandatory. For this reason, combining IGCC and CCS is promising for the future. As part of this, for example, GE participates in the Advanced Energy Systems/ Hydrogen Turbine program of the U.S. Department of Energy (DOE) and is engaged in research and development of hydrogen gas turbines. The project is planning to develop and demonstrate a 1700°C class hydrogen turbine by around 2020. In addition, GE has been developing combustors using the premixed combustion method for a long time, and although it is at the laboratory level, GE has developed a combustor that can handle up to 90% hydrogen concentration in a gas turbine that is supposed to combine CCS with IGCC, and has been proven that it is possible to cofire up to 30% hydrogen concentration in the mixed fire test with natural gas. In addition, in the diffusion combustion system, GE's gas turbine delivered to a Korean petrochemical plant has been shown to have been operating for 17 years (over 100,000 h) with a hydrogen concentration of 95% [76–78].

## Japan

An example of power generation using hydrogen mixed-fired fuel is also conducted in Japan. Until now, about 150,000 kW of electricity is generated, but in the future research on mixed combustion of coal, biomass, and ammonia will be in progress, and it is expected that this type of power generation will become popular. This situation is summarized in Table 5.22.

## HYDROGEN AND FOSSIL FUELS

Finally, we will describe mixed combustion of hydrogen and fossil fuel.

In the transition to the Great East Japan Earthquake, thermal power generation occupies a very large proportion as the energy crisis is being emphasized. This indicates that the emission amount of carbon dioxide that is the global warming gas is large. For this reason, the thought is to reduce the carbon dioxide emissions by introducing hydrogen or ammonia into fossil fuels.

Since fossil fuels emit $CO_2$, they have been regarded as a problem in recent years. Mixed combustion with hydrogen or hydrogen compounds is studied to reduce $CO_2$ emissions.

**TABLE 5.22** Examples of Power Generation by Hydrogen Cofired Fuel in Japan

| Power Plant | Power Generation Method | Combustion Method | Electric Terminal Power | Power Generation Efficiency, Low Heat Generation Efficiency (%) | Feedstock | Hydrogen Concentration (%) |
|---|---|---|---|---|---|---|
| Kashima Co operative thermal power units 3 and 4 | Steam power generation | Premixed combustion | 700 MW (350 MW × 2) | 40 | Heavy oil, blast furnace gas, coke oven gas | 5 |
| Kashima Cooperative thermal power units 5 | Gas turbine combined cycle GTCC | Diffusion combustion | 300 MW | 50 | Blast furnace gas, coke oven gas | 10 |
| Kimitsu cofired thermal power unit 4 | Steam power generation | Premixed combustion | 350 MW | 40 | Heavy oil, blast furnace gas, coke oven gas | 1 |

Based on an modified the environmental assessment deliberations created by the Agency for Natural Resources and Energy. It is based on the website of JX and Mitsubishi Heavy Industries. (http://www.nedo.go.jp/english/publications_reports_index.html; http://www.mhi.co.jp/technology/review/index-54-3.html).

Research on mixed fuels of hydrogen and methane and fossil fuels such as coal has been conducted already. Therefore, here we will describe research on methane and hydrogen, gasification of coal, methane and ammonia, and mixed combustion of pulverized coal and ammonia.

## Combustion of Hydrogen and Methane Mixed Fuel

A fundamental study on the combustion characteristics of hydrogen/methane mixed fuel will be the basis of mixed gas combustion. With the continuous use of fossil fuels such as the emission of greenhouse gases and threatening life have been demonstrated. In recent years, however, hydrogen has been placed as the primary energy to replace those fossil fuels. It is becoming clear that a mixed gas containing hydrogen mixed with fossil fuel not only contributes to the reduction of carbon dioxide but also has high combustion characteristics and low NOx emissions [81]. The movement to shift from the present fossil-fuel society to a hydrogen society using hydrogen as an energy source has already begun, but there are still many problems in terms of technology, cost, institutional aspects, and infrastructure to make a complete transition. As a breakthrough to solve the infrastructure issue, hydrogen is mixed in the existing methane pipeline and transported, and it is introduced to compatible equipment as methane gas, hydrogen, and hydrogen mixed according to the application at the destination. Although the use of a mixed fuel of hydrogen and methane gas is being studied, it is thought that it can be used for existing combustion equipment in a mixing amount to such an extent that the combustion speed does not change abruptly. The mixing ratio with hydrogen is at most 15%. Indeed, it is clear that if the mixing amount of hydrogen is of such a level, the burning rate for all equivalent ratios will be only about 10% or less compared with the case where pure methane is used as fuel (Fig. 5.36). This phenomenon is greatly different from 100% hydrogen gas, and it is necessary to solve in the future. As described earlier, combustion research on a mixed gas of hydrogen and methane has been actively studied [82].

Experimental investigations of hydrogen/methane/air flame propagation have been conducted for stoichiometric gases with different concentrations of hydrogen— 0%, 25%, 50%, 75%, and 100%. Experimental results showed that the flame propagation behavior accompanying pressure dynamics was investigated, and it was shown that the flame shape changes as the aspect ratio and hydrogen fraction change.

In the duct with the smallest aspect ratio, only the first four stages of the tulip flame were observed. As the aspect ratio increased, the flame structure evolved differently. When the hydrogen fraction is low, the flame face becomes asymmetric, and in the classical tulip flame reversal, the lower side lip propagates later than the upper part [83,84].

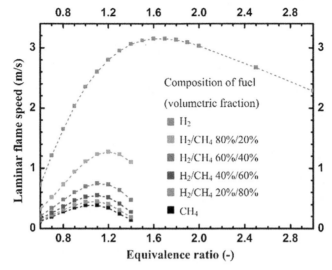

**FIGURE 5.36** Flame propagation speed when hydrogen concentration is changed. The propagation speed increases as the hydrogen concentration increases, but it differs greatly between 80% and 100%.

As the partial pressure of hydrogen increased, the flame propagated while maintaining its tulip shape with small vibrations. In addition, a cellular flame appeared in the duct having the maximum aspect ratio before the end of the propagation. For these ducts, distorted tulip flames with pitched tulip lips were observed for pure hydrogen. In all the configurations, the dynamics of the flame velocity are closely related to the pressure and both results are significantly increased with increasing hydrogen.

As the aspect ratio and hydrogen fraction increase, the flame shape develops differently. It was also found that the maximum flame velocity and maximum pressure greatly increased with hydrogenation [85].

Furthermore, we confirmed the fact that the dynamics of flame velocity are closely related to pressure. Research on such combustion is examined. This combustion study is also being studied in the catalyst field. An example of this research follows.

Catalytic microcombustors have been developed via deposition of active layers of $Pt/\gamma-Al_2O_3$, $LaMnO_3/\gamma-Al_2O_3$, and $Pt-LaMnO_3/\gamma-Al_2O_3$ on Fe-Cr alloy and $\alpha-Al_2O_3$ slabs. Diffusional resistance in the gas phase has been experimentally excluded, even if a catalyst is present on only one side of the reactor. $Pt/\gamma-Al_2O_3$ gave best performances in the combustion of $H_2$ while $LaMnO_3/\gamma-Al_2O_3$ revealed to be the best catalyst for methane combustion. The mixture $Pt-LaMnO_3$ supported on $\gamma-Al_2O_3$ showed intermediate properties for both fuels.

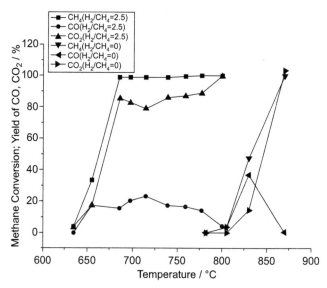

**FIGURE 5.37** Result of catalytic combustion of methane with hydrogen introduced. It has been shown that the combustion initiation temperature can be reduced by 100°C or more. Such a method has a wide range of applications, and application to the next generation of thermal power generation is considered [86].

The combustion of $H_2$–$CH_4$ mixtures showed that the presence of hydrogen promotes the oxidation of methane, since above 700°C radical reactions involving $CH_4$ are activated by $H_2$ ignition. Fig. 5.37 contains a graph in which the temperature is shown on the horizontal axis and the combustion of $CH_4$ (the conversion) is plotted on the vertical axis. It shows that introduction of hydrogen lowers the combustion initiation temperature. In this way, research is being conducted to burn at a lower temperature as a fuel that facilitates combustion, and by combining it with a combustion catalyst (Fig. 5.37).

## Coal Gas: A Mixed Gas of Hydrogen and Carbon Monoxide

IGCC is an integrated coal gasification combined cycle that combines coal at a high temperature to produce gaseous fuel, which is burned in a gas turbine. Steam is generated by the heat of the exhaust gas to move the steam turbine system, which requires a combustion process to use a gas turbine.

The gas produced by the coal gasification technology is mainly composed of carbon monoxide (CO) and hydrogen ($H_2$), and it is a widely available technology such as hydrogen production, chemical raw material production, synthetic fuel production, as well as power generation. After long-term research, the IGCC began operating gas turbines in recent years.

This coal gas is a mixed gas of hydrogen and carbon monoxide called synthesis gas. Gas synthesis research has been conducted for quite some time, and the basic reaction formula of general gasification is as follows.

$$Coal \rightarrow H_2 + CmHn + Char(C) \tag{5.21}$$

$$C + O_2 \rightarrow CO_2 \tag{5.22}$$

$$C + 1/2CO_2 \rightarrow CO \tag{5.23}$$

$$C + CO_2 \rightarrow 2CO \tag{5.24}$$

$$C + H\rho \rightarrow CO + H_2 \tag{5.25}$$

$$C + 2H_2O \rightarrow CO_2 + 2H_2 \tag{5.26}$$

$$CO + H_2O \rightarrow CO_2 + H_2 \tag{5.27}$$

$$C + 2H_2 \rightarrow CH_4 \tag{5.28}$$

Eq. (5.21) represents the thermal decomposition of coal; it decomposes into volatile matter such as hydrogen and hydrocarbon gas and char (residual solid content composed of fixed carbon and ash). Eqs. (5.22) and (5.23) indicate combustion and partial combustion by reaction with oxygen, and supply heat of reaction required for gasification. Eqs. (5.24) and (5.25) are main gasification reaction equations and are endothermic reactions. With these reactions, the reactions shown in Eqs. (5.26)–(5.28) also proceed.

The present study will be a stepping stone to further work concerning understanding the effects of hydrogen gas on hydrocarbon flames. An important continuation is the determination of flame properties at pressures relevant to gas turbine combustion.

Hydrogen production by coal gasification is also capable of separating and recovering $CO_2$ at the same time as necessary hydrogen production and is attracting attention as one of key technologies for achieving a hydrogen society.

This technology is currently under consideration for adapting to gas turbines, and it is estimated that this technology will be put into practical use after several years [64].

## Combustion of Mixed Fuel of Methane and Ammonia

Ammonia is attracting attention as a hydrogen carrier, having a large hydrogen content, and it is expected particularly to be used as a power generation fuel. This time we succeeded in gas turbine power generation using a mixed gas of methane and ammonia and showed the possibility of power generation by ammonia mixed firing in a large thermal power plant using natural gas as a fuel. In addition, we succeeded in power generation by 100% ammonia combustion (ammonia dedicated firing) leading to large-scale thermal power generation free of carbon dioxide ($CO_2$). These results can be expected to be put to practical use as a technology, contributing to a significant reduction of greenhouse gas emissions in the power generation field.

Research and development of hydrogen carriers that support massive introduction of renewable energy has been promoted, but ammonia has problems such as igniting less easily than general fuel, and its burning speed is slow. Gas turbine power generation using ammonia as a fuel has not been done so far. Therefore, gas turbine power generation with this 100% ammonia raw material is a very high achievement, and the height of the technology for countering global warming at the Renewable Energy Research Center at AIST Fukushima Center is outstanding [87].

An overview of the micro gas-turbine power generator used for the combustion test is shown in Fig. 5.38.

The power generator is a gas turbine with a rated output of 50 kW. We succeeded in generating 41.8 kW with about 80% power output by cofiring methane-ammonia and ammonia-exclusive firing. In addition, NOx, which is the nitrogen oxide after combustion can be suppressed to less than 10 ppm, was obtained (Fig. 5.39).

Power generation using ammonia as a fuel was a very clean power generation that does not emit carbon dioxide, and it showed an extremely advanced result, demonstrating the possibility of putting it into practical use.

In the ammonia monofuel combustion experiment, the gas turbine was started by feeding kerosene. After that the fuel was transitioned to ammonia monofuel combustion by increasing the introduction amount of ammonia.

When confirming the output at this point, at the rated rotation speed of 80,000 rpm, the power generation output of 41.8 kW was confirmed (Fig. 5.40).

These experimental results indicate that it is possible that the main fuel for thermal power generation can be transferred from natural gas to ammonia.

**FIGURE 5.38** Micro gas-turbine generator that can burn ammonia directly.

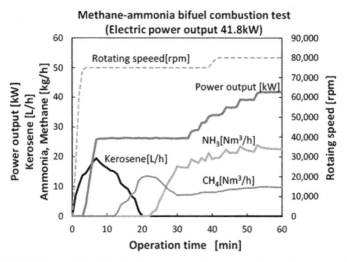

**FIGURE 5.39** Changes in fuel feed and power generation output in the methane-ammonia bifuel combustion experiment.

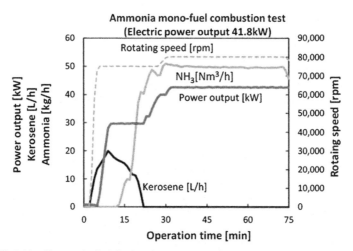

**FIGURE 5.40** Changes in fuel feed and power generation output in the ammonia monofuel combustion experiment.

It suggests that ammonia may be introduced as part of the fuel, or there is a possibility of ammonia burning alone [88].

It shows the possibility that ammonia can become $CO_2$-free power generation fuel, which can greatly reduce greenhouse gas emissions.

## Mixed Combustion of Pulverized Coal and Ammonia

Development of mixed combustion technology of pulverized coal and ammonia that suppresses the emission of nitrogen oxides has been advanced toward the reality of ammonia use in coal-fired power plants.

The Central Research Institute of Electric Power Industry is developing a mixed combustion with pulverized coal, assuming that ammonia is introduced into thermal power generation [89].

When ammonia is used as fuel, NOx as an air pollutant is generated. In order to mix and combust ammonia with pulverized coal, it is necessary to reduce the NOx concentration of the boiler exhaust gas without causing a new cost by large-scale remodeling of the denitration apparatus and the like.

By optimizing the method of introducing ammonia, we succeeded in increasing the fuel input rate of ammonia to 20%. In addition, even when burning in a high-concentration ammonia atmosphere of 20%, the NOx concentration does not rise and only the concentration of about 200 ppm is produced [90] (Fig. 5.41).

In order to clarify the detailed burning behavior of pulverized coal and ammonia mixed fire, spatial resolution photography was realized by high-speed camera and long-range microscope. This is research that can be expected to apply to renewable energy origin hydrogen energy utilization technology.

A research group such as Professor Akamatsu of Osaka University succeeded for the first time in the world to visualize the state of cofiring pulverized coal and ammonia with high spatial resolution. Ammonia is attracting attention as an energy carrier that transports and stores hydrogen efficiently, but research on techniques related to direct use as fuel instead of reconverting to hydrogen is being promoted. Due to research results, the elucidation of the burning process of pulverized coal with a size of several tens of microns, which is less than the thickness of human hair, which was not well understood in the field of combustion engineering, has greatly advanced. The results of this research are to improve combustion efficiency in current thermal power plants and industrial furnaces utilizing pulverized coal. To clean the exhaust gas, combusting the hydrogen produced from ammonia bring hope for the future of pulverized coal. It is also exploited to elucidate the combustion phenomenon.

In the future, if knowledge of large-scale demonstration tests can be obtained, by fusing them with the results of this research we will contribute greatly to the reduction of $CO_2$, a global issue, and the realization of a hydrogen society that determines Japan's future of small energy resources is expected.

In addition, the results of this research are utilized as databases (basic data for verification) for combustion simulation in pulverized coal-fired thermal power boiler combustors and industrial furnaces of actual machines, contributing to

**FIGURE 5.41** Overview of the equipment used for the test. We devised the position where ammonia is introduced and succeeded in reducing the NOx concentration in the exhaust gas. When introducing ammonia from the introduction part of the pulverized coal in the vicinity of 1.0 m, the NOx concentration did not rise, and the NOx concentration was almost the same as in the case of the coal cooking.

reduction of actual machine verification opportunities. It is expected to accelerate development and lower cost. Furthermore, the result of this research is positioned as the first step to clarify the mechanism of reduction of NOx at the time of cocombustion of pulverized coal and ammonia. This technology is expected to greatly contribute to the development not only of the future combustion technology of pulverized coal and ammonia but also the development of the current pulverized coal thermal power generation when adopting various types of coals [91].

## CONCLUSION

Hydrogen energy is indispensable for building a sustainable energy society. Hydrogen power generation uses hydrogen as a fuel for thermal power generation as described earlier. Currently, thermal power generation uses fossil fuels such as petroleum, coal, and natural gas as fuels, but the occurrence of $CO_2$ is inevitable and it is inevitable to become a factor in global warming in the future. Hydrogen power generation has attracted attention in order to avoid this problem.

Hydrogen power generation does not emit $CO_2$ during power generation. However, since hydrogen power generation requires a large amount of hydrogen as a fuel, a large amount of hydrogen production sources are indispensable, and this problem must be solved. In the production from fossil fuels, a large amount of $CO_2$ is discharged at the time of hydrogen production. With hydrogen production using CCS, pure hydrogen can be obtained without discharging $CO_2$ to the atmosphere, but since the area where CCS can be implemented is limited, electricity generation using fossil fuel as a raw material will generate $CO_2$. It is inevitable, and it is necessary to introduce hydrogen in order to promote low carbon emission of exhaust gas. Currently, several thermal power plants remain in the area of mixed combustion called hydrogen introduction into natural gas or coal gas, but the development of a combustor capable of technically difficult hydrogen exclusive firing has been conducted.

Regarding the supply of hydrogen, hydrogen production from renewable energy such as hydroelectric power generation and wind power generation is envisioned in the future, and a region with a huge potential is regarded as a hydrogen production destination candidate. This is thought to contribute greatly from the aspect of energy security, and is expected. Hydrogen power generation and large amounts of hydrogen import from overseas are rare in the world, and if it leads to the massive use of hydrogen, it also affects economic efficiency, and has a great driving force for realizing a hydrogen energy society.

Finally, a triple combined cycle system is possible. This system is a power generation system combining a solid oxide fuel cell, a gas turbine, and a steam turbine. Despite being developed at the moment, expectations are increasing for this system. The reason for this is that the theoretical possibility of the efficiency exceeding 60% is shown. Ultimately, it is expected to be put into practical use as a zero-emission generator by docking with hydrogen power generation as shown in this chapter.

# Application of Hydrogen by Use of Chemical Reactions of Hydrogen and Carbon Dioxide

## SIGNIFICANCE OF CHEMICAL REACTION USING HYDROGEN

It is already clear that there is a high possibility that fossil fuel will not be depleted during this century.

However, it is thought that it is very difficult to prevent the reduction of carbon dioxide ($CO_2$) caused by combustion of fossil fuel, and it is thought that technology to reduce $CO_2$ concentration by using reaction with hydrogen will be important in the future [92].

There are various theories about the rise in temperature due to the increase in $CO_2$ concentration, but in the worst case scenario, it is predicted that the temperature will rise by 5°C or more in 2100. It is hoped that supply by renewable energy will be realized by the middle of the century, but it is also necessary to find a way to reduce $CO_2$.

Hydrogen is one of a few elements that react well with $CO_2$. Therefore, synthesizing a useful chemical substance from the reaction between hydrogen and $CO_2$ is a useful method for this problem [93,94].

From a global viewpoint, research in this field encompasses the following.

- To counter global warming, the reaction between hydrogen and $CO_2$ has attracted attention. This method is effective for $CO_2$ reduction measures.
- The method of synthesizing methanol from hydrogen and $CO_2$ was industrially established, and a demonstration project for producing olefins in a chemical plant was conducted.
- The reaction of synthesizing methane from hydrogen and $CO_2$ has been recently called Power to Gas. Research to produce hydrogen using electricity from renewable energy and to synthesize methane by reaction of hydrogen with $CO_2$ has been conducted.

In this chapter, we will mainly explain methanol synthesis and methane synthesis and show the usefulness of hydrogen.

Science and Engineering of Hydrogen-Based Energy Technologies. https://doi.org/10.1016/B978-0-12-814251-6.00013-7

## METHANOL SYNTHESIS FROM HYDROGEN AND CARBON DIOXIDE

### Methanol Synthesis Reaction Formula From Methane, Water, and Carbon Dioxide

Catalytic steam reforming reaction produces carbon monoxide and hydrogen from methane and water. This reaction is a large endothermic reaction. Industrial hydrogen production is mainly produced by this reaction. This reaction is also used for hydrogen, which is a feedstock of methanol [95].

This steam reforming reaction is shown by Eq. (5.29).

$$CH_4 + 2H_2O \rightarrow CO_2 + 4H_2 - 165 \text{ kJ/mol} \qquad (5.29)$$

The reaction called dry reforming reaction produces carbon monoxide and hydrogen from methane and $CO_2$. Actually, a small amount of steam is introduced into the process, but this reaction is a bigger endothermic reaction than the catalytic steam reforming reaction. This reaction is shown in Eq. (5.30).

$$CH_4 + CO_2 \rightarrow 2CO + 2H_2 - 247.7 \text{ kJ/mol} \qquad (5.30)$$

These two hydrogen production reactions are respectively carried out and combined to give the following reaction. That is, adding Eqs. (5.29) and (5.30) results in Eq. (5.31):

$$3CH_4 + 2H_2O + CO_2 \rightarrow 4CO + 8H_2 \qquad (5.31)$$

This reaction formula shows that the molar ratio of carbon monoxide to hydrogen is 1:2. This molar ratio is the equivalent ratio of the methanol production reaction, which is very convenient for synthesizing methanol.

Methanol is synthesized by applying a Cu/ZnO catalyst to a gas having a molar ratio of carbon monoxide to hydrogen of 1:2. This reaction is shown in Eq. (5.32).

$$CO + 2H_2 \rightarrow CH_3OH \qquad (5.32)$$

This is an overview of methanol synthesis reaction.

### Methanol Synthesis Catalyst and Yield

Catalysts used for synthesizing methanol from hydrogen and carbon monoxide have already been established industrially. The catalyst $Cu/ZnO/Al_2O_3$ is widely used industrially.

Metals with high activity as methanol synthesis catalysts are Cu and Pd. Previous studies have clarified that oxides carrying this metal have high suitability for ZnO, $ZrO_2$, and the like, resulting in high activity.

The reaction of synthesizing methanol from hydrogen and $CO_2$ is represented by Eq. (5.33), but in reality, the reverse shift reaction shown in

Eq. (5.34) also occurs. It is considered that a competitive reaction is occurring, and selectivity of reaction is important.

$$CO_2 + 3H_2 \rightarrow CH_3OH + H_2O + 49 \text{ kJ/mol} \tag{5.33}$$

$$CO_2 + H_2 \rightarrow CO + H_2O - 41.2 \text{ kJ/mol} \tag{5.34}$$

Fig. 5.42 shows the equilibrium yield of methanol and carbon monoxide produced in Eqs. (5.33) and (5.34).

At low temperature and at high pressure, the equilibrium yield of CO decreases, which is a good condition for the purpose of methanol synthesis. This is clear from the graph of the equilibrium yield of methanol shown below Fig. 5.42.

In order to improve the equilibrium yield, it is obvious from the viewpoint of Eq. (5.32) that it is advantageous to operate at low temperature and at a higher pressure.

In fact, low temperatures and high pressures are advantageous to suppress the concentration of carbon monoxide and increase the yield of methanol. However, in order to proceed with the synthesis reaction, it is necessary to increase the temperature, and the temperature of the industrial catalyst layer is set to 250–400°C.

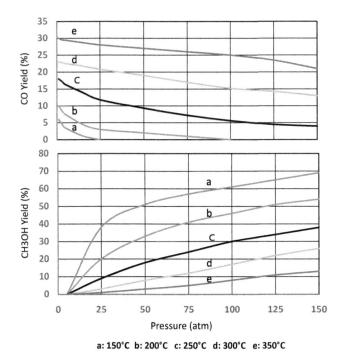

a: 150°C  b: 200°C  c: 250°C  d: 300°C  e: 350°C

**FIGURE 5.42** Equilibrium yield of methanol and carbon monoxide in chemical reactions from Eqs. (5.33) and (5.34).

It is known that the method of synthesizing methanol from $CO_2$ and hydrogen has a low yield with respect to the method from carbon monoxide and hydrogen. Therefore, development of a highly active catalyst is desired. The impregnation method and coprecipitation method are used for preparing methanol synthesis catalyst, but the coprecipitation method is generally used to prepare highly activated catalyst. The reason for choosing this adjustment method is that the content of Cu serving as the active metal can be increased and the Cu metal particles can be made into fine particles.

In the coprecipitation method, sodium carbonate is added to an aqueous solution in which copper nitrate, zinc nitrate, or aluminum nitrate is dissolved to cause precipitation as mixed carbonate, followed by washing, drying, and calcining at 400°C for about 2 h to produce a catalyst.

The activity of the methanol synthesis reaction depends on the surface area of Cu. Therefore, sintering of Cu causes a decrease in activity. When $Al_2O_3$ and $ZrO_2$ are added, the surface area of Cu increases and the activity improves. In order to further increase the surface area, $Ga_2O_3$ or $Cr_2O_3$ is added to the catalyst. It has been found that good activity appears when this method is adopted.

Fig. 5.43 shows data on the life of $Cu/ZnO/Al_2O_3$ and $Cu/ZnO/ZrO_2/Al_2O_3/Ga_2O_3$. This result shows that by complexing the oxide, a catalyst

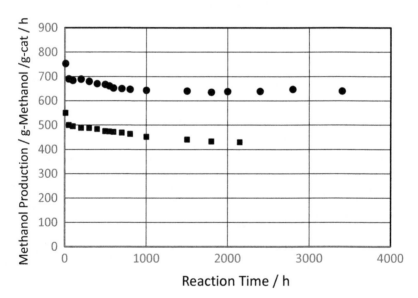

**FIGURE 5.43** Comparison of activity between conventional catalyst and novel composite catalyst. The activity of the conventional $Cu/ZnO/Al_2O_3$ catalyst gradually declines, but the novel $Cu/ZnO/ZrO_2/Al_2O_3/Ga_2O_3$ catalyst has excellent heat resistance and surface area increase. For this reason, the activity is high and stable over a long period of time. This result shows that it is promising as a practical catalyst [97].

having high activity and high heat resistance and long life was developed. As shown in the figure, it shows stable activity and lifespan over 3000 h or more.

## Pilot Plant and Its Results

The demonstration pilot plant process is methanol synthesis from natural gas. The process consists of four sections (pretreatment, synthesis gas production, methanol synthesis, and separation) and a utility section. Each section has the following functions [98].

## Pretreatment

Natural gas contains impurities typified by sulfur components. For this reason impurities must be removed. This method is called hydrodesulfurization. Normally, hydrogen is added to remove the sulfur component, which is called the desulfurization operation.

## Hydrogen Production

Water vapor is added to natural gas to produce supply reformatted gas (CO, $CO_2$, $H_2$) to the methanol synthesis section.

## Methanol Synthesis

After pressurizing the feed gas from the previous step, methanol is synthesized in the reactor. After cooling, crude methanol (methanol and by-product water) and unreacted feed stocks are separated, and unreacted gases are recycled to the reactor.

## Separation

Impurities are removed from crude methanol to obtain a methanol product of a given concentration. Distillation is generally used for separation.

In synthesis gas production, large amounts of steam and fuel are required, and high-temperature waste heat is also generated. Methanol synthesis also requires high pressure conditions and requires power. Therefore, utility equipment that covers these is necessary.

Among these, an example of a process block diagram when combining four section units excluding the utility section is shown in Fig. 5.44.

The reactor was filled with the $Cu/ZnO/ZrO_2/Al_2O_3/Ga_2O_3$ catalyst as described earlier. Methanol production of 50 kg/day was confirmed under the condition of SV=5000 (Space Velocity [L $kg^{-1}h^{-1}$]), pressure 5 MPa, and reaction temperature 250°C. In this examination, stable and satisfactory

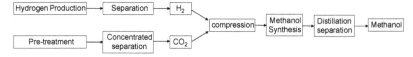

**FIGURE 5.44** Process block diagram of methanol synthesis [99].

results with a methanol conversion rate of 80% or more were obtained over 3000 h [100].

## METHANE SYNTHESIS FROM HYDROGEN AND CARBON DIOXIDE

### Significance of Methanation Reaction

Because excess power of renewable energy is frequently available in several countries, there is method for hydrogen production that uses such secondary energy to store energy as hydrogen. Furthermore, attention has been drawn to a method for producing methane by using hydrogen for reducing $CO_2$, which is a greenhouse gas causative of the global warming problem. This reaction, called Sabatier reaction, is an industrially important reaction for the synthesis of gas, methane production as a hydrogen carrier, and purification of hydrogen for ammonia synthesis.

### Methane Synthesis Reaction

The methanation reaction of $CO_2$ is an exothermic catalytic reaction and is typically operated at temperatures between 150°C and 550°C depending on the catalyst used:

$$CO_2 + 4H_2 \rightarrow CH_4 + 2H_2O + 165 \text{ kJ/mol} \qquad (5.35)$$

A two-step reaction mechanism is assumed for this reaction.

In the first step, $CO_2$ and hydrogen are converted to carbon monoxide and steam via the inverse water gas shift reaction:

$$CO_2 + H_2 \rightarrow CO + H_2O - 41 \text{ kJ/mol} \qquad (5.36)$$

In the subsequent reaction, methane is synthesized from carbon monoxide and hydrogen:

$$CO + 3H_2 \rightarrow CH_4 + H_2O + 206 \text{ kJ/mol} \qquad (5.37)$$

Besides methane and water, also higher saturated hydrocarbons can be found in the products. The most stable $C_{2+}$ hydrocarbon is ethane and it is synthesized according to

$$2CO_2 + 7H_2 \rightarrow C_2H_6 + 4H_2O + 264 \text{ kJ/mol} \qquad (5.38)$$

Under certain process conditions, carbon deposit can occur according to the following reaction mechanism. Such thermodynamics situation must be avoided in industrial operation. It is because the activity of catalyst declines when carbon precipitates.

$$CO_2 + 2H_2 \rightarrow C + 2H_2O + 90 \text{ kJ/mol} \tag{5.39}$$

$$2CO \rightarrow C + CO_2 - 172.5 \text{ kJ/mol} \tag{5.40}$$

Since carbon deposition destroys the catalyst and causes catalyst deterioration, it should be avoided as much as possible.

## Methanation Catalyst

The suitability and effectiveness of a catalyst is determined by the catalyst metal and the support material. The support material consists of a metal oxide because of its large specific surface. A support with a large specific surface area (m$^2$/g-cat) generally promotes the reaction. Nickel is generally used for the methanation of $CO_2$ and hydrogen. It is also possible to use ruthenium (Ru), rhodium (Rh), platinum (Pt), iridium (Ir), cobalt (Co), iron (Fe), or palladium (Pd).

On the basis of the information in this chapter, the following order of selection is used for the methanation of hydrogen and $CO_2$, where the catalyst with the highest specific activity is listed first.

$$Ru > Fe > Ni > Co > Rh > Pd > Pt > Ir$$

However, literature study also presents that the results are slightly different for the methanation for $CO_2$.

The selectivity not only depends on the catalyst material, but is also strongly affected by the support material. A recent literature review proved that the higher methane production is expected from the following precious metal catalysts:

$$Ru > Rh > Pt > Ir$$

Nickel is less selective for hydrogen/ $CO_2$ than ruthenium or rhodium, but it is economic.

The disadvantage of nickel is that at low temperatures ($<250°C$) carbon deposition occurs sooner than precious metal, which reduces the effective surface. Although ruthenium has less carbon deposition, it produces higher hydrocarbon [101]. In general, Ni is industrially used as a catalyst.

Fig. 5.45 shows the equilibrium curve of the methanation conversion. The horizontal axis shows temperature and the vertical axis shows methane concentration. Practice shows what happens in one reactor until equilibrium is achieved. The generated gas is returned to the atmospheric temperature, superheated again, and introduced into the next reactor, and similar equilibrium

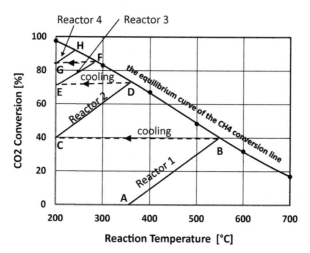

**FIGURE 5.45** Equilibrium curve of the methanation. Example using four reactors. The first reactor is set to the high temperature region, and after the second reactor, the methane concentration is raised by gradually reacting at low temperature. This reaction is an advantageous reaction in terms of equilibrium in the high pressure reaction. Industrially, many plants have four or more reactors installed in series [102].

is reached. In normal design, the methane concentration is increased in four stages.

There are also systems that add further reactors to further improve the methane concentration.

Fig. 5.45 is the operation line of this methanation system. The first reactor, Reactor 1, is located from point A to point B and is a reactor having a temperature difference of 350°C at the inlet temperature and 550°C at the outlet temperature. The outlet temperature becomes high because methanation reaction occurs intensely. At the outlet of the reactor, the concentration of methane is about 40%. Next, cooling is carried out between points B and C, the methanation reaction is carried out again at the Reactor 2 positioned at points C and D, and then cooled. Repeat this process to raise the methane conversion rate. From Fig. 5.45, we can see that the operating temperature of the reactor varies, so the composition of the catalyst packed in the reactor varies for each reactor. This catalyst composition is shown in Table 5.23.

Since Reactor 1 reaches a high temperature of 550°C, the Ni content of the catalyst to be charged is sufficient at about 15%. Also, if the Ni content is more than this, sintering occurs and the activity decreases. Since the maximum temperature of the second positioned Reactor 2 is about 350°C, the Ni content can be increased. For this reason, about 20% of catalyst is used. Further, in the

**TABLE 5.23** An Example of Metal Content of the Methanation Catalyst (wt%)

| Catalyst[a] | Ni/$\gamma$-Al$_2$O$_3$ | Ru/$\gamma$-Al$_2$O$_3$ | Rh/$\gamma$-Al$_2$O$_3$ | Pt/$\gamma$-Al$_2$O$_3$ |
|---|---|---|---|---|
| Reactor-1 | 10–15 | 0.5 | 1 | 1 |
| Reactor-2 | 20–25 | 1 | 2 | 2 |
| Reactor-3 | 30 | 3 | 3 | 3 |
| Reactor-4 | 40 | 5 | 5 | 5 |

[a]Usually the catalyst is pelletized. The noble metal catalyst has an eggshell-type structure [103].

third and fourth reactors, since the reaction temperature is 300°C or lower, the influence of sintering is not considered. In order to operate at low temperature, a catalyst that can maximize activity can be required. In the case of a Ni catalyst, a catalyst having a content of 30%–35% is necessary. Also, as described later, a noble metal catalyst may be used. Table 5.24 lists the temperature condition, inlet gas condition, and outlet gas concentration of this reaction operation curve. It is important to improve the methane concentration by reducing the hydrogen concentration as much as possible.

## Safety and Efficiency for Synthesis System

For construction of industrial systems, the influence of the various process parameters on the conversion from hydrogen into methane and energetic efficiency were studied. The safety risks, requirements for the outgoing gas mixture, and environmental aspects were also considered.

The various safety provisions were also implemented on the system. The required process and gas-quality measurements were also included. Risk decrease took place by means of continuous monitoring of the threshold values of the measured parameters and by switching off the system in a controlled manner when they were not met or exceeded.

The technical feasibility of the complete system and the individual system elements were obtained during the implementation of the practical development. For example, the conversion of electrical energy to hydrogen and oxygen was performed with an energetic efficiency of 50%. The remainder of the energy could not be consumed in the synthesis plant. Heat is also released during the catalytic conversion of hydrogen and $CO_2$, and this could not be consumed either. The energetic efficiency of this methanation process was set at about 70% or more. The energy balance of the entire Power-to-Gas system demonstrated an energetic efficiency of 35%, as represented in Fig. 5.46.

**TABLE 5.24** An Example of Mass Balance of Ideal Four-Stage Series Reactors [104]

|  | Reactor 1 | | Reactor 2 | | Reactor 3 | | Reactor 4 | |
|---|---|---|---|---|---|---|---|---|
|  | Inlet | Outlet | Inlet | Outlet | Inlet | Outlet | Inlet | Outlet |
| $H_2$ (mol/h) | 12 | 2 | 2 | 1 | 1 | 0.5 | 0.5 | 0.1 |
| $CO_2$ (mol/h) | 50 | 10 | 10 | 6 | 6 | 4 | 4 | 3.6 |
| $CH_4$ (mol/h) | 0 | 4 | 4 | 6 | 6 | 7 | 7 | 7.8 |
| $H_2O$ (mol/h) | 0 | 2 | 0 | 1 | 0 | 0.5 | 0 | 0.4 |
| Temperature (°C) | 350 | 550 | 200 | 350 | 200 | 275 | 200 | 230 |
| Pressure (MPa) | 1 | 1 | 1 | 1 | 1 | 1 | 1 | 1 |

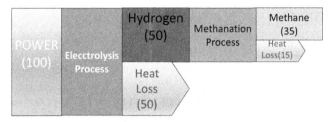

**FIGURE 5.46** Energy balance of power to gas. The conversion of electrical energy to hydrogen and oxygen was performed with an energetic efficiency of 50%. The efficiency of the methanation process was set at about 70% or more. The energy balance of the entire Power-to-Gas system demonstrated an energetic efficiency of 35% [105].

## CONCLUSIONS

In order to reduce $CO_2$, reaction with hydrogen was explained. Technologies for reducing $CO_2$ by methanol and methane synthesis were outlined. As a measure against global warming, effective utilization of $CO_2$ using chemical reactions has been established technically, but it is a field that has not been put to practical use.

For this reason, development will continue in the future. Since methanol synthesis and methane synthesis are restricted in chemical equilibrium, multiple reactors are required to improve the conversion ratio. Especially in methane synthesis, various systems have been devised, but the conventional multistage reaction system is considered to be the most efficient. These processes are considered to be areas that need further development from the viewpoint of energy conservation, and we hope to see further progress in research and development.

Chapter 5.3.4

# Application of Hydrogen Storage Alloys

Hydrogen storage alloys have unique and interesting characteristics such as a higher hydrogen storage density than that of the liquid hydrogen state, and reversible exothermic hydriding and endothermic dehydriding with fast reaction rates. Cyclic hydriding-dehydriding treatment induces pulverization of the alloy because of relatively high volume changes of the alloy by the treatment [106]. The pulverization means the reduction of the alloy particle, resulting in increasing surface area and reaction rate while decreasing heat conductivity among the particles. These phenomena should be taken into account in application [107]. Many interesting phenomena can be observed as surface and volume effects of the alloy reacting with hydrogen [107–112].

The high hydrogen storage capacity of the alloy is applied to the nickel-metal hydride (Ni-MH) rechargeable battery as commercial products. The Ni-MH battery is widely used in various scales. A small size of the Ni-MH battery is used for daily electric commodities, and the larger Ni-MH batteries are used for hybrid vehicles such as the Toyota Prius and Honda Insight. Much larger sizes of the Ni-MH batteries with megawatt (MW) capacity are used for power peak-cut and/or peak-shift of power grid, and indispensable for the efficient use of unstable renewable energy [113].

As is well known, the density of electric capacity of the Ni-MH battery is lower than the lithium (Li) ion battery. However, the Ni-MH battery is not flammable and has a high chemical stability compared with the Li ion battery. This is the main reason that many hybrid vehicles are adopting the Ni-MH battery in order to avoid firing of vehicles in accidents. Use of the Li ion battery is strictly controlled or even inhibited by mailing or air transport, as is well known.

The reversible heat reactions of the alloy can be applied as a metal hydride (MH) heat pump. Unique MH freezer/water cooling systems are applied to the cultivation of strawberry in agriculture and water temperature control in fish breeding. The MH freezer technology is found markedly effective in cutting carbon emission and in saving energy compared with conventional freezing technologies using Freon gas [114].

Science and Engineering of Hydrogen-Based Energy Technologies. https://doi.org/10.1016/B978-0-12-814251-6.00014-9

In this section, typical examples of application of hydrogen storage alloys are demonstrated in terms of the Ni-MH battery and an MH freezer system combined with waste heat.

## NICKEL-METAL HYDRIDE RECHARGEABLE BATTERY

Typical rechargeable batteries for power storage are sodium-sulfur (NaS), redox flow (RF), Li ion and Ni-MH, and lead (Pb) cells. Typical costs per kWh of each battery are about US$400/kWh for NaS, US$ 2000/kWh for Li ion, US$ 1000/kWh for Ni-MH, and US$ 500/kWh for Pb batteries, respectively. No cost for the RF battery is given because large-scale RF batteries are still at the test stage. The RF battery, using vanadium ions for positive and negative electrodes, is attracting a strong battery at US$ 600−700/kWh. NGK Insulators, Ltd. has applied the NaS battery to MW class power storage. The NaS battery is used rather for industrial use because the battery is operated at 573K. For residential use, the Li ion battery with 7 kWh is commercialized. Typical features of the Ni-MH battery, such as high chemical stability and high cyclic charge-discharge durability under high current and load use, are confirmed by many commercialized hybrid vehicles in and outside Japan. The high diffusivity of H atoms inside the negative electrode of hydrogen storage alloys enables the high current density, and the rare earth−based hydrogen storage alloys used for the negative electrode are responsible for high chemical and cyclic stabilities. Large-scale Ni-MH batteries are used for power storage and control in transportation and residential use because of their high safety and reliability under high load. Kawasaki Heavy Industries, Ltd. has developed Ni-MH rechargeable batteries with high capacities for power storage and control (Fig. 5.47).

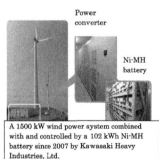

Power converter

Ni-MH battery

A 1500 kW wind power system combined with and controlled by a 102 kWh Ni-MH battery since 2007 by Kawasaki Heavy Industries, Ltd.

**FIGURE 5.47** Applications of the nickel-metal hydride (Ni-MH) rechargeable batteries GIG-ACELL (Kawasaki Heavy Industries, Ltd.) to power storage and control systems for monorail trains of Tokyo Monorail Co., Ltd. (left), and to power smoothing of a wind power system in Akita Prefecture, Japan (right) [115].

Many applications using the Ni-MH rechargeable battery are demonstrating prominent results in transportation use such as light rail vehicles, trains, and hybrid vehicles; and in residential use, such as smoothing of fluctuating Renewable Energy (RE) supply. Many companies are constructing and demonstrating smart communities or towns where the Li ion and the Ni-MH battery are widely used for power storage and grid control.

## APPLICATIONS OF METAL HYDRIDE AS A FREEZER SYSTEM [114,116]

### Operating Principle of a Metal Hydride Freezer [116]

In an MH freezer system, a pair of hydrogen storage alloys for high and low temperatures are used as shown in Fig. 5.48. In this system, waste heat from outside is needed to drive hydrogen in hydrides. A hydride of a high temperature alloy, Ma, desorbs $H_2$ gas by absorbing waste heat Qa at $T_H$. The desorbed $H_2$ gas is transported to another low temperature alloy, Mb, to form a hydride MHb at room temperature $T_M$. This exothermic hydriding reaction releases heat Qb, which is not used in this case. Then MHb desorbs $H_2$ gas by absorbing heat Qb from the surroundings, resulting in reducing temperature of air or water from $T_M$ to a lower temperature. The $H_2$ gas desorbed from MHb is transported back to Ma to form a hydride. By implementing this cyclic $H_2$ gas reaction between the alloys Ma and Mb, the temperature of the surroundings of Mb cools.

Using this system, different types of the MH freezer can be operated in temperature ranges below 243 K, or at 273−278 K, respectively. The temperatures can be selected and adjusted for freezing foods (meats/fishes) or for

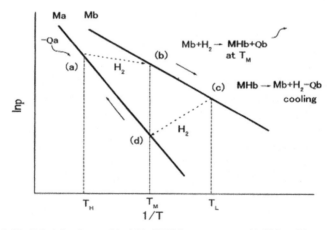

**FIGURE 5.48** Principle of a metal hydride (MH) freezer system with high and low temperature hydrogen storage alloys, Ma and Mb, respectively.

cooling agricultural products such as vegetables and fruits above freezing temperatures. An MH freezer system can be used not only for freezing, but for the production of cold water at temperatures above freezing. The latter temperatures are important to control water temperature for agriculture and fish cultivation. These technologies have been examined in various ways over 10 years at the city of Saijo, Ehime Prefecture, Japan, by the cooperation among Tokai University, Japan Steel Works (JSW), Ltd., the Ministry of Economy, Trade and Industry (METI), and Saijo City, Ehime Prefecture.

Fig. 5.49 shows a plan of an MH freezer system using a high-temperature heat source of waste heat from an industry or an incinerator, and a low-temperature heat source of ground water.

## Hydrogen Storage Alloys for a Metal Hydride Freezer

Fig. 5.50(A) and (B) show the pressure isotherms of the developed hydrogen storage alloys, Ti−Zr−Mn−V−Fe and Ti−Zr−Cr−Fe−Ni−Mn−Cu for high- and low-temperature alloys, respectively. The amount of hydrogen transferred to drive heat reactions was over 100 cc/g-alloy at these conditions: dehydriding at 433 K and 1.0 MPa, and hydriding at 313 K and 0.075 MPa, respectively. The reduction of the maximum hydrogen storage capacity was found to be less than 2% after 10,000 hydriding−dehydriding cycles.

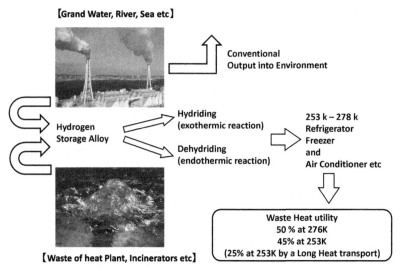

**FIGURE 5.49**  Utilization of high-temperature waste heat sources of industry and ground water, respectively, for the operation of a metal hydride (MH) freezer.

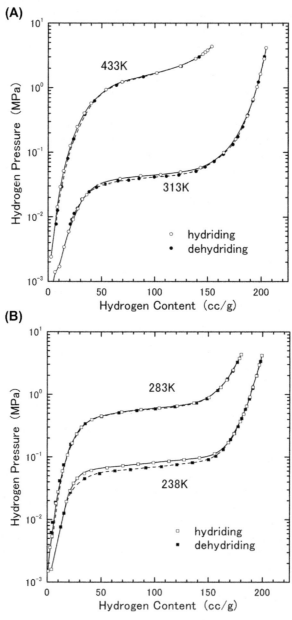

**FIGURE 5.50** Pressure isotherms of the Ti–Zr–Mn–V–Fe (A) and the Ti–Zr–Cr–Fe–Ni–Mn–Cu and (B) for high- and low-temperature hydrogen storage alloys, respectively (JSW, Ltd.).

 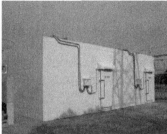

**FIGURE 5.51** A metal hydride (MH) freezer system (left) with control unit and two freezer rooms. Two freezer rooms (right) for 243 K and for 273–278 K with a volume of 67 m$^3$, respectively.

Fig. 5.51 show a whole MH freezer system and two freezer rooms, respectively. Each room has a volume inside 67 m$^3$. The temperature of each room was designed for food preservation.

The amounts of hydrogen storage alloys in MH tanks were 137 kg × 2 and 120 kg × 2 for high- and low-temperature heat reactions, respectively. The time for a cyclic operation of the alloys was around 2000 s. At temperature 283 K outside, the temperature inside a room fell to below 243K within 4 h. Realizing a temperature as low as 243 K, the low temperature alloy is cooled until 290 K using ground water. And then the alloy is cooled further using a hydrocarbon cooler (5.5 kW) until 283 K. Endothermic reactions by dehydriding of the low temperature alloy realizes a freezer temperature of 243 K after these two steps. If a groundwater or cold heat source has a temperature less than 290 K, such an additional cooler is not needed to realize 243 K. The durability of the alloys is very stable at cycles over 30,000–50,000 over 3 years' use.

## Energy Consumption and CO$_2$ Reduction

The energy consumption is less than 30% that of a conventional chlorofluorocarbon-type freezer, because the MH system utilizes high-temperature waste heat from industrial facilities and low temperature of groundwater. The extra energy is required for an additional hydrocarbon cooler and for electric fans to circulate air inside the rooms. This system reduces the CO$_2$ emission over 70% compared with a conventional CFC system. The MH freezer system is a prominent result of eco-technology, or environment-friendly green technology.

This system was also examined to produce cold water for strawberry cultivation and fish breeding. In that system, the effects of CO$_2$ reduction and energy saving were over 80%. Thus the combination of an MH freezer and high temperature waste heat is a prominent way for energy saving and CO$_2$ reduction for freezing/cold food preservation, and environmentally friendly agriculture and fish breeding.

## CONCLUSION

Hydrogen storage alloy has highly interesting features as shown here. Many types of the alloy have been developed so far, however only limited numbers of the alloy are practically used such as rare earth base type $MmNi_5$, magnesium-base type $Mg_2Ni$, and titanium base type FeTi. Some alloys with specific compositions have been developed and used for a certain purpose as shown for the MH freezer system. The cost of the alloys was relatively high, however mass production of the nanostructured FeTi alloy was accomplished, and this reduced the cost by a third compared with those of conventional alloys [117]. This may accelerate use of the alloy for safe hydrogen storage in a large scale; for example, hydrogen fuel stations for fuel cell vehicles and effective control of fluctuating electric generation by renewable energy. Many more extended applications of the alloy are expected.

## CONCLUDING REMARKS

Traditionally, hydrogen production has been using fossil fuels, but in recent years technologies that can be manufactured with renewable energy have been developed and expected to be high. Fossil fuels are used for most of the hydrogen produced, but hydrogen production by renewable energy is expected to develop in the future. Hydrogen production that does not emit carbon dioxide is an important development item because it can reduce the impact on global warming. Hydrogen production including electrolysis will become mainstream in the future.

In addition, demonstration projects of fuel cell vehicles have been carried out many times in Europe, America, and Asia in the 21st century. As a result, fuel cell vehicles were sold and fuel cell buses also began to run on ordinary roads. Production volume of these fuel cell vehicles surely grows, and it seems that it will occupy a part of mainstream automobiles in the future.

In addition, hydrogen refueling stations have been under construction since 2015 in Japan, and now more than 100 hydrogen stations have been built. It is planned that 160 places will be built in 2020 and about 900 stations will be constructed by 2030. Hydrogen energy technology is definitely beginning to penetrate society.

Because concern about global warming due to carbon dioxide has increased, a technique of reacting hydrogen with carbon dioxide is attracting attention as hydrogen utilization technology. Although this is a challenging study, it will be developed as a technology to be incorporated into the chemical process.

Furthermore, the use of hydrogen storage alloy is promoted as an energy stockpile. Hydrogen energy plants using storage alloys have been commercialized in recent years. Hydrogen storage is a technology that enables clean energy on a yearly basis, and it can be regarded as a distinguishing technology.

In this way, hydrogen energy technology is beginning to penetrate into society and has made remarkable progress. In the future, further growth in this field will continue.

## REFERENCES

[1]  M. Asif, T. Muneer, Energy supply, its demand and security issues for developed and emerging economies, Renew. Sustain. Energy Rev. 11 (2007) 1388−1413.

[2]  T. Muneer, M. Asif, J. Kubie, Generation and transmission prospects for solar electricity: UK and global markets, Energy Convers. Manag. 44 (2003) 35−52.

[3]  T.N. Veziroglu, Hydrogen movement and the next action: fossil fuels industry and sustainability economics, Int. J. Hydrogen 22 (1997) 551−556.

[4]  Hydrogen Scaling Up - A sustainable pathway for the global energy transition, Hydrogen Council, November 2017. http://hydrogencouncil.com/wp-content/uploads/2017/11/Hydrogen-scaling-up-Hydrogen-Council.pdf.

[5]  M.A. Pena, J.P. Gómez, J.L.G. Fierro, New catalytic routes for syngas and hydrogen production, Appl. Catal. Gen. 144 (1996) 7−57.

[6]  I. Dincer, C. Zamfirescu, Sustainable hydrogen production, Elsevier (2016).

[7]  J.W. Chun, R.G. Anthony, Partial oxidation of methane, methanol, and mixtures of methane and methanol, methane and ethane, and methane, carbon dioxide, and carbon monoxide, Ind. Eng. Chem. Res. 32 (5) (1993) 788−795.

[8]  S.H.D. Lee, D.V. Applegate, S. Ahmed, S.G. Calderone, T.L. Harvey, Hydrogen from natural gas: part I—autothermal reforming in an integrated fuel processor, Int. J. Hydrogen 30 (2005) 829−842.

[9]  S. Ahmed, D.D. Papadias, S.H.D. Lee, R.K. Ahluwalia, Autothermal and Partial Oxidation Reformer-Based Fuel Processor, Method for Improving Catalyst Function in Autothermal and Partial Oxidation Reformer-Based Processors, United States Patent 20100095590, 2010, p. 487.

[10]  C.H. Bartholomew, Mechanisms of catalyst deactivation, Appl. Catal. 212 (2001) 17−60.

[11]  R.T.K. Baker, Catalytic growth of carbon filaments, Carbon 27 (3) (1989) 315−323.

[12]  M.S. Kim, N.M. Rodriguez, R.T.K. Baker, The interaction of hydrocarbons with copper nickel and nickel in the formation of carbon filaments, J. Catal. 131 (1) (September 1991) 60−73.

[13]  F. Abild-Pedersen, J.K. Nørskov, J.R. Rostrup-Nielsen, J. Sehested, S. Helveg, Mechanisms for catalytic carbon nanofiber growth studied by ab initio density functional theory calculations, Phys. Rev. B 73 (20 March 2006) 115419.

[14]  C.P. Forzatti, L. Lietti, Catalyst deactivation, Catal. Today 52 (1999) 165−181.

[15]  J.R. Rostrup-Nielsen, New aspects of syngas production and use, Catal. Today Vol. 63 (2−4) (2000) 159.

[16]  J.R. Rostrup-Nielsen, T. Rostrup-Nielsen, Large-scale hydrogen production, CATTECH 6 (4) (August 2002) 150−159.

[17]  (a) http://www.jfecon.jp/product/hpc/fcv.html.
      (b) http://www.hydrogencarsnow.com/index.php/hydrogen-fuel-tanks/.

[18]  T. Nejat, Transportation fuel hydrogen, Energy Technol. Environ. 4 (1995) P2712−P2730.

[19]  Grace Odaz et al., Hydrogen Storage: 2005 Annual DOE Hydrogen.

[20]  K.T. Møllera, T.R. Jensen, E. Akiba, H.-W. Li, Prog. Nat. Sci. Mater. Int. 27 (2017) 34−40.

[21]   (a) http://www.jari.or.jp/portals/0/jhfc/station/index.html.
       (b) www.airliquide.com/united-states-america/air-liquide-announces-locations-several-hydrogen-fueling-stations-northeast.
       (c) California Fuel Cell Partnership OEM Advisory Group. Letter to Hydrogen Infrastructure Stakeholders. June 11, 2014. TS. http://cafcp.org/getinvolved/stayconnected/blog/automakers_release_list_station_priority_locations.
       (d) http://media.daimler.com/marsMediaSite/en/instance/ko/First-hydrogen-filling-station-opens-in-Bremen.xhtml?oid=29897600.

[22]   H2-Mobility Hydrogen Station. http://h2-mobility.de/en/h2-stations/.

[23]   https://www.iea.org/publications/freepublications/publication/TechnologyRoadmapHydrogen andFuelCells.pdf.

[24]   K. Mazloomi, C. Gomes, Hydrogen as an energy Carrier: prospects and challenges, Renew. Sustain. Energy Rev. 16 (5) (June 2012) 3024–3033.

[25]   J. Gretz, J.P. Baselt, O. Ullmann, H. Wendt, The 100 MW euro-Quebec hydro-hydrogen pilot project, Int. J. Hydrogen Energy 15 (6) (1990) 419.

[26]   J. Gretz, J.P. Baselt, D. Kluyskens, F. Sandmann, O. Ullmann, Status of the hydro-hydrogen pilot project (EQHHPP), Int. J. Hydrogen Energy 19 (2) (1994) 169.

[27]   K. Jouthimurugesan, A.K. Nayak, G.K. Metha, K.N. Rai, S. Bhatia, R.D. Srivastava, Role of rhenium in Pt-Re-$Al_2O_3$ reforming catalysis—an integrated study, AIChE J. 31 (12) (1985) 1997–2007.

[28]   K. Jouthimurugesan, S. Bhatla, R.D. Srivastava, Kinetics of dehydrogenation of methylcyclohexane over a platinum-rhenium-alumina catalyst in the presence of added hydrogen, Ind.. Eng. Chem. Fundam. 24 (4) (1985) 433.

[29]   R.W. Coughlin, K. Kawakami, A. Hasan, Activity, yield patterns, and coking behavior of Pt and PtRe catalysts during dehydrogenation of methylcyclohexane: I. In the absence of sulfur, J. Catal. 88 (1984) 150.

[30]   W.J. Doolittle, N.D. Skoularikis, R.W. Coughlin, Reactions of methylcyclohexane and n-heptane over supported Pt and PtRe catalysts, J. Catal. 107 (2) (1987) 490.

[31]   M. Markiewicz, Y.Q. Zhang, A. Bösmann, N. Brückner, J. Thöming, P. Wasserscheid, S. Stolte, Environmental and health impact assessment of Liquid Organic Hydrogen Carrier (LOHC) systems – challenges and preliminary results, Energy Environ. Sci. 8 (2015) p1035–1045.

[32]   A. Ozaki, K. Aika, Catalysis, science and technology, in: J.R. Anderson, M. Boudart (Eds.) Vol. 1, Springer Verlag, Berlin, 1981, p. 88 (Chapter 3).

[33]   https://minerals.usgs.gov/minerals/pubs/mcs/2018/mcs2018.pdf.

[34]   S.R. Tennison, Catalytic ammonia synthesis: fundamentals and practice, in: J.R. Jennings (Ed.), Fundamental and Applied Catalysis, Plenum Press, New York, 1991.

[35]   M. Boudart, G. Djéga-Mariadassou, Kinetics of Heterogeneous Catalytic Reactions, Princeton University Press, Princeton, 1984.

[36]   S.R. Tennison, Catalytic ammonia synthesis: fundamentals and practice, in: J.R. Jennings (Ed.), Fundamental and Applied Catalysis, Plenum Press, New York, 1991.

[37]   F.M. Hoffmann, D.J. Dwyer, Surface science of catalysis: in situ probes and reaction kinetics, in: D.J. Dwyer, F.M. Hoffmann (Eds.) Vol. 482, American Chemical Society, Washington, DC, 1991, p. 1 (Chapter 1).

[38]   https://ammoniaindustry.com/download-salient-ammonia-statistics-usgs/.

[39]   H. Yan, Y.-J. Xu, Y.-Q. Gu, H. Li, X. Wang, Z. Jin, S. Shi, R. Si, C.-J. Jia, C.-H. Yan, Promoted multimetal oxide catalysts for the generation of hydrogen via ammonia decomposition, J. Phys. Chem. C 120 (2016) 7685–7696.

[40] A. Boisen, S. Dahl, J.K. Nørskov, C.H. Christensen, Why the optimal ammonia synthesis catalyst is not the optimal ammonia decomposition catalyst, J. Catal. 230 (2) (March 10, 2005) 309–312.

[41] C. Jacobsen, S. Dahl, B.S. Clausen, S. Bahn, A. Logadottir, J.K. Nørskov, Catalyst design by interpolation in the periodic Table: bimetallic ammonia synthesis catalysts, J. Am. Chem. Soc. 123 (34) (2001) 8404–8405.

[42] JHFC was a Demonstration Project Regarding FCVs Implemented From FY2002 to FY2010. http://www.jari.or.jp/Portals/0/jhfc/e/jhfc/index.html.

[43] Toyota Envisions the Fuel Cell Bus of The Future. https://newatlas.com/toyota-sora-fuel-cell-bus-tokyo/51825/.

[44] Toyota Mirai Fuel Cell Vehicle, The Future, 2017. Available from: https://ssl.toyota.com/mirai/fcv.html.

[45] Policy Research Institute for Land, Infrastructure, Transport and Tourism. Ministry of Land, Infrastructure, Transport and Tourism Search Japanese. http://www.mlit.go.jp/pri/english/houkoku/gaiyou/english_kkk59.html.

[46] L. Eudy, M. Post, M. Jeffers, Fuel Cell Buses in U.S. Transit Fleets: Current Status 2016, National Renewable Energy Laboratory, November 2016. Prepared under Task No. HT12.8210, Technical Report, NREL/TP-5400-67097, Contract No. DE-AC36-08GO28308.

[47] The Association of Hydrogen Supply and Utilization Technology HySUT. http://hysut.or.jp/en/index.html.

[48] Hydrogen Station. http://fccj.jp/hystation/.

[49] Hydrogen Station. https://www.navitime.co.jp/category/0818/.

[50] Hydrogen Station. http://toyota.jp/mirai/station/.

[51] Y. Matsuda, et al., JARI Res. J. 32 (7) (2010) 345–348.

[52] ISO FDIS 14687-2. International Standardization for Hydrogen Fuel for Fuel Cell Vehicles.

[53] Hydrogen Fueling Infrastructure Analysis. https://www.nrel.gov/hydrogen/hydrogen-infrastructure-analysis.html.

[54] A California Road Map. http://www.fuelcellpartnership.org/carsandbuses/caroadmap/Overview-of-FCH2-Developments-in-Germany.pdf.

[55] Towards sustainable energy systems —Overview of German FC-Developments http://www.iphe.net/docs/Meetings/SC21/Educational%20Event%20Presentations/Overview-of-FCH2-Developments-in-Germany.pdf.

[56] Scenario for FCV and Hydrogen Station Dissemination. http://fccj.jp/pdf/22_cse.pdf.

[57] UK H2 Mobile "UK H2 Mobile Phase1 Result". http://www.iphe.net/docs/Meetings/SC21/Educational%20Event%20Presentations/Overview-of-FCH2-Developments-in-Germany.pdf.

[58] H2USA Home-Page. http://h2usa.org/.

[59] H2 Mobility Press Release. http://www.now-gmbh.de/en/presse-aktuelles/2013/h2-mobility-initiative.html.

[60] Hydrogen Society in Japan Plane. http://www.meti.go.jp/english/press/2014/0624_04.html.

[61] A.H. Kakaee, A. Paykani, M. Ghajar, The influence of fuel composition on the combustion and emission characteristics of natural gas fueled engines, Renew. Sus. Energ. Rev. 38 (2014) 64–78.

[62] G.A. Richards, M.M. McMillian, R.S. Gemmen, W.A. Rogers, S.R. Cully, Issues for low-emission, fuel-flexible power systems, Prog. Energy Combust. Sci. 27 (2) (2001) 141–169.

[63] R.L. Bannister, R.A. Newby, W.C. Yang, Final report on the development of a hydrogen-fuelde combustion turbine cycle for power generation, J. Eng. Gas Turbines Power 121 (1999) 38–45.

[64] K.J. Bosschaart, L.P.H. de Goey, The laminar burning velocity of flames propagating in mixtures of hydrocarbons and air measured with the heat flux method, Combust. Flame 136 (3) (2004) 261−269.

[65] W. Lowry, J. de Vries, M. Krejci, E. Petersen, Z. Serinyel, W. Metcalfe, H. Curran, G. Bourque, Laminar flame speed measurements and modeling of pure alkanes and alkane blends at elevated pressures, J. Eng. Gas Turbines Power Trans. ASME 133 (9) (2011).

[66] E. Bancalari, I.S. Diakunchak, G. McQuiggan, A review of W501G engine design, development and field operating experience, in: ASME Paper GT 2003−38843, 2003.

[67] MITI, Comprehensive approach to the new sunshine program which supports the 21st century (Agency of Industrial Science and Technology in Ministry of International Trade and Industry [MITI]), Sunshine J. 4 (1993) 1−6.

[68] NEDO, International Clean Energy Network Using Hydrogen Conversion (WE-NET), 1993 Annual Summary Report on Results, New Energy and Industrial Technology Development Organization [NEDO], 1994.

[69] NEDO, Subtask 8 − development of hydrogen-combustion turbine, study for an optimum system for hydrogen combustion turbine, in: 1995 Annual Technical Results Report, New Energy and Industrial Technology Development Organization [NEDO], 1996.

[70] G.A. Geoffroy, D.J. Amos, Four years operating experience update on a coal gasification combined cycle plant with two 100 MW class gas turbines, in: Presented at Combined Heat and Power Independent Producers Conference, Birmingham, England, 1991.

[71] J. Karg, G. Haupt, B. Wiant, IGCC plants provide clean and efficient power using refinery residues and coal, in: Electric Power Conference, Cincinnati, Ohio, 2000.

[72] a N. Kizuka, et al., Conceptual design of the cooling system for 1700°C-class, hydrogen-fueled combustion gas turbines, Trans. ASME 121 (1999) 108−115;
b H. Jericha, O. Starzer, M. Theissing, Towards a Solar-Hydrogen System, in: ASME Cogen Turbo, vol. 6, IGTI, 1991, pp. 435−442.

[73] F. Hannemann, B. Koestlin, G. Zimmermann, H. Morehead, F.G. Peña, Pushing forward IGCC technology at siemens, in: 2003 Gasification Technologies Conference, San Francisco, California, 2003.

[74] http://www.nedo.go.jp/news/press/AA5_100596.html

[75] https://green.blogs.nytimes.com/2009/08/17/a-hydrogen-power-plant-in-italy/.

[76] GE Internal Study, 2007 and EPRI IGCC Design Considerations for $CO_2$ Capture: Engineering and Economic Assessment of IGCC Coal Power Plants for near-term Deployment, 2008.

[77] https://ja.scribd.com/document/46732183/GE-Hydrogen-Fueled-Turbines.

[78] http://www.iea-coal.org.uk/documents/82227/7239/Gas-turbine-technology-for-syngas/hydrogen-in-coal-based-IGCC.

[79] http://www.nedo.go.jp/english/publications_reports_index.html.

[80] http://www.mhi.co.jp/technology/review/indexj-54-3.html.

[81] J. Gao, A. Hossain, Y. Nakamura, Flame base structures of micro-jet hydrogen/methane diffusion flames, Proc. Combust. Inst. 36 (2017) 4209−4216.

[82] K. Kuwana, S. Kato, A. Kosugi, T. Hirasawa, Y. Nakamura, Experimental and theoretical study on the interaction between two identical micro-slot diffusion flames: burner pitch effects, Combust. Flame 165 (2016) 346−353.

[83] K. Zhao, D. Cui, T. Xu, Q. Zhou, S. Hui, H. Hu, Effects of hydrogen addition on methane combustion, Fuel Proces. Technol. 89 (11) (November 2008) 1142−1147.

[84]  Q. Ma, Q. Zhang, Jiachen Chen, Ying Huang, Yuantong Shi, Effects of hydrogen on combustion characteristics of methane in air, Int. J. Hydrogen Energy 39 (21) (July 15, 2014) 11291–11298.

[85]  M. Klelle, H. Eichlseder, M. Sartory, Mixtures of hydrogen and methane in the internal combustion engine – synergies, potential and regulations, Int. J. Hydrogen Energy 37 (15) (August 2012) 11531–11540.

[86]  M.C. Lee, S.B. Seo, J.H. Chung, Si M. Kim, Y.J. Joo, D.H. Ahn, Gas turbine combustion performance test of hydrogen and carbon monoxide synthetic gas, Fuel 89 (7) (July 2010) 1485–1491.

[87]  O. Kurata, N. Iki, T. Matsunuma, T. Inoue, T. Tsujimura, H. Furutani, H. Kobayashi, A. Hayakawa, Performances and emission characteristics of NH$_3$-air and NH$_3$-CH$_4$-air combustion gas-turbine power generations, Proc. Combust. Inst. 36 (2017) 3351–3359.

[88]  Y. Arakawa, A. Hayakawa, K.D.K.A. Somarathne, T. Kudo, H. Kobayashi, Flame characteristics of ammonia and methane flames in a swirl combustor, in: Proceedings of 12th International Conference on Flow Dynamics (12th ICFD), Sendai, Japan, October 27–29, 2015.

[89]  R. Murai, R. Omori, R. Kano, Y. Tada, H. Higashino, N. Nakatsuka, J. Hayashi, F. Akamatsu, K. Iino, Y. Yamamoto, The radiative characteristics of NH$_3$/N$_2$/O$_2$ non-premixed flame in 10 kW test furnace, Energy Procedia 120 (2017) 325–332.

[90]  Mixed combustion of pulverized coal and ammonia, <http://www.denken.or.jp/en/index.html>.

[91]  N. Nakatsuka, J. Fukui, K. Tainaka, H. Higashino, J. Hayashi, F. Akamatsu, Detailed observation of coal-ammonia Co-combustion processes, in: AIChE Annual Meeting, Accepted, 2017.

[92]  J. Nakamura, Ghokubai no Jiten ISBN978-4-254-25242-2 C3558 P545 (2000).

[93]  https://energy.gov/sites/prod/files/Elem_Coal_Studyguide.pdf.

[94]  Canonico, S., Sellman, R., Preist, C., As published in the Proceedings of the 2009 IEEE International Symposium on Sustainable Systems and Technology (ISSST), ISBN 978-1-244-3456-5, IEEE, 2009 Reducing the Greenhouse Gas Emissions of Commercial Print With Digital Technologies.

[95]  J.H. Shinfelt, in: J.R. Anderson, M. Boudart (Eds.), Catalysis, Science and Technology vol. 1, Springer, Berlin, Heidelberg, New York, 1981, p. 257.

[96]  G.C. Chinchen, P.J. Denny, et al., Review: synthesis of ethanol, Part 1 catalysts and kinetics, Appl. Catal. 36 (1988) 1–65.

[97]  T. Chang, R.W. Rousseau, P.K. Kilpatrick, Methanol synthesis reactions: calculations of equilibrium conversions using equtions of states, Ind. Chem. Process Des. Dev. 25 (1986) 477–481.

[98]  K. Klier, et al., J. Catal. 74 (1982) 343.

[99]  H.F. Rase, Hand Book of Commercial Catalysts, Heterogeneous Catalysts, CRC Press, Boca Raton, FL, 2000.

[100]  Patent WO2013183577 A1 PCT/JP2013/065325, Mitsui Chemical. Inc.

[101]  B.B. Pearce, M.V. Twigg, C. Woodwrd, Catalyst Handbook Twigg, Wolfe, London, 1989, pp. 340–378 (Chapter 7).

[102]  T. Inui, Methanation, in: I.T. Horvath (Ed.), Encyclopedia of Catalysis, vol. 5, John Wiley & Sons, New Jersey, 2003, pp. 51–61.

[103]  E. Iglesia, Isotopic and kinetic assessment of the mechanism of reactions of CH$_4$ with CO or H$_2$O to form synthesis gas and carbon on Ni catalysis, J. Catal. 244 (2004) 370–383.

[104] Catalyst for Production of Hydrocarbons, U.S. Patent 4,801,573.

[105] S. Rittmann, A. Seifert, C. Herwig, Essential prerequisites for successful bioprocess development of biological $CH_4$ production from $CO_2$ and $H_2$, Crit. Rev. Biotechnol 35 (2) (Jun 2015) 141–151.

[106] H. Uchida, H.H. Uchida, Y.C. Huang, J. Less-Common Met. 101 (1984) 459–468.

[107] H. Uchida, K. Terao, Y.C. Huang, Z. Phys. Chem. N.F. 164 (1989) 1275–1284.

[108] H. Uchida, Y. Ohtani, M. Ozawa, T. Kawahata, T. Suzuki, J. Less-Common Met. 172–174 (1991) 983–996.

[109] Y. Ohtani, S. Hashimoto, H. Uchida, J. Less-Common Met. 172–174 (1991) 841–850.

[110] S. Seta, H. Uchida, J. Less-Common Met. 231 (1995) 448–453.

[111] H. Uchida, S. Seki, S. Seta, J. Less-Common Met. 231 (1995) 403–410.

[112] H. Uchida, Int. J. Hydrogen Energy 24 (1999) 861–869.

[113] H. Uchida, Japan's energy policy after the 3.11 natural and nuclear disasters – from the viewpoint of the R&D of renewable energy and its current state (chapter 2), in: D. Stolten, V. Scherer (Eds.), Energy Transition to Renewable Energy, Wiley VCH, Germany, June 2013, pp. 109–125. ISBN: 978-7-527-67390-2.

[114] K. Ochi, K. Itoh, H. Uchida, Proc.. The18th WHEC 2010, Essen Germany vols. 78–4, Forschnungszentrum Juelich GmbH, Germany, 2010, pp. 377–384. ISBN: 978-3-89336-654-5.

[115] Kawasaki Heavy Industry, GIGACELL, Railway Battery Power System for Tokyo Monorail, 2012.

[116] H. Uchida, An metal hydride refrigerator using hydrogen storage alloys, Refrigerator (REITOU) 79 (2004) 837–840.

[117] T. Haraki, K. Oishi, H. Uchida, Y. Miyamoto, M. Abe, T. Kokaji, S. Uchida, Int. J. Mater. Res. 99 (2008) 507–512.

Chapter 6

# Regulatory Framework, Safety Aspects, and Social Acceptance of Hydrogen Energy Technologies

Andrei V. Tchouvelev[1], Sergio P. de Oliveira[2], Newton P. Neves Jr.[3]
[1]A.V. Tchouvelev & Associates, Mississauga, Canada; [2]National Institute of Metrology, Quality and Technology, Duque de Caxias, Brazil; [3]H2 Technical Analyses and Expertise in Gases, Capivari, Brazil

## PREAMBLE

The actual engineering utilization of hydrogen energy technologies requires a series of best practices and regulations, codes, and standards (RCS) to be properly followed for safety, efficiency, and sustainability reasons. Metrology is a foundation and guarantor of high-quality technical RCS and is critically essential for conformity assessment of products and services that ensure public safety. In addition to that, the introduction of a new form of energy in the world energy matrix is very much dependent on the social acceptance of new technologies from safety and risk perspective. These themes should be reviewed and specified and have their introduction in the society evaluated by time-independent analyses.

## STAGE SETTING

### Hierarchy of Regulatory Framework—Pyramid of RCS

Visually and simplistically the hierarchy of the regulatory framework of any country on the globe could be presented as a pyramid consisting of three main segments as shown on Fig. 6.1 [1].

So-called "component standards" are the foundation of any national standards system, which in turn is a foundation for a regulatory framework.

Science and Engineering of Hydrogen-Based Energy Technologies. https://doi.org/10.1016/B978-0-12-814251-6.00006-X

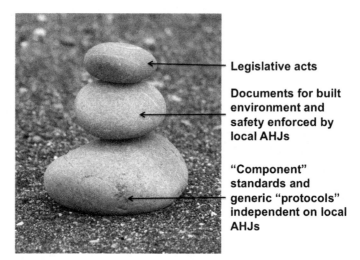

Legislative acts

Documents for built environment and safety enforced by local AHJs

"Component" standards and generic "protocols" independent on local AHJs

**FIGURE 6.1** Hierarchical pyramid of regulations, codes, and standards. *AHJs*, authorities having jurisdiction. *A.V. Tchouvelev, Presentation at the ISO/TC 197 Plenary Meeting in Montreal, February 28, 2013. Stone Pyramid Image, 2013. https://www.123rf.com/profile_snolligoster.*

"Component" is a cumulative generic term for individual components (such as valves and hoses), more complex components (such as a dispenser), and component systems (such as an electrolyzer, reformer, storage unit, and pressure swing adsorption, PSA, for hydrogen purification). "Protocol" is a cumulative generic term for protocols and procedures.

Documents for built environment and safety that are enforced by local authorities having jurisdiction (AHJs) occupy the middle segment of the pyramid. Those normally include installation codes and building and fire codes (sometimes called "model codes") adopted by local regulations and thus have the force of law. More details on the relationship between codes and standards are given in the Best Practices and RCS section.

The top of the pyramid is taken by "legislative acts." Those include a wide variety of legally binding documents issued by local (provincial), national (federal), or regional (such as in the European Union) acts of various levels of government. Such documents may be called "Regulations," "Code of Regulations," "Directives," "Acts," or similar. Such documents may and often do refer to codes and/or standards in regard to specific requirements. Once those codes and standards are explicitly mentioned in a legislative act they are themselves part of the legislative act and have the force of law. Otherwise standards are voluntary compliance documents.

As an illustration of the complexity of interaction between local, regional, and international aspects related to RCS and safety, Fig. 6.2 is a web diagram representing key North American and global entities involved in

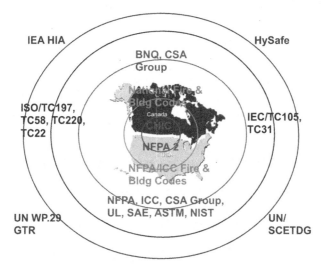

**FIGURE 6.2** Safety regulations, codes, and standards web diagram from North American perspective. Note: Acronyms used in diagram above are as follows:

- CHIC—Canadian Hydrogen Installation Code (Canada)
- NFPA 2—Hydrogen Technologies Code (United States)
- NFPA—National Fire Protection Association (United States)
- ICC—International Code Council (United States)
- BNQ—Bureau de normalisation du Québec (Canada)
- CSA Group—Canadian Standards Association (Canada and United States)
- UL—United Laboratories (United States)
- SAE—Society of Automotive Engineers (United States)
- ASTM—American Society for Testing and Materials (United States)
- NIST—National Institute of Standards and Technology (United States)
- ISO—International Organization for Standardization
- IEC—International Electrotechnical Commission
- TC—Technical Committee (most relevant TCs from ISO and IEC are shown)
- IEA HIA—International Energy Agency Hydrogen Implementing Agreement
- HySafe—International Association for Hydrogen Safety
- UN WP.29 United Nations Working Party 29 that developed GTR #13 (Global Technical Regulation) for hydrogen and fuel cell vehicles (FCVs)
- UN SCETDG—United Nations Sub-Committee on Transportation of Dangerous Goods

*A.V. Tchouvelev, Presentation at the International Energy Agency (IEA) Hydrogen Roadmap North America Workshop in Washington in January 2014, 2014.*

hydrogen and transportation fuel cell safety RCS (including R&D support) development [2].

## BEST PRACTICES AND REGULATIONS, CODES, AND STANDARDS

Best practices and RCS play a critical role in ensuring public safety. It is thus useful to examine the meaning and purpose of each one in the society.

## Best Practices

Here is a sample of available original definitions for "best practices":

*Merriam-Webster Dictionary* (Generic Perspective): "*Best practices: **a procedure that has been shown by research and experience to produce optimal results** and that is established or proposed as a standard suitable for widespread adoption.*"

*Wikipedia* (Generic Perspective): "**A best practice is a method or technique that has been generally accepted as superior to any alternatives because it produces results** *that are superior to those achieved by other means or because it has become a standard way of doing things, e.g., a standard way of complying with legal or ethical requirements.*"

*Investopedia* (Finance Management Perspective): "**Best practices are a set of guidelines, ethics or ideas that represent the most efficient or prudent course of action.** *Best practices are often set forth by an authority, such as a governing body or management, depending on the circumstances. While best practices generally dictate the recommended course of action, some situations require that industry best practices be followed.*"

*Technopedia* (IT Perspective): "**A best practice is an industry-wide agreement that standardizes the most efficient and effective way to accomplish a desired outcome.** *A best practice generally consists of a technique, method, or process. The concept implies that if an organization follows best practices, a delivered outcome with minimal problems or complications will be ensured. Best practices are often used for benchmarking and represent an outcome of repeated and contextual user actions.*"

*US DOE Hydrogen Tools* (Hydrogen Perspective): "**A best practice is a technique or methodology that has reliably led to a desired result.** *Using best practices is a commitment to utilizing available knowledge and technology to achieve success.*"

All above original definitions are quite consistent with each other.

We should just add here that key attributes for safety best practices include safety culture, safety planning, incident procedures, and communications. These attributes will be reviewed in detail in the Safety Best Practices Attributes section.

## Codes and Standards

There are a significant number of generic definitions of codes and standards. To reference the most suitable ones for compressed hydrogen storage and distribution infrastructure application, it is best to check with the organizations that have been developing codes and standards for over a century for pressure vessels (such as American Society of Mechanical Engineers [ASME]) and built environment and fire protection (such as NFPA).

## American Society of Mechanical Engineers Perspective [3].

ASME provides the following definitions for codes and standards.

A standard consists of technical definitions and guidelines that function as instructions for designers/manufacturers and operators/users of equipment. Standards can run from a few pages to a few hundred pages and are written by professionals who serve on ASME committees. *Standards are considered voluntary because they are guidelines and not enforceable by law.*

A code is a standard that has been adopted by one or more governmental bodies and is enforceable by law.

## National Fire Protection Association Perspective [4].

With a reference to NFPA's "A Reporter's Guide to Fire and the NFPA," the earliest building code is thought to have been developed sometime between 1955 BC and 1913 BC, during the reign of King Hammurabi of Babylon. The code did not specify how to build a building—but laid out the consequences of not building well. If a house fell and killed the owner or his child, then the builder, or his child, would be slain in retaliation.

Today's codes are more elaborate and less punitive. However, like Hammurabi's code, they express society's will on a particular technical issue, *specifying a desired outcome*:

- A *code* is a model, a set of rules that knowledgeable people recommend for others to follow. It is not a law but can be adopted into law:
  - Note from the author: that is why often in the United States, codes for built environment such as building and fire codes are called "model codes."
- A *standard* tends to be a more detailed elaboration, the "nuts and bolts" of meeting a code; in other words, it comprises practical aspects rather than general rules.

One way of looking at the differences between codes and standards is that a *code tells you what you need to do* and a *standard tells you how to do it*. A code may say that a building must have a fire alarm system. The standard will spell out what kind of system and how it must work. The NFPA has few codes; most of its documents are standards.

Interesting that ASME emphasizes adoption into law as the main difference between a code and a standard, whereas NFPA underscores the purpose and level of detail: what versus how.

In reality, both organizations are correct: Codes are adopted into laws and regulations, whereas standards are referenced (using US terminology—listing standards) by codes. A sort of paradox is when a standard is mentioned in a code that has been adopted into a regulation, its requirements become mandatory, i.e., enforceable by law, which erases the difference between the two mentioned by ASME above.

The point here is that it is sometimes hard to determine which document is a code or a standard, which often leads to situations when installation codes (such as NFPA 2 or CHIC) get into the territory of component standards and specify requirements that should only be contained in a component standard.

## Requirements for Codes and Standards

The following text (slightly modified and trimmed) from the ASME Codes and Standards Writing Guide referred to earlier is equally applicable to NFPA and any other code or standard within hydrogen and fuel cell space.

To be considered a valid (ASME or any other hydrogen and fuel cell) code or standard, a document should have the following features:

1. *Suitable for repetitive use.* A major requirement of a code or standard is that its set of requirements is not too specialized and thus can be used repeatedly.
2. *Enforceable.* A standard's requirements should be worded so that a person auditing its use or application can point out where it has or has not been followed.
3. *Definite.* Requirements should be expressed as specific instructions and never as explanations.
4. *Realistic.* Requirements that are unrelated, excessive, or more restrictive than necessary should not be included. A standard that is too restrictive or detailed imposes a burden on both the administrator and user. Increasing the severity or detail of a requirement does not automatically increase quality or safety but will nearly always increase cost.
5. *Authoritative.* Requirements should be technically correct and accurate and cover only those properties that are subject to control or are of legitimate use. Requirements should be reasonably consistent with current practices and capabilities in the industry and be attainable for the user.
6. *Complete.* All areas open to question or interpretation (or misinterpretation) should be covered.
7. *Clear.* Express the requirements in easily understood language that is not ambiguous.
8. *Consistent.* Requirements should not be contradictory or incompatible with one another; similarly, the requirements of related and dependent standards should also be consistent.
9. *Focused.* When too much is covered by one standard, its requirements become confusing and the standard becomes less effective; users may be left wondering which parts of the standard apply to their work. When the standard applies to a variety of users, with different requirements, it is often more desirable to provide separate standards.

For completeness of the above picture, it is worth mentioning that SAE (Society of Automotive Engineers) that develops standards for hydrogen fuel cell electric vehicles (FCEVs) often calls standard documents "recommended practices" such as SAE J2578 Recommended Practice for General Fuel Cell Vehicle Safety.

## Safety Best Practices Attributes

The last point in the section above as well as mentioned earlier NFPA's point that codes "express society's will on a particular technical issue, *specifying a desired outcome*" suggest there is a connection between codes and standards and best practices. Often best practices and lessons learned become foundations and important sources of information for codes and standards.

Thus, borrowing NFPA's "nuts and bolts" expression above, best practices could be described as *nonformalized and voluntary field-tested and proven-to-work nuts and bolts of installing and establishing safe operation of equipment or process*.

In the context of this chapter, it is important to highlight safety best practices and provide more details of their key attributes because those are very relevant to the design and operation of the hydrogen and fuel cell facilities. From this perspective, the body of knowledge developed by the US DOE-supported Hydrogen Safety Panel provides the timeless insight into those key attributes [5].

### Safety Culture

An organization's safety culture can be observed in the beliefs and behaviors of its staff members regarding the importance of eliminating or minimizing workplace hazards. Safety culture encompasses the following elements:

- Conducting work safely and responsibly to protect:
  - the health, safety, and welfare of the organization's staff and
  - its equipment and property
- Protecting the health, safety, and welfare of the general public
- Protecting the environment

Only trained and qualified/certified personnel should be allowed to handle compressed hydrogen gas or cryogenic liquid hydrogen.

### Safety Planning

Safety planning should be an integral part of the design and operation of a system. Safety approvals should not be after thoughts, or final hurdles to be overcome before a system can become operational. Initial safety approvals are just that, initial. Safety can only be assured if researchers and users are vigilant in the maintenance of safety.

Safe practices in the production, storage, distribution, and use of hydrogen are essential to protect people from injury or death. A catastrophic failure in any hydrogen project could negatively impact the public's perception of hydrogen systems as viable, safe, and clean alternatives to conventional energy systems.

### Incident Procedures and Communications

An emergency action plan covering incident procedures for the event of a hydrogen release should be developed and implemented to ensure employee safety. Personnel who work with hydrogen should be knowledgeable regarding the properties of hydrogen, the materials and equipment used in hydrogen service, and the correct procedures for responding to potential incidents involving hydrogen.

All safety-related information should be maintained and should be made available to employees. This information can be communicated in a variety of media including staff handbooks, electronic newsletters, and hallway posters.

## KEY RELEVANT GLOBAL STANDARDS DEVELOPMENT ORGANIZATIONS

In the context of this chapter, we will examine the roles and relevance of key global Standards Development Organizations that develop international standards and guidelines for hydrogen technologies, fuel cell technologies, and metrology.

## Importance of Global Standardization and Harmonization—Role of ISO and IEC

Because of their nature, global standard development organizations such as ISO (International Organization for Standardization) and IEC (International Electrotechnical Commission) focus their activities on the development of component standards and generic protocols.

International (ISO and IEC) component standards are being developed to eliminate global barriers to trade. So that a hydrogen component (such as a hose or breakaway device) or an assembly (such as electrolyzer or reformer or dispenser) can meet the same design and testing criteria and thus can be sold across the globe without any such additional requirements. It is important to stress here that installation requirements of those components or assemblies (such as separation distances) can differ from jurisdiction to jurisdiction, but their design and testing requirements should not.

Because ISO and IEC standards are developed by the broadest spectrum of international stakeholders, they become "super" standards and thus should replace any existing similar/analogous national component standards. This consideration leads to at least three following implications:

- National component standards including those that served as seed documents for the development of international standards must be prepared to harmonize their design and testing requirements with the international standards. Essentially, national standards should become harmonized with adopted international standards, where the only deviations are references to specific relevant national standards and regulations and climatic conditions where justified.
- National legislation and installation codes should recognize international standards or their national harmonized adoptions as the only/preferred listing or certification components standards.
- National installation codes should remove any design and testing requirements related to components and assemblies and focus solely on their installation requirements. They should also explicitly reference available international component standards or their national harmonized adoptions for design and testing requirements [6].

## ISO/TC 197 Hydrogen Technologies

The scope of this TC is *"Standardization in the field of systems and devices for the production, storage, transportation, measurement and use of hydrogen"*.

The secretariat of this TC is held by the Standards Council of Canada.

ISO/TC 197 is composed of 20 participating countries, including active participation from all of the G7 countries, as well as participation from China, Korea, India, Russia, etc. as shown in Fig. 6.3 [7]. In combination with observing members (13), ISO/TC 197 global participation covers most of the biggest world economies.

Driven by the need for global relevance to support the deployment of FCEVs, ISO/TC 197 has been developing many international standards related to the hydrogen refueling infrastructure as well as critical components of the compressed hydrogen storage system on-board hydrogen-fueled vehicles. The standards being developed by ISO/TC 197 will contribute to the harmonization of safety aspects and equipment requirements around the globe.

The centerpiece of its work program shown in Fig. 6.4 [8] is the so-called fueling family that includes all standards directly related to hydrogen fueling stations. Many of the standards that are being developed by ISO/TC 197 have direct relevance to the GTR #13, a global technical regulation on hydrogen and FCVs.

## ISO/TC 197 Global Participation

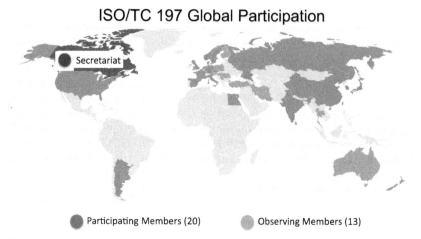

FIGURE 6.3   ISO/TC 197 global participation. *Compiled by A.V. Tchouvelev from the material on the ISO website.*

## ISO/TC 197 Work Program

| Fuel Quality WG 27 ISO 14687 Rev | Active Collaboration CEN/CENELEC TC 268/WG5, TC 6 |
| --- | --- |

**Vehicle Components**

**Fueling Connectors** WG 5 **ISO 17268 Rev**

**On-board Storage** WG 18 **ISO 19881**

**TPRD** WG 18 **ISO 19882**

**Fueling Family ISO 19880**

-1: **HFS Gen'l Req'ts** WG 24
-2: **Dispensers** WG 19
-3: **Valves** WG 20
-4: **Compressors** WG 21
-5: **Hoses** WG 22
-6: **Fittings** WG 23
-7: **Reserved**
-8: **Fuel Quality Control** WG 28

**Storage Technologies**

**GH₂ Ground Storage** WG 15 **ISO 19884**

**Me-Hy Portable Storage** WG 25 **ISO 16111 Rev**

**Electrolysers** WG 26 **ISO 22734 Rev**

FIGURE 6.4   ISO/TC 197 work program. *A.V. Tchouvelev, Presentation at the ISO/TC 197 Strategic Planning Meeting, December 6, 2017, 2017.*

An essential component of ISO/TC 197 activities is the collaboration with European CEN/CENELEC TCs under the umbrella of the Vienna Agreement (see box below), which achieves two important objectives. The first one is a streamlined mechanism for ISO/TC 197 standards adoption within the European Union (via CEN/TC268/WG5 that acts as a CEN ISO/TC 197 mirror committee). The most recent example of such collaboration was the adoption of ISO/TC 197 work products onto European standards related to the deployment of hydrogen fueling infrastructure per EU Alternative Fuel Infrastructure Directive that came in full force in November 2017.

The second objective is the ability to actively participate (via CEN/ CENELEC TC 6 Hydrogen in energy systems) in the international standardization activities that do not directly fit into the scope of ISO/TC 197 yet are of significant interest to ISO/TC 197 stakeholders. Such collaboration includes, but is not limited to, such areas as power-to-gas/power-to-hydrogen, multifuel stations, broad technical terminology for energy systems, certification of green hydrogen, and others.

**Vienna Agreement**

The Agreement on technical cooperation between ISO and CEN (Vienna Agreement) is an agreement on technical cooperation between ISO and the European Committee for Standardization (CEN). Formally approved on June 27, 1991, in Vienna by the CEN Administrative Board following its approval by the ISO Executive Board at its meeting on May 16 and 17, 1991, in Geneva, it replaced the Agreement on exchange of technical information between ISO and CEN (Lisbon Agreement) concluded in 1989. The "codified" Vienna Agreement was approved by ISO Council and the CEN Administrative Board in 2001.

The main objective of the Vienna Agreement is to ensure we make the best use of the resources available for standardization. It helps ISO and CEN exchange information and increases the transparency of CEN work to ISO members as well as helping to make sure work does not have to happen twice at the regional or international level.

Where an International Standard is simultaneously approved as a European Standard, it automatically becomes a national standard for all CEN members.

However, the Agreement also recognizes that the Single European Market may have particular needs, for example:
- standards for which there is no international need currently recognized;
- standards that are required urgently in the European Union but that have a lower priority at the international level.

In these cases the Agreement therefore permits ISO committees to request that work being carried out within CEN, which answers the specific needs of the Single European Market, be made available for voting and comment by all ISO member bodies at the enquiry and formal approval stages. This allows non-European ISO members to influence the content of European Standards and where appropriate to approve those standards as International Standards [9].

## *Standardization of Fuel Cells at IEC*

Founded in 1906, the IEC is the world's leading organization for the preparation and publication of International Standards for all electrical, electronic, and related technologies. These are known collectively as "electrotechnology." Millions of devices that contain electronics, and use or produce electricity, rely on IEC International Standards and Conformity Assessment Systems to perform, fit, and work safely together.

One of the IEC's TC has a relevant role in the hydrogen energy and fuel cell technologies: IEC/TC 105 Fuel Cells. Its stated scope is: "*to prepare international standards regarding fuel cell (FC) technologies for all FCs and various associated applications such as stationary FC power systems, FCs for transportation such as propulsion systems, range extenders, auxiliary power units, portable FC power systems, micro FC power systems, reverse operating FC power systems, and general electrochemical flow systems and processes.*"

By the end of 2017, it had 17 participating members and 15 observer members [10].

Some insight on the IEC/TC 105 work program can be gained from Table 6.1. The standardization interests of individual projects can be grouped into the following areas: all types of fuel cells safety and performance, use of reversible fuel cells for energy storage, and fuel cells environmental performance—based life cycle thinking.

## Role of Metrology for RCS Quality and Public Safety

As the reader is well aware, the field of hydrogen and fuel cell technologies involves a significant number and variety of measurements. Hence, RCS in this highly scientific and technical field are technical documents containing a variety of numbers and quantities, i.e., variables in measurable units. Their (RCS) quality thus relies on the quality and reliability of metrology, which is defined as "*the science of measurement, embracing both experimental and theoretical determinations at any level of uncertainty in any field of science and technology*" [11]. It establishes a common understanding of units, crucial to human activity [12].

### Metrology Facts

Metrology has become a natural and vital part of everyday life in modern society. In Europe, for example, the cost of doing measure and weigh operations is equivalent to 6% of EU-combined GNP.

In industry, processes are regulated and alarms are set off because of measurements. Systematic measurement with known degrees of uncertainty is one of the foundations of industrial quality control and, generally speaking, in most modern industries, the costs bound up in taking measurements constitute 10% −15% of production costs. Good measurements can, however, significantly increase the value, effectiveness, and quality of a product.

**TABLE 6.1** IEC/TC 105 Work Program

| # | IEC Project | IEC Title of Project |
|---|---|---|
| 1 | IEC 62282-2-100 ED1 | Fuel cell technologies—Part 2-100: Fuel cell modules—Safety |
| 2 | IEC 62282-2-201 ED1 | Fuel cell technologies—Part 2-201: Fuel cell modules—Performance (PEFC) |
| 3 | IEC 62282-3-100 ED2 | Fuel cell technologies—Part 3-100: Stationary fuel cell power systems—Safety |
| 4 | IEC 62282-5-100 ED1 | Fuel cell technologies—Part 5-100: Portable fuel cell power systems—Safety |
| 5 | IEC 62282-6-101 ED1 | Fuel cell technologies—Part 6-101: Micro fuel cell power systems—Safety—General requirements |
| 6 | IEC 62282-6-400 ED1 | Fuel cell technologies—Part 6-400: Micro fuel cell power systems—Power and data interchangeability |
| 7 | IEC 62282-8-101 ED1 | Fuel cell technologies—Part 8-101: Energy storage systems using fuel cell modules in reverse mode—Test procedures for solid oxide single cell and stack performance including reversing operation |
| 8 | IEC 62282-8-102 ED1 | Fuel cell technologies—Part 8–102: Energy storage systems using fuel cell modules in reverse mode—PEM single cell and stack performance including reversing operation |
| 9 | IEC 62282-8-201 ED1 | Fuel cell technologies—Part 8-201: Energy storage systems using fuel cell modules in reverse mode—Power-to-power systems performance |
| 10 | IEC TS 62282-9-101 ED1 | Fuel cell technologies—Part 101: Evaluation methodology for the environmental performance of fuel cell power systems based on life cycle thinking—Streamlined life cycle considered environmental performance characterization of stationary fuel cell power systems for residential applications |
| 11 | IEC TS 62282-9-102 ED1 | Fuel cell technologies—Part 102: Evaluation methodology for the environmental performance of fuel cell power systems based on life cycle thinking—Product category rules for environmental product declarations of stationary fuel cell power systems and alternative systems for residential applications |

IEC website/TC 105.

## Categories of Metrology

Metrology is separated into three categories with different levels of complexity and accuracy

*Scientific metrology* deals with the organization and development of measurement standards and with their maintenance (highest level, SI basic and derived units). Scientific metrology is divided into nine technical subject fields by BIPM (The International Bureau of Weights and Measures): acoustics, ultrasound, and vibration; amount of substance (metrology in chemistry and biology); electricity and magnetism; ionizing radiation; length; mass and related quantities; photometry and radiometry; thermometry; time; and frequency:

1. All of these fields are relevant to hydrogen and fuel cell technologies.
2. *Industrial metrology* encompasses the adequate functioning of measurement instruments used in industry and academic research for ensuring quality of many industrial, research, and quality of life-related activities and processes. This includes the need to demonstrate traceability, which is becoming just as important as the measurement itself.
3. *Legal metrology* "concerns regulatory requirements of measurements and measuring instruments for the protection of health, public safety, the environment, enabling taxation, protection of consumers, and fair trade" [13]. One practical example of legal metrology is the assurance that a hydrogen fuel dispenser at a public fueling station accurately tracks the hydrogen fuel mass transferred to a customer's hydrogen-fueled vehicle, thus ensuring proper customer monetary charging.

## Key Relevant Attributes of Metrology

A number of attributes of metrology are particularly important for the hydrogen and fuel field and thus needed to be mentioned specifically: uncertainty, testing, traceability, and calibration [14].

*Uncertainty* is a quantitative measure of the quality of a measurement result, enabling the measurement results to be compared with other results, references, specifications, or standards.

All measurements are subject to error, in which the result of a measurement differs from the true value of the measured object. Given time and resources, most sources of measurement errors can be identified: Systematic measurement errors can be quantified and corrected for, for instance, through repair, adjustment, and/or calibration, whereas random errors are taken into account in the estimation of measurement uncertainty. There is, however, seldom time or resources to determine and correct completely for these measurement errors.

Measurement uncertainty can be determined in different ways. A widely used and accepted method, e.g., accepted by the accreditation bodies, is originally published by ISO GUM method and then republished with minor modifications by the Joint Committee for Guides in Metrology [15]. Monte Carlo numerical simulations have also been used on the estimation of uncertainties in measurement including the application to fuel cell efficiency [16].

*Testing* is the determination of the characteristics of a product, a process, or a service, according to certain procedures, methodologies, or requirements. The aim of testing may be to check whether a product fulfills specifications such as safety or performance requirements. Testing in general is carried out widely, covers a range of fields, and takes place at different levels and at different requirements of accuracy.

A necessary approach to performing experimental or conformance test measurements with quality and reliability can be highlighted by a diagram sequence shown in Fig. 6.5 [17].

Metrology delivers the basis for the comparability of test results, e.g., by defining the units of measurement and by providing *traceability* and associated uncertainty of the measurement results.

A *traceability* chain shown in Fig. 6.6 [16] is an unbroken chain of comparisons, all having stated *uncertainties*. This ensures that a measurement result or the value of a standard is related to references at the higher levels, ending at the primary standard.

A basic tool in ensuring the traceability of a measurement is the *calibration* of a measuring instrument, measuring system, or reference material. Calibration determines the performance characteristics of an instrument, system, or reference material. It is usually achieved by means of a direct

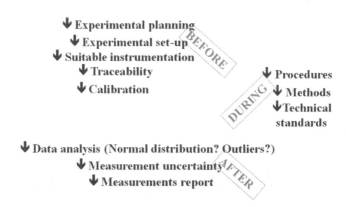

**FIGURE 6.5** Experimental or conformance testing sequence. *From S.P. Oliveira, Metrology and quality in measurements of fuel cells variables, Presentation at 1st Conference Cycle: Hydrogen and Sustainable Energetic Future of Ceara State. Fortaleza, Brazil, March 15–16, 2011.*

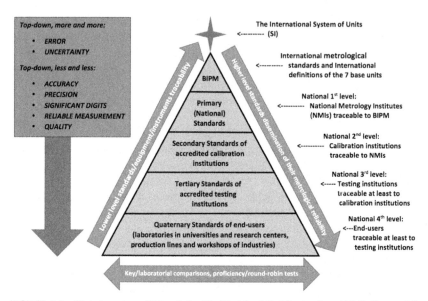

**FIGURE 6.6** Metrology traceability chain. *Modified by S.P. Oliveira from P.R.G. Couto, J.C. Damasceno, S.P. Oliveira, Chapter 2 Monte Carlo simulations applied to uncertainty in measurement, in: V.(W.K.) Chan (Ed.), Theory and Applications of Monte Carlo Simulations, INTECH, (March 6, 2013) (978-953-51-1012-5 under CC BY 3.0 license), https://doi.org/10.5772/53014.*

comparison against measurement standards or certified reference materials (CRMs). A calibration certificate is issued, and in most cases, a sticker is provided for the instrument.

Four main reasons for having an instrument calibrated:

1. to establish and demonstrate traceability;
2. to ensure readings from the instrument are consistent with other measurements;
3. to determine the accuracy of the instrument readings;
4. to establish the reliability of the instrument, i.e., that it can be trusted.

## Metrological International Infrastructure

### The Metre Convention

In the middle of the 19th century, the need for a universal decimal metric system became very apparent, particularly during the first universal industrial exhibitions. In 1875, a diplomatic conference on the meter took place in Paris where 17 governments signed the diplomatic treaty "*the Metre Convention.*" The signatories decided to create and finance a permanent scientific institute:

the "Bureau International des Poids et Mesures" (BIPM). The Metre Convention was slightly modified in 1921.

Representatives of the governments of the member states meet every fourth year for the "Conférence Générale des Poids et Mesures" (CGPM) or General Conference on Weights and Measures. The CGPM discusses and examines the work performed by National Metrology Institutes (NMIs) and the BIPM and makes recommendations on new fundamental metrological determinations and all major issues of concern to the BIPM.

As of August 17, 2016, the General Conference membership was made up of 58 Member States, 41 Associate States and Economies, and 4 international organizations [18].

CGPM elects up to 18 representatives to the "Comité International des Poids et Mesures" (CIPM) or International Committee for Weights and Measures, which meets annually. The CIPM supervises BIPM on behalf of the CGPM and cooperates with other international metrology organizations.

## Key Relevant Organizations

The *CIPM Mutual Recognition Arrangement*, CIPM MRA, is an agreement between NMIs (see below). It was signed in 1999, slightly revised on some technical points in 2003, and has two parts. One part relates to the establishment of the degree of equivalence of national measurement standards, where the second part concerns the mutual recognition of calibration and measurement certificates issued by participating institutes. The CIPM MRA does not extend or replace any part of the Metre Convention and is a technical arrangement between the directors of the NMIs, not a diplomatic treaty.

The objectives of the CIPM MRA are:

- to establish the degree of equivalence of national measurement standards maintained by NMIs;
- to provide for the mutual recognition of calibration and measurement certificates issued by NMIs;
- thereby to provide governments and other parties with a secure technical foundation for wider agreements related to international trade, commerce, and regulatory affairs.

An NMI is an institute designated by national decision to develop and maintain national measurement standards for one or more quantities. Examples of NMIs include, but not limited to, NIST (USA), NPL (UK), LNE (France), and PTB (Germany).

An NMI's participation in the CIPM MRA enables national accreditation bodies and others to be assured of the international credibility and acceptance of the measurements the NMI disseminates. It therefore also provides the basis for the international recognition of the measurements made by accredited testing and calibration laboratories, provided that these laboratories can demonstrate competent traceability of their measurements to a participating NMI.

As of 2018, the CIPM MRA has been signed by the representatives of 102 institutes—from 57 Member States, 41 Associates of the CGPM, and 4 international organizations—and covers further 156 institutes designated by the signatory bodies [19]. Over 90% of world trade in merchandise exports is between CIPM MRA participant nations.

The *International Organization of Legal Metrology* (OIML) is an intergovernmental treaty organization established in 1955 on the basis of a convention. The purpose of OIML is to promote the global harmonization of legal metrology procedures that underpin and facilitate international trade. Such harmonization ensures that certification of measuring devices in one country is compatible with certification in another, thereby facilitating trade in the measuring devices and in products that rely on the measuring devices.

The OIML issues international recommendations that provide members with an internationally agreed basis for the establishment of national legislation on various categories of measuring instruments. The OIML works closely with other international organizations such as the BIPM and ISO to ensure compatibility between each organization's work.

OIML draft recommendations and documents are developed by TCs and subcommittees composed of representatives from member countries. In 2018, OIML had 18 TCs. Most relevant to the hydrogen technologies is TC 8 Measurement of quantities of fluids, SC 7 Gas metering. This subcommittee has published and is currently revising two critical documents that will include requirements for hydrogen dispensing to hydrogen-fueled vehicles: R139:2014 Compressed gaseous fuel measuring systems for vehicles and R140:2007 Measuring systems for gaseous fuel.

In 2018, OIML had 62 member countries and 66 corresponding member countries that joined the OIML as observers.

---

**Metrology Impact—Example of Natural Gas Measurements [12]**

Measuring the value of natural gas must be uniform and reliable throughout the world to protect consumers and fiscal revenue. For example, North America and EU have over half a billion consumers of natural gas combined, supplied from millions kilometers of pipelines. Thus their annual consumption is worth many hundreds of billions of dollar and euros, respectively.

Natural gas is an expensive commodity that is traded across the globe and is subject to fiscal charges, so it is important that consumers, importing/exporting countries, and tax authorities can trust that the measurements made are fair, consistent, and reliable.

Payment of gas is made according to the volume and calorific value of the gas, which is determined by the composition of the gas. Gas chromatography is used to measure the composition of the gas, and the measurements are complex. Measurements are made in many places on the gas grid on a daily, weekly, monthly, and annual basis using a gas chromatograph. Calculation of the calorific value is

**Metrology Impact—Example of Natural Gas Measurements [12]—cont'd**

done automatically in the gas chromatograph according to international technical standards.

The calibration of the gas chromatograph is performed using a gas CRM, which is traceable to a CRM calibrated by an NMI. Under the CIPM MRA, all participating NMIs and designated institutes are obliged to submit their calibration and measurement capabilities and quality systems for peer review and to participate in appropriate key comparisons.

These arrangements and the reviews, and practical measurements and comparisons that underpin them provide confidence in the accuracy and transparency of this commodity trading and transfer across borders.

## Metrological Units

The idea behind the metric system—a system of units based on the *meter* and the *kilogram*—arose during the French Revolution when two platinum artifact reference standards for the meter and the kilogram were constructed and deposited in the French National Archives in Paris in 1799—later to be known as the Metre of the Archives and the Kilogram of the Archives. The French Academy of Science was commissioned by the National Assembly to design a new system of units for use throughout the world, and in 1946 the MKSA system (*meter, kilogram, second, ampere*) was accepted by the Metre Convention countries. In 1954, the MKSA was extended to include the *kelvin* and *candela*. The system then assumed the name the International System of Units, SI (Le Système International d'Unités).

The SI system was established in 1960 by the 11th General Conference on Weights and Measures, CGPM.

At the 14th CGPM in 1971, the SI was again extended by the addition of the *mole* as base unit for amount of substance. The SI system now comprises *seven base units*, which together with derived units make up a coherent system of units. In addition, certain other units outside the SI system are accepted for use with SI units [12].

### SI Units and Prefixes to Express Concentrations of Gases

Table 6.2 [20] presents the units and common SI prefixes used to express concentration of gases. In general, the calculations are facilitated with those units. A useful format to work in extensive concentration ranges, from traces to percentage values, is 0.0000% mol/mol or 0.0000% molar fraction, where the last decimal figure represents $\mu$mol/mol.

Although $Nm^3$ and Nl, normal cubic meter and normal liter, appear regularly in literature as standard units to express gas volume, their use should be discontinued because of strong arguments. First, in SI the symbol N is reserved for the unit of force, newton. Originally, normal temperature and pressure refer to

**TABLE 6.2** Recommended Units to Express Concentrations of Gases

| Not Recommended | | | Recommended | |
|---|---|---|---|---|
| Name | Symbol | Value | Symbol | Name |
| percent | %, %v/v | $10^{-2}$ | %mol/mol | percent mol per mol |
| parts per million | ppm, ppmv, ppv | $10^{-6}$ | µmol/mol | micromol per mol |
| parts per billion | ppb | $10^{-9}$ | ηmol/mol | nanomol per mol |
| parts per trillion | ppt | $10^{-12}$ | pmol/mol | picomol per mol |

N.P. Neves Jr., International Conference on Hydrogen Safety 2011, 2011. http://conference.ing.unipi.it/ichs2011/.

**FIGURE 6.7** Metrology is precursor of high-quality RCS and guarantor of public safety. *Modified by A.V. Tchouvelev from S.P. Oliveira, Metrology and quality in measurements of fuel cells variables. Presentation at 1st Conference Cycle: Hydrogen and Sustainable Energetic Future of Ceara State. Fortaleza, Brazil, March 15–16, 2011.*

273.15 K and 101 325 Pa. However, today IUPAC (the International Union of Pure and Applied Chemistry) adopts 273.15 K and 100 kPa as the "standard conditions for gases," and there are many institutions defining "normal" or "standard" conditions differently. To overcome this difficulty, we recommend to use $m^3$ (or L) and inform in the text, just once, that the volumes were measured at 273.15 K and 100 kPa; otherwise one should inform the actual conditions. Additionally, a good practice to avoid problems with quantification of gases is to use units of mass (g, kg, t) instead of units of volume.

## Metrology Conclusion

In conclusion, metrology is an essential precursor and guarantor of high-quality RCS as well as conformity assessment of products and services that ensure public safety. Graphically this thought can best be expressed as shown in Fig. 6.7 [17].

## SAFETY, RISK, AND PUBLIC ACCEPTANCE

### Safety and Risk Concepts and Definitions

Explaining a connection between safety and risk can be tricky. The reader may be aware from the personal experience that when people are asked to explain what the terms "safety" and "safe" actually mean, they are often lost for the right words. The answer comes the ISO/IEC Guide 51 that contains a definition for "safety" stating *"safety is freedom from unacceptable risk"* [21]. This definition became an internationally accepted paradigm, which effectively means that:

- risk is the measure of safety;
- society accepts the fact that there is neither absolute (i.e., 100%) safety nor zero risk;
- society, de facto, establishes acceptable levels of risk [22].

Following this thread of thought, we arrive to a conclusion that "safety" is a *societal construct* by nature and thus cannot be calculated. As an example below will show, it varies per societal needs. "Safety" thus can only be measured through "risk," which is a *technical construct* and can be calculated.

---

### Definitions of Risk

Risk's quantitative nature "is said to be rooted in the Hindu-Arabic numbering system" [23].

Generically speaking, risk is a quantitative evaluation of *potential losses and damages* that can affect individuals (*individual risk*), populations (*societal risk*), equipment, buildings, businesses (*business and financial risks*), and the environment (*environmental risk*) in the case of an event such as accidents, equipment or operational failures, natural causes, or specific human actions, each of those having their own *probabilities*.

That is why informally, risk is understood as the *probability of realization of a hazard* (which in turn is defined as the *potential source of harm* [21]).

There are two official definitions of risk offered by ISO/IEC Guides:
- Combination of the probability of occurrence of harm and the severity of that harm [21], and
- The effect of uncertainty on objectives [24].

The first one is more technical and comes from a safety perspective; the second attempts to reflect a more generic societal perspective.

In view of the above, ISO/TC 197, in its standard for hydrogen fueling stations, adopted the following definition of risk: *"combination of the probability of occurrence of harm and the severity of that harm; encompassing both the uncertainty about and severity of the harm"* [25], which is essentially the combination of the two ISO/IEC definitions with the emphasis on public safety from technical standpoint.

In practice, risk is expressed in terms of a ratio between quantities, or a frequency, although in this case there are always "hidden units." For example, lightning kills 132 people per year, on average, in Brazil from 2000 to 2009 (number of deaths per year) or $7 \times 10^{-6}$ deaths per inhabitant per year [26]; on average, 48 people were hurt by a gas station fire each year, and there were two fatalities per year by the same cause, in the United States between 2004 and 2008 [27]; the fatality risk of a pedestrian crossing a road is 1 in 300 million per crossing, or $3.3 \times 10^{-9}$, estimated for Great Britain for 2013 [28].

Symbolically the relation between safety and risk can be visualized as the mythical Roman Janus Bifrons [29], a god of beginnings on one side and a demon on the other.

A technical visualization of this relation can be presented as communicating vessels: A higher established level of safety means a lower acceptable level of risk. When the level of safety drops, the risk gets higher and vice versa [30]. Both images are shown side by side in Fig. 6.8.

Historically, the term "risk" comes from Italian "*risicare*," which means "*to dare*." In this sense, *risk is a choice rather than a fate* [23]. This illustrates the societal side of risk (or rather safety). An RSA Risk Commission report investigating the risks associated with childhood points out that "*risk and uncertainty are key features of contemporary society*" adding that "*being at risk unintentionally is not the same as knowingly taking risks*" [31]. The last thesis is of particular importance for setting risk criteria for various stakeholder groups (or so-called *parties*) within the society.

### Risk Acceptance Criteria

The above discussion leads us right to the second important for this topic paradigm and definition that comes from ISO/IEC Guide 73: "*risk criteria are the terms of reference by which the significance of risk is assessed*" [24]. This internationally accepted paradigm effectively means that

- society, de facto, establishes terms of reference for acceptable levels of risk or *risk acceptance criteria*.

One can then go further and suggest that should the world share the same risk acceptance criteria, then any product acceptance from a safety perspective would be much easier: Just prove the compliance of your product to acceptable risk levels in your home country and ship it anywhere around the world.

 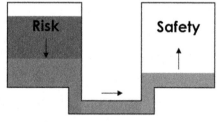

**FIGURE 6.8** Visualization of the relation between safety and risk. Left: symbolical via the image of Janus Bifrons. Right: technical via the concept of communicating vessels. *Left: Head of Janus, Vatican museum, Rome. Author: Loudon Dodd, commons wiki, 27 July 2009, under Creative Commons licence. Right: modified from A.V. Tchouvelev presentation at the ICHS2011, September 12, 2011, San Francisco.*

**FIGURE 6.9** Example of different risk acceptance criteria for the same worker occupation. *A.V. Tchouvelev, Presentation at the ICHS2011, September 12, 2011, San Francisco. 2011.*

The reality, however, is that risk acceptance or tolerance (which is another term that is often used in the context) varies greatly across the globe. Fig. 6.9 [30] shows two images of steel workers' protective equipment from two different parts of the world. The difference is stark. Clearly what seems to be acceptable in one part of the world would not be acceptable in the other. In addition, clearly the workers on the left are less protected from the hazards associated with metal forging than the worker on the right. Hence, the workers on the left are exposed to a higher risk than the worker on the right. Yet, both situations are accepted as "safe" by their respective societies. The conclusion from the above example is that "safety" and "acceptable level of risk" are determined solely by societal values and priorities. Developed nations tend to establish more stringent risk acceptance criteria than developing nations. Level of safety tends to increase with the increase of the average societal quality of life.

## Risk Criteria and Public Acceptance

The public acceptance of a hydrogen facility design and operation should be determined by whether the associated risk meets established risk criteria. Establishment of risk criteria is a key element required to utilize a risk-informed approach for developing hydrogen RCS. Because the primary concern is the potential for personnel injury, risk criteria can be established for all the people exposed to the consequences of facility-related accidents, which could include the public located outside the boundaries of the facility (so-called third party), users of the facility or customers (so-called second party), and the facility workers (so-called first party) [32]. Graphical visualization of risk parties or exposures is presented in Fig. 6.10 [33].

# Exposures or Parties

❑ **First party:**
- ✓ Workers / employees – most trained and have most knowledge about hazards and related risks – *highest acceptable risk*

Attendant –
1st Party
Customer
watching –
2nd Party

❑ **Second party:**
- ✓ Customers – **knowingly** take risk
- ✓ Note: customer becomes 1st party when self-servicing

Customer becomes
1st Party when self-
refueling

❑ **Third party:**
- ✓ Public – may not be aware of hazards and risks, i.e. **unknowingly** takes risk – *lowest acceptable risk*

Third
Party –
Public

**FIGURE 6.10** Typical risk exposures or parties considered for risk criteria. *A.V. Tchouvelev, Presentation at the IEA HIA Hydrogen Safety Stakeholder Workshop "Sharing Knowledge, Identifying Needs, Celebrating Progress" Bethesda, October 2–3, 2012. Top and bottom images: Shutterstock 8443399 and 509722159 under Standard Licence. Mid image: Stuart Energy (2003).*

Risk guidelines can be specified with regard to individuals or the society at large. Individual risk reflects the frequency that an average person located at a certain location is harmed. Generally, *individual risk* is evaluated for the most exposed individual who can be a person at an actual location or a person assumed to be constantly at the facility boundary. Characterization of the population surrounding a facility is thus not required to evaluate individual risk. *Societal risk* reflects the relationship between the frequency (F) and the number (N) of people harmed and is usually expressed in the form of an FN curve. The slope of the FN curve is defined by a *risk aversion factor* that is designed to reflect the society's aversion to single accidents with multiple fatalities as opposed to several accidents with few fatalities. Evaluation of societal risk requires determination of the population surrounding a facility.

**ALARP (As Low As Reasonably Practicable) Principle**

Principle of ALARP is fundamental in achieving risk acceptance criteria in real life. It was introduced by Det Norske Veritas (DNV) GL for maritime industry, later was used for offshore oil and gas industry, and finally was brought into the hydrogen

**ALARP (As Low As Reasonably Practicable) Principle—cont'd**

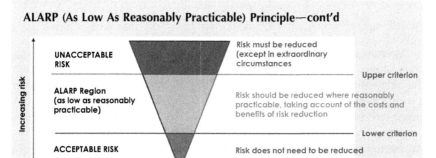

FIGURE 6.11   ALARP principle. *DNV GL, Risk Acceptance Criteria and Risk Based Damage Stability. Final Report, Part 1: Risk Acceptance Criteria. Report No: 2015-0165, Rev. 1. February 24, 2015 (Note: Text on the right is colored for emphasis by A.V. Tchouvelev).*

domain via HySafe in the mid-2000s. It is based on the following assumptions: (1) there are no zero-risk situations; (2) managing risk to a reasonable level is achievable; (3) acceptable risk represents the level below which an investment should be made to further reduce risk via cost–benefit analysis; (4) acceptable risk represents the *minimum risk* level that must *be obtained, regardless of cost*; (5) the ALARP principle is that the residual risk should be **A**s **L**ow **A**s **R**easonably **P**racticable—risk-reducing measures are feasible and their costs are not larger than the benefits.

Graphically the ALARP principle is best explained in the DNV GL recent report to European Maritime Safety Agency [34] and shown in Fig. 6.11. An earlier version of ALARP diagram can be found in DNV presentation at the 21st International Conference on Offshore Mechanics and Arctic Engineering (OMAE) in 2002 [35].

Risk acceptance criteria for individual and societal risk, although de facto exist everywhere, are not always obvious. In most Western European countries, they are incorporated into law. In the United States and Canada, to the contrary, as in many other jurisdictions around the world, they are not publicly defined in any way and are, thus, subject to interpretation. However, a common theme when introducing new technologies is that no new technology shall impose a greater societal risk than an incumbent similar technology; in other words, they have to be at least at par with each other in terms of societal risk (i.e., they must satisfy the same risk acceptance criteria). Hence, a reasonable approach for establishing risk criteria for hydrogen fueling stations is to equate it to the equivalent risk level presented by hydrocarbon fuels.

There are, thus, several options for selecting risk criteria. The first option is to specify that the risk associated with hydrogen refueling stations be at par with the risk associated with gasoline or compressed natural gas (CNG) stations. A second option is to specify that the risk from hydrogen accidents be some fraction of the total risk to individuals from all unintentional injuries.

Several groups have adopted the second option for hydrogen safety applications. The European Integrated Hydrogen Project (EIHP) has specified the value to be 1% of the average fatality death rate of $1 \times 10^{-4}$/year or $1 \times 10^{-6}$/year [36]. The European Industrial Gas Association (EIGA) has suggested an individual risk value of $3.5 \times 10^{-5}$/year (approximately one-sixth the average fatality risk) [37]. The EIHP has also established individual fatality risk value of $1 \times 10^{-4}$/year for fueling station customers (second party) and workers (first party).

A third option is to use the practice of oil and gas and process industries where risk criteria are expressed as the statistical *expected number of fatalities per 100 million exposed hours* called *fatal accident rate* (FAR). The advantage of using FAR versus conventional number of fatalities per year (common for public risk) is that FAR accounts for actual exposure (number of exposed hours) regardless of how long it takes to accumulate them and does not require any further normalization.

## Guidance on Risk Criteria for Public Acceptance of Hydrogen Fueling Stations

Based on the above analysis, IEA HIA Task 19 Hydrogen Safety experts developed the following guidance on risk criteria in 2008

*Individual risk (third party)*—ALARP with the following criteria:

- 24/7 exposure (24 h/day, 7 days/week)—site independent—generic and more conservative guideline
- Most exposed individual—site-specific guideline
- Acceptable risk level (fatality) $< 1 \times 10^{-5}$/year
  - o Basis—comparative risk to gasoline stations, 10% of risk to society from all other accidents, representative value used by most countries
- Cost—benefit analysis limit—$1 \times 10^{-7}$/year
  - o Basis—representative of most countries

*Societal risk*—adopt EIHP ALARP FN curve:

- Basis—risk aversion factor of 2 with a pivot point for 10 fatalities of $1 \times 10^{-5}$/year for acceptable risk curve and $1 \times 10^{-7}$/year for cost—benefit analysis limit curve

An example of FN curve for societal risk showing the ALARP region including the cost—benefit risk reduction zone as well as risk boundaries for societal risk is presented in Fig. 6.12 [38].

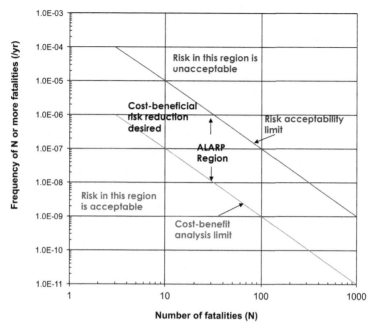

**FIGURE 6.12**   An example of FN curve for societal risk showing ALARP region. *J. LaChance, A.V. Tchouvelev, Development of risk-informed approach to safety. Presentation at US DOE-hysafe Workshop on QRA Tools. Washington DC, June 11−12, 2013.*

*Customer (second party) and worker risk (first party)*:

- Conventional approach: use traditional frequency of fatality per year (like in individual risk). Suggested acceptable risk for both second and first parties $< 1 \times 10^{-4}$/year
  - Basis—order of magnitude higher than the individual acceptable risk
  - Both customers and workers accept higher risk versus general public not using the refueling facility
- **Alternative approach:** use FAR similar to oil and gas/process industry approach (per 100 million h).
  - Option 1: FAR can be calculated from gasoline station statistics (e.g., *NFPA data*) and adopted for hydrogen stations
  - Option 2: use existing statistics for gasoline cars: e.g., *FAR for drivers is 25 and for passengers is 29 per 100 million h (UK)*

    ✔ Both drivers and passengers should accept *at least the same level of risk* for vehicle refueling as they accept while using their vehicles

The above guidance has no expiry date, so to speak, at least not in the foreseeable future while the human race continues making and driving cars and other means of transportation and, hence, continues using fueling stations and related infrastructure. The guidance is not rigid, and variations within the range of numbers proposed by EIHP ($1 \times 10^{-6}$/year) and EIGA ($3.5 \times 10^{-5}$/year) mentioned above are reasonable and acceptable depending on each country's situation.

For example and as a useful reference, NFPA used an individual risk of $2 \times 10^{-5}$/year for the development of risk-informed separation distances requirements for hydrogen facilities. This number was selected mostly for the reason of maintaining the individual risk at an equivalent level to gasoline stations, which in turn was assessed based on NFPA statistical data on fires, injuries, and fatalities at gasoline stations in the United States [39]. It also represents 10% of an average accidental fatality rate in the United States of $2 \times 10^{-4}$/year. Thus, the individual risk criterion of $2 \times 10^{-5}$/year selected by NFPA is in the midrange of numbers proposed earlier by EIHP and EIGA and fits well into the proposed IEA HIA Task 19 generic guideline.

### Risk-Informed Approach

As the reader may be aware, risk, despite being a technical construct, cannot be measured in a way the physical objects are measured. It can only be analyzed, estimated, evaluated, or assessed. That is why there are terms such as "risk analysis," "risk estimation," "risk evaluation," and "risk assessment" in risk terminology. A handy diagram [40] (Fig. 6.13) shows the exact meaning of the above terms.

Hence, we rely on models to estimate and quantify or qualify risk. However, models and their assumptions do not always (or rather never) address all contributors and failure mechanisms. In addition, input data, particularly on failure frequencies and probabilities, is often (or rather always)

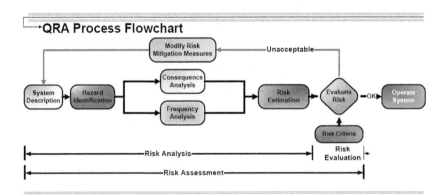

**FIGURE 6.13** Steps of quantitative risk assessment. *A.V. Tchouvelev, Based on ISO/IEC Guide 73:2002. Topical Lecture at ICHS2007 "Risk Management and Hydrogen Safety". San Sebastian, September 11–13, 2007, 2006.*

sparse. As a result, uncertainties of any risk assessment can be quite large. Therefore, evaluated risk should be used in conjunction with other information when making decisions! This is called a *risk-informed approach*.

**Risk-Informed Approach Facts**

- Risk-informed approach originated in nuclear industry:
  - Terms such as RIDM (*risk-informed decision making*) and IRIDM (*integrated risk-informed decision making*) were introduced in the late 1990s.
- US Nuclear Regulatory Commission's (NRC) definition for RIDM [41] reads:
  - A "risk-informed" approach to regulatory decision-making represents a philosophy whereby *risk insights* are considered together with *other factors* to establish requirements that better focus licensee and regulatory *attention on design and operational issues* commensurate with their *importance to health and safety*.
- Today NRC defines *risk-informed* approach as "... a decision-making approach that uses risk insights, engineering judgment, safety limits, and other factors ...."

Sandia National Laboratories (Sandia) in the United States has been instrumental in introducing the risk-informed approach for hydrogen energy applications in the mid-2000s as part of international collaborative effort within IEA HIA Task 19 Hydrogen Safety. This approach has been generally embraced by participating members of Task 19 and later implemented within ISO/TC 197 standard development work.

Fig. 6.14 [38] summarizes hydrogen safety community vision of risk-informed approach for hydrogen energy applications: use of validated models (both computational 3D and engineering 1D), field data, and expert input to determine risk through quantitative risk assessment (QRA).

As noted above, a risk-informed process utilizes risk insights obtained from QRAs combined with other considerations to establish code requirements. Further expanding this premise, the QRAs are used to identify and quantify scenarios for the unintended release of hydrogen, identify the significant risk contributors at different types of hydrogen facilities, and identify potential accident prevention and mitigation strategies to reduce the risk to acceptable levels. Examples of other considerations used in this risk-informed process can include the results of deterministic analyses of selected accidents scenarios, the need for defense-in-depth for certain safety features (e.g., combination of overpressure protection, continuous ventilation, hydrogen leak, and fire detection), the use of safety margins in the design of high-pressure components, and requirements identified from the actual occurrences at hydrogen facilities. A key component of this process is that *both accident prevention and mitigation features are included in the codes and standards requirements*.

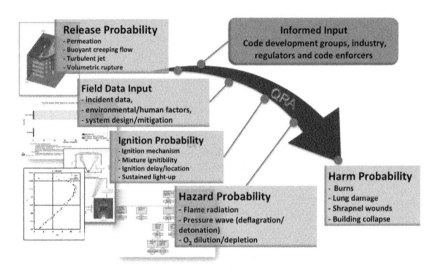

**FIGURE 6.14** Visualization of risk-informed approach developed by Sandia. *J. LaChance, A.V. Tchouvelev, Development of risk-informed approach to safety. Presentation at US DOE-Hysafe Workshop on QRA Tools. Washington DC, June 11−12, 2013.*

Once this has been achieved, i.e., hydrogen installation codes had become risk-informed, a risk analysis of "standard design" fueling facilities should not be required. However, some facilities may require special design or operational considerations that may require exemptions from the codes and standard requirements. For example, most of the public hydrogen dispensing stations deployed today are located in existing stations for other fuels such as gasoline, diesel, and CNG. Thus, limited space is normally available for locating the hydrogen-related generation, storage, and dispensing equipment, and the potential for additional accident initiators is introduced. Similarly, locations for new, dedicated hydrogen fueling stations particularly within urban centers may also be space-limited. Thus, meeting specified separation distances might not always be possible for some hybrid and new hydrogen fueling stations. A facility operator may thus be required by the AHJ to address the risk implications of the exemption and include additional safety features or take other steps to address an increase in risk associated with the facility design and operation.

## Risk Assessment Tools

The above discussion brings us to a very critical topic of risk assessment tools.

Most hazard assessment tools available to stakeholders, authorities having jurisdiction, and insurance firms to assess the safety of hydrogen systems are based on correlations that have been established for hydrocarbon fuels.

At the same time, hydrogen differs from commonly used hydrocarbon fuels through its wider flammability range, larger diffusion constant, higher buoyancy, and its large laminar flame velocity. In addition, the absence of soot as combustion product changes its heat radiant properties with respect to other fuels. Its ignitability and self-ignition properties also differ markedly from hydrocarbon fuels. Hydrogen is also usually stored under unique conditions: as a highly pressurized gas, as a cryogenic liquid, or through sorption into or onto solids. These specific properties of hydrogen, which may improve or hinder safety depending on the situation, must be taken into account to provide realistic assessments of the risks associated with a facility [42].

For these very reasons, the development of proper risk assessment tools for hydrogen applications has been a priority for International Association for Hydrogen Safety (HySafe) and IEA HIA Task 37 (successor of Tasks 19 and 31 on hydrogen safety) in the recent years.

**International Association for Hydrogen Safety, HySafe**

The International Association for Hydrogen Safety (HySafe) strives to be the main global forum for hydrogen safety–related issues. Founded in 2009, it is an international nonprofit organization registered under Belgian Law that currently has 38 members from industry, research organizations, and universities representing 14 countries worldwide. Its mission is to promote the safe use of hydrogen as a sustainable energy carrier. The Association facilitates the networking for the further development and dissemination of knowledge and for the coordination of research activities in the field of hydrogen safety. HySafe experts collaborate to assess the state of the art in hydrogen safety approaches and assessments and to identify and prioritize topics for further hydrogen safety research to be fed into the strategic agenda of hydrogen technology research and innovation programs worldwide.

HySafe regularly organizes unique marquee events on hydrogen safety: on the odd years (since 2005) the *International Conferences on Hydrogen Safety*, and on the even years (since 2010) the *Research Priorities Workshops* (RPW). The RPWs are coorganized together with US Department of Energy (DOE) and the Joint Research Center of the European Commission (EC JRC). The purpose of these internationally open activities is to share and update the state of the art in hydrogen safety with respect to hydrogen safety knowledge and to prioritize the research activities to address corresponding gaps in the short and medium term.

From the hydrogen safety research community perspective, one of the key tasks of scientific work is to translate fundamental scientific findings into practical formulas, which are easily applied in daily work. Although Computational Fluid Dynamics (CFD) is capable to producing high-resolution results for complex physical phenomena, the substantial financial and computational resources required for CFD make it unusable for daily safety decisions. For many safety decisions or for analyses with large numbers of relevant variables (i.e., QRA), validated simplified models, correlations, and statistics data can be used to provide robust results with significantly fewer resources than required by simulations [43].

In June 2013, Sandia and HySafe organized an expert workshop to identify limitations of existing QRA tools [44]. Since then, Sandia and HySafe have been working to build simplified toolkits to facilitate the use of QRA within the hydrogen fuel cell industry. The primary objective has been to develop a library of modern hazard assessment tools that contains best available models and data relevant to understanding and quantifying risk in hydrogen fuel cell infrastructure. The toolkit must:

- contain the latest available data and models (ideally, validated for hydrogen infrastructure use) relevant to quantifying the probability of progression of various hazard scenarios;
- contain the latest available data and models (ideally, validated for hydrogen infrastructure use) relevant to prediction of physical properties of hydrogen releases and ignition events, and the consequences of those events;
- contain risk metrics that represent observable quantities (e.g., physical parameters, losses, number of fatalities) relevant to decision-making for safety, codes, and standards;
- facilitate relative risk comparison, sensitivity analysis, and treatment of uncertainty;
- be built in a modular configuration;
- contain user-friendly, graphical interfaces;
- provide default models, values, and assumptions and provide transparency about those defaults; furthermore, it must allow modification of these defaults to reflect different systems and new knowledge.

At the HySafe RPW2014 [45] in Washington DC, the topical research areas directly relevant to Integrated Computational Tools (highlighted bold for convenience) were ranked at the top of research priorities as follows:

1. **QRA Tools (23%)**
2. **Reduced Model Tools (15%)**
3. Indoor (13%)
4. Unintended Release—Liquid (11%)
5. Unintended Release—Gas (8%)
6. Storage (8%)
7. **Integration Platforms (7%)**
8. Hydrogen Safety Training (7%)
9. Materials Compatibility/Sensors (7%)
10. Applications (2%)

To better understand the context and purpose of Integrated Computational Tools, the following definition has been proposed.

*Integrated Computational Tools for hydrogen safety — a suite of* **engineering** *probabilistic and/or physical effects (consequence) validated models integrated into a user-friendly interface allowing the user to input user-specific information and boundary conditions and capable of generating risk and/or hazard assessment data within reasonably short time (in sec and min).*

The word "engineering" is a key word here and is highlighted to differentiate the tools from more sophisticated ones, such as CFD, that normally require hours and days to get a result.

Responding to the challenge, in 2014, Sandia developed the HyRAM (*Hydrogen Risk Assessment Models*) tool, which fits perfectly the above definition of an integrated computational tool. HyRAM is a comprehensive methodology and accompanying software toolkit for assessing the safety of hydrogen fueling and storage infrastructure via QRA with integrated consequence analysis (discussed in this section) and/or stand-alone use of deterministic consequence models. The HyRAM software toolkit provides a consistent, documented methodology for QRA with integrated reduced-order physical models that have been validated for use in hydrogen systems. HyRAM also contains probabilistic data and models that have been vetted by the hydrogen research community. HyRAM is intended to facilitate evidence-based decision-making to support codes and standards development and compliance.

HyRAM is a model integration platform for comprehensive QRA, providing a unified language and architecture for models and data relevant to hydrogen safety. The development of a unified software framework also facilitates completeness and usability: Experts from across the hydrogen safety research community can contribute validated models from their domain of expertise, and the hydrogen industry benefits from a "one-stop shop" for those models.

HyRAM core functionality includes:

- documented QRA methodology
- generic data for gaseous hydrogen ($GH_2$) systems
  - ○ component leak frequencies, ignition probability; modifiable by users
- fast-running, experimentally validated models of gaseous hydrogen physical effects for consequence modeling
  - ○ release characteristics (plumes, accumulation)
  - ○ flame properties (jet fires, deflagration within enclosures)
- probabilistic models for human harm from thermal and overpressure hazards

HyRAM key features are:

- GUI and mathematics middleware
- Documented approach, models, algorithms
- Fast running: to accommodate rapid iteration
- Flexible and expandable framework; supported by active R&D

The HyRAM QRA methodology follows the general QRA approach. The HyRAM toolkit contains two user interfaces—one that allows stand-alone implementation of the physical effects models for flames and overpressures and one for a QRA with those models. In general the QRA approach uses a combination of probabilistic and deterministic models to evaluate the risk for a given system. The methodology uses traditional QRA probabilistic model approaches to assess the likelihood of various hydrogen release and ignition scenarios, which can lead to thermal and overpressure hazards. Several deterministic models are used together to characterize the physical effects for the scenarios. Information from the physical effect models is passed into probit functions that calculate consequences in terms of number of fatalities.

A significant value of HyRAM is that it supports calculation of the following key fatality risk metrics (expected value):

- FAR (fatal accident rate)—number of fatalities per 100 million exposed hours
- AIR (average individual risk)—number of fatalities per exposed individual
- PLL (potential loss of life)—number of fatalities per system-year

It also can calculate the following accident scenario metrics (expected value):

- Number of hydrogen releases per system-year (unignited and ignited cases)
- Number of jet fires per system-year (immediate ignition cases)
- Number of deflagrations/explosions per system-year (delayed ignition cases)

In summary, HyRAM provides a platform, which integrates state-of-the-art, validated science and engineering models and data relevant to hydrogen safety into a comprehensive, industry-focused decision support system. The use of a standard platform for conducting hydrogen QRA ensures that various industry stakeholders can produce metrics for safety from defensible, traceable calculations. The physical models underlying the HyRAM platform have been experimentally validated with hydrogen in the parameter (e.g., pressure, temperature) range of interest for hydrogen systems. The probability data included in HyRAM have been developed by reference to systems using hydrogen as much as possible. The software architecture of HyRAM is modular, with the anticipated addition and revision of modules and data as the state-of-the-art advances [46]. Harm criteria come from the earlier work within IEA HIA Task 19 Hydrogen Safety [47].

Thanks to these unique features of HyRAM, its approach has been adopted by ISO/TC 197 and incorporated into ISO 19880-1 draft international standard (DIS, publication expected in 2018). Fig. 6.15 illustrates an approach to risk-informed safety distances based on HyRAM and recommended by ISO 19880-1 [48].

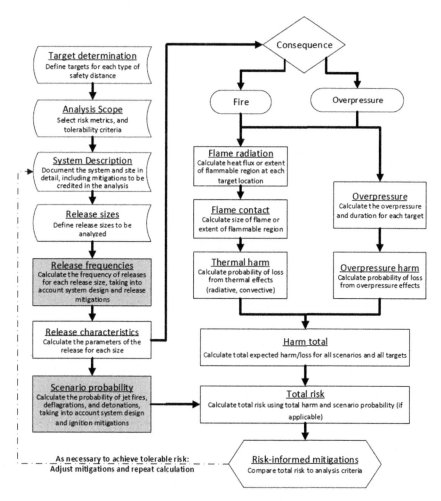

- Grey shading denotes an analysis step that is used only in full-QRA approach.
- Concave rectangle denotes analysis step
- Rectangle denotes calculation step
- Diamond denotes branching

**FIGURE 6.15** Example of a risk-informed approach to safety distances (based on HyRAM). *ISO/DIS 19880-1, Gaseous Hydrogen — Fueling Stations — Part 1: General Requirements, December 2017.*

## Public Engagement and Acceptance

Public engagement is absolutely essential in achieving public acceptance of a new technology. Speaking specifically of hydrogen and fuel cell technologies, public engagement has traditionally included a range of events from small and regional awareness and more technical workshops for various AHJs and fire fighters/emergency responders to broad open forums such as Fuel Cell Expo in Japan and Hannover Fair in Germany. Lately, such events included popular ride and drive of FCEVs that help public to familiarize themselves with hydrogen-fueled vehicles and appreciate their appeal, which includes no noise, no pollution, and fun to drive.

It must be noted here that engagement, in itself, does not necessarily result in social acceptance or endorsement—and arguably, nor should it, if it is open-ended and enables alternative options to be debated [49].

The research quoted above also reveals that *innovations where the scientific knowledge base is still evolving, pose particular problems for engaging the public.* A number of studies of public perceptions of emerging technologies, such as biotechnology and its long-term effects and nano-technologies, reveal a deep-rooted ambivalence. Such ambivalence is not simply a reflection of a lack of knowledge or information. Rather, it signifies reluctance by people to make unequivocal statements about these innovations, and the wider sociotechnical systems in which they might be embedded, *until they are directly experienced and/or it is shown how they might connect with their everyday lives.* Thus, the cautious and provisional nature of public attitudes toward emerging technologies is to be expected.

In Japan, the Association of Hydrogen Supply and Utilization Technology (HySUT) has been conducting surveys of public opinion on FCEVs and hydrogen infrastructure safety and performance at the Fuel Cell Expo in Tokyo since 2014. HySUT exhibition in 2017 included exhibits of FCEVs from Honda and Toyota as well as ride and drive of FCEVs as shown in Fig. 6.16 [50].

Fig. 6.17 shows the summary of the public opinion survey conducted at the FC Expo between 2014 and 2017 [50].

The above summary, although showing generally a favorable attitude in regard to safety of hydrogen fuel cell technologies, confirms the public's ambivalence noted earlier in regard to other novel technologies. The society just needs more time to get used to FCEVs and hydrogen stations to become more comfortable with them. As the quoted research explains, this is a natural process and we should not be discouraged with what may seem like a luke-warm initial response to these great technologies.

A great example of passive unobtrusive public engagement and education about hydrogen and fuel cell technologies and their safety has been

**FIGURE 6.16**  HySUT exhibition at FC Expo in Tokyo, 2017. *T. Ikeda, T. Abe. Research and development for safety improvement of hydrogen refueling stations in Japan. Presentation at ICHS2017, Hamburg, September 11—13, 2017.*

| Date | Survey respondents | Number of respondents |
|---|---|---|
| 2/27-28/2014 | FC Expo visitors | 208 (Male: 194 / Female: 14) |
| 2/25-26/2015 | FC Expo visitors | 327 (Male: 246 / Female: 81) |
| 3/3-4/2016 | FC Expo visitors | 329 (Male: 246 / Female: 83) |
| 3/2-3/2017 | FC Expo visitors | 332 (Male: 289 / Female: 43) |

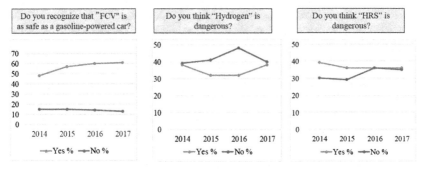

**FIGURE 6.17**  Summary of HySUT public opinion survey at FC Expo in Tokyo, 2014—17. *FCV, fuel cell vehicle; HRS, hydrogen refueling station. T. Ikeda, T. Abe, Research and development for safety improvement of hydrogen refueling stations in Japan. Presentation at ICHS2017, Hamburg, September 11—13, 2017.*

**FIGURE 6.18** Sui Sophia begins hydrogen story at hydrogen-navi portal. *T. Ikeda, T. Abe, Research and development for safety improvement of hydrogen refueling stations in Japan. Presentation at ICHS2017, Hamburg, September 11–13, 2017.*

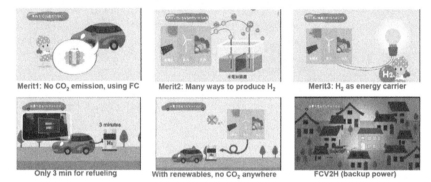

**FIGURE 6.19** Sui Sophia explains merits of hydrogen and fuel cell technologies. *FC*, fuel cell; *FCV2H*, fuel cell vehicle to hydrogen. *A. Maruta, Hydrogen information portal site "hydrogen energy navi". Presentation at US DOE and NEDO Meeting, June 2017, 2017.*

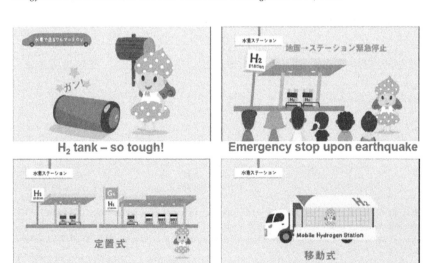

**FIGURE 6.20** Sui Sophia explains key safety features and fueling options. *A. Maruta, Hydrogen information portal site "hydrogen energy navi". Presentation at US DOE and NEDO Meeting, June 2017, 2017.*

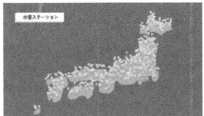

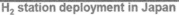

**FIGURE 6.21** Conclusion of the virtual hydrogen story at hydrogen-navi portal emphasizing clean and green future. *A. Maruta, Hydrogen information portal site "hydrogen energy navi". Presentation at US DOE and NEDO Meeting, June 2017, 2017.*

demonstrated in Japan via a project commissioned by the New Energy and Industrial Technology Development Organization (NEDO) to HySUT and Technova Inc [51]. In May 2015, Technova opened a "One-stop info portal for public: *'hydrogen-energy-navi'*" (http://www.hydrogen-navi.jp/). Since the portal opening, it has had over 500 visits per day, two-thirds of which were new visitors. The portal tells a hydrogen story illustrated by Figs. 6.18−6.21 through a character called Sui Sophie, which means "*$H_2$ + Sophia*." It appeals to younger generations who will most likely drive fuel cell cars.

## SOME PRACTICAL EXAMPLES

The examples below are based on many years of experience in risk and safety assessment of hydrogen installations and facilities and in research and modeling for hydrogen safety RCS. They address four key areas from safety and RCS perspective: selection of credible hydrogen leak orifice, hydrogen flammability and lean limits of combustion, defense-in-depth and ventilation approach, and gas purging.

## Selection of Credible Leak Orifice for Risk Assessment and Safety Engineering

Selecting a representative leak hole size and thus a representative leak rate has always been a tricky endeavor in hydrogen release modeling. Until recently, published guidances on this matter (such as TNO Purple Book) referred to the oil and gas industry experience, which inevitably led to overconservative approach when applied to significantly smaller hydrogen installations. Modeling to determine hazardous areas is likely the best reference for the purposes of this discussion.

From this perspective, IEC 60079-10-1 standard may be a good source of information. Previous editions of IEC 60079-10-1 did not address this issue although mass flow rate of a flammable gas was prominently displayed in recommended algorithms to determine the size of hazardous areas.

### EC 60079-10-1:2015 [52].

The recent edition of IEC 60079-10-1:2015 corrected this situation by providing suggested hole cross sections for secondary grade releases (see Fig. 6.22).

This should be viewed as a significant improvement of the standard because the suggested leak orifices now can be applied to relatively small hydrogen installations, piping, and equipment.

| Type of item | Item | Leak Considerations | | |
|---|---|---|---|---|
| | | Typical values for the conditions at which the release opening will not expand | Typical values for the conditions at which the release opening may expand, e.g erosion | Typical values for the conditions at which the release opening may expand up to a severe failure, e.g blow out |
| | | $S$ (mm$^2$) | $S$ (mm$^2$) | $S$ (mm$^2$) |
| Sealing elements on fixed parts | Flanges with compressed fibre gasket or similar | ≥ 0,025 up to 0,25 | > 0,25 up to 2,5 | (sector between two bolts) × (gasket thickness) usually ≥ 1 mm |
| | Flanges with spiral wound gasket or similar | 0,025 | 0,25 | (sector between two bolts) × (gasket thickness) usually ≥ 0,5 mm |
| | Ring type joint connections | 0,1 | 0,25 | 0,5 |
| | Small bore connections up to 50 mm [a] | ≥ 0,025 up to 0,1 | > 0,1 up to 0,25 | 1,0 |
| Sealing elements on moving parts at low speed | Valve stem packings | 0,25 | 2,5 | To be defined according to Equipment Manufacturer's Data but not less than 2,5 mm$^2$ [d] |
| | Pressure relief valves [b] | 0,1 × (orifice section) | NA | NA |
| Sealing elements on moving parts at high speed | Pumps and compre- ssors [c] | NA | ≥ 1 up to 5 | To be defined according to Equipment Manufacturer's Data and/or Process Unit Configuration but not less than 5 mm$^2$ [d and e] |

[a]   Hole cross sections suggested for ring joints, threaded connections, compression joints (e.g.,metallic compression fittings) and rapid joints on small bore piping.

**FIGURE 6.22**   Table B1 from IEC 60079-10-1:2015 showing suggested hole cross sections for secondary grade of releases. *IEC 60079-10-1, Explosive Atmospheres − Part 10: Classification of Hazardous Areas − Explosive Gas Atmospheres, second ed., 2015−09.*

Of particular interest is the suggestion for small pipes and fittings with internal diameter (ID) up to 50 mm, marked by red arrows on Fig. 6.22. The suggested size of the hole of 0.025 mm$^2$ translates into **0.18 mm ID**. What is particularly significant is that this pipe ID would apply to *all piping and fittings up to 50 mm*.

The above approach is comparable with the recent industry practice as well as advanced approach to hydrogen leak rate data analysis performed by Sandia for the development of NFPA 55 and NFPA 2 standards, which later incorporated into ISO/TC 197 work on safety distances.

### Sandia National Labs Analysis for NFPA 2/55 Separation Distances

Sandia conducted extensive statistical analysis of available data on gas leaks including hydrogen. Based on their analysis shown in Fig. 6.23, leaks between 0.1 and 0.2 mm, ranging from 0.03% to 0.1% of pipe cross-sectional area, contribute to between 90% and 96% of all leaks [53].

### Recent Industry Practice

It is useful to review the recent industry practice in establishing criteria for "small" and "major" leaks for the purpose of establishing standard requirements for hydrogen installations.

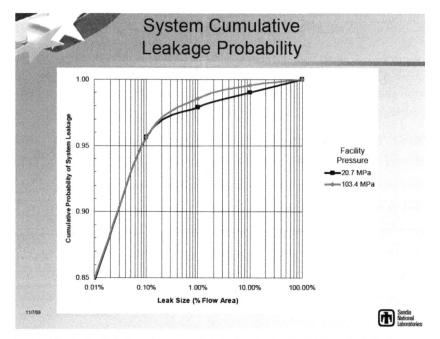

**FIGURE 6.23** Sandia leaks probability analysis for NFPA 55 and NFPA 2 standards development. *J. LaChance, J. Brown, B. Middleton, D. Robinson, Leakage data for the use in quantitative risk analysis of hydrogen refueling stations. ISO/TC 197, WG11, the Hague, Netherlands, November 2007.*

## HyApproval

HyApproval was high-profile project funded by the FCH JU (Fuel Cell Hydrogen Joint Undertaking) between 2005 and 2007. It included significant industry presence (Air Liquide, Air Products, BP, GM/Opel, Linde, Shell) as well as big research centers (CEA, NCSRD, DNV, INERIS, EC-JRC, HSL). Of particular interest from the point of this analysis is the report of WP4 Safety on "Agreement on required modelling tools & techniques for risk assessments and simulations, accident scenarios, credible leak rates" [54]. The members of the group "customer safety" that included all industry partners agreed on the following sizes for "small" and "major" leaks that might happen at hydrogen refueling sites (see Fig. 6.24).

As it can be seen from the above data, 0.1 mm leak orifice was selected as credible leak size for small leaks, which could be reasonable expected under normal upset conditions.

## Other Examples

***Canadian Hydrogen Airport Project.*** This project, in part funded by the Canadian government and in part by the industry, assessed the equipment readiness for deployment at two Canadian airports—Montreal and Vancouver between 2008 and 2011. For every major piece of equipment for hydrogen use, an FMECA (Failure Mode Effects and Criticality Analysis) was conducted. The approved methodology included the use of 0.1 mm leak orifices for risk assessment and establishment of hazardous areas [55].

***HyQRA by NoE HySafe.*** Network of Excellence HySafe in 2008 conducted an internal project HyQRA to develop a uniform methodology for QRA of hydrogen facilities. The applied methodology included the use of leak sizes starting with 0.1 mm being identified as "small leak" [56].

***Stuart Energy Experience.*** Stuart Energy Systems (Canada) in 2005 published a paper summarizing its experience and research in determining

| Scenario | Pressure | Leak size | Leak rate (g/s) |
|---|---|---|---|
| D5 minor leak | 350 bar | 0.1 mm | 0.10 g/s |
| D5 major leak | 350 bar | 0.2 mm | 0.41 g/s |
| D6 minor leak | 700 bar | 0.1 mm | 0.17 g/s |
| D6 major leak | 700 bar | 0.2 mm | 0.70 g/s |
| DL10 minor leak | 8 bar | 0.1 mm | 0.003 g/s |
| DL10 major leak | 8 bar | 0.2 mm | 0.013 g/s |

**FIGURE 6.24** A table from HyApproval WP4 Safety final report on selection of modeling scenarios and leak sizes for hydrogen gas. *HyApproval WP4 Safety, Agreement on required modelling tools & techniques for risk assessments and simulations, accident scenarios, credible leak rates, V1.2 Final (May 15, 2007).*

hazardous areas. The sizes of credible leaks for hydrogen determined through engineering correlations and assessed by CFD modeling were 0.1 and 0.2 mm [57].

***Japan High-Pressure Gas Safety Law.*** Requirements for installation of hydrogen equipment were developed with the consideration of 0.1 and 0.2 mm leaks [58].

## Summary

Based on the above evidence, the "true value" lies between these two numbers. Depending on the nature of the project, i.e., commercial or research, and its sensitivity in terms of public exposure, a number of approaches to determining the exact value could be taken:

- Take the highest value of 0.2 mm.
- Select a recommended value from IEC 60079-10-1:2015 of 0.18 mm.
- Prorate the leak ID based on pipe size—use a certain percentage value of pipe ID (NFPA approach for calculation of separation distances, was 3%, now 1% of pipe ID).
- Use the average between IEC and NFPA values.

Either method has merits. Important in this discussion is that there is significant evidence on the size range of a credible leak for the risk assessment and hazardous areas estimation purposes for hydrogen applications.

## On Hydrogen Flammability and Lean Limits of Combustion

Hazards associated with flammable gas are directly dependent on the probability of creating an ***explosive*** atmosphere. IEC 60079-10-1 mentioned above gives the following definition of an explosive gas atmosphere:

**explosive gas atmosphere**
mixture with air, under atmospheric conditions, of flammable substances in the form of gas, or
vapor, which, after ignition, ***permits self-sustaining flame propagation***.

The meaning of this definition is unmistakable: An explosive gas atmosphere is such that if ignited it self-sustains flame propagation through the entire gas—air mixture until the gas (fuel) or air (oxidant) is consumed. It also means that if the mixture is unable to sustain flame propagation through the entire volume, this gas atmosphere is not explosive.

This is a very important conclusion for *hydrogen* because it *is the only gas for which flammability limits and explosion limits are not the same.* In other words, hydrogen flammability limits and lean limits of combustion—when fuel can light up and sustain the flame—are not equal.

Hydrogen low flammability limit (LFL) of 4% molar fraction in air at STP (standard temperature and pressure) is a theoretical value at which an ignition and flame propagation is possible only in one direction—vertical and only under specific laboratory test conditions: fully homogeneous (premixed) and quiescent (not moving) mixture. Even under these ideal laboratory conditions, flame propagation in horizontal direction, which is often the case, is not possible.

The US Bureau of Mines published a report [59] that summarized more than 78 laboratory tests to determine hydrogen flammability and lean limits for combustion.

So, it is well documented that hydrogen flammability is geometry dependent as follows:

- For vertical flame propagation, LFL is around 4% molar fraction in air.
- For horizontal flame propagation, LFL is between 6.2% and 7.15% molar fraction in air,
- For downward flame propagation, LFL is between 8.5% and 9.45% molar fraction in air.

It must be noted that all the above concentrations are for ideal conditions of homogeneous and quiescent mixtures. Under real release conditions where mixing of hydrogen with air occurs during the release, the above numbers should be increased.

Those findings were later confirmed by the work and publications of Sandia [60] and Dr. Mike Swain of University of Miami [61].

Mike Swain reported experiments with various sparking electrical appliances with deliberately exposed electrical contacts trying to ignite flammable hydrogen concentrations in air with the following results:

- toggle light switch—unable to ignite 4.0%–10.0% molar fraction hydrogen concentrations;
- shop vacuum motor—able to ignite 6.0%–10.0% molar fraction hydrogen concentration;
  - Note: in this case the motor was able to ignite the premixed $H_2$–air mixture (like in the case below) inside the test box. There was no downward propagation.
- pull chain ceiling light—able to ignite 8.0%–10.0% molar fraction hydrogen concentration;
- garage door opener motor—unable to ignite 4.0%–10.0% molar fraction hydrogen concentrations.

This work indicates that real world electrical appliances may be safe in atmospheres significantly higher than 4.0% molar fraction hydrogen concentration in air.

Ballard Power Systems during exhaustive testing of potential releases from a hydrogen FCV showed that for horizontal releases sustained combustion does not happen until hydrogen concentration is increased to approximately 8% molar fraction in air. Downward flame propagation and hence flame propagation in all directions is only possible when hydrogen concentration in air reaches 10% molar fraction. Ballard concluded *"that typically about 8 to 10% H₂ by volume (mol/mol) in air was required before sustainable combustion would occur. The testing therefore demonstrated that use of the LFL criteria is design restrictive for flowing discharges and that it can be replaced with performance-based criteria for ensuring that discharges are not ignitable."* [62].

NFPA 55 in its 2013 revision noted that *"Although the generally accepted value for the upward-propagating lower flammability limit of hydrogen in air is 0.04 mole fraction, experimental data in the literature indicate that the limit may be as high as 0.072 mole fraction for horizontal-propagating flames and 0.095 mole fraction for downward-propagating flames (Zebetakis, 1965, Coward and Jones, 1952)"* [63].

Finally, a group of experts from HySafe, ISO, and NFPA after an exhaustive analysis of *"What is an Explosion?"* concluded that *"While 4% is recognized as the lower flammability limit for hydrogen air flames this is not the lower **hazardous** flammability limit which can be defensibly defined to be 8%"* [64].

This last consideration is particularly valid for an outdoor environment, large unconfined indoor spaces, and ventilated enclosures where accumulation of hydrogen and development of homogeneous mixtures with air is unlikely. Hence, there is significant scientific evidence for reasonably conservatively using the following hydrogen concentrations in air in the true meaning of an "explosive gas atmosphere" for the establishment of hazardous areas around vent stacks and ventilated enclosures or separation distances around hydrogen-containing equipment at STP conditions:

- 4% molar fraction—in vertical direction, and
- 8% molar fraction—in all other directions.

## Defense-in-Depth Approach to Safety

One of the key best practices for ensuring safety of operation is the so-called *defense-in-depth* approach. The term, originating from a military strategy of using several independent defense methods against any enemy attack, when applied to a hydrogen system means that there is a combination of several independent safety features engaged simultaneously. At the same time, failure of any of the key safety features leads to a safe shutdown of the main processes such as hydrogen production, compression, and/or dispensing depending on the location and severity of the failure.

Such safety features normally include continuous ventilation and flame and/or smoke and/or hydrogen (concentration) detection. Each feature has its own circuit; however, those are normally linked together via an integrated safety circuit controlled by a PLC (a programmable logic controller) and connected to a system "fire panel." The name "fire panel" is used here as an example and an indication that the primary objective here is achieving an acceptable level of fire safety of the hydrogen installation or facility. Hydrogen "fire panel" can then be connected to a generic fire panel for the cases when a hydrogen facility is an integrated component of a larger facility or operation. This way hydrogen fire safety is seamlessly integrated into the overall fire safety and emergency response plan.

The use of defense-in-depth approach guarantees the safe operation of a hydrogen system by significantly reducing the probability for escalation of small events and thus mitigating the associated risk.

### Definitions Important for Ventilation and Gas Purging

**Dead Volume** Space within a reservoir or system that lies outside the main flow of a gas introduced into the system, or flowing through it, making gas exchange more difficult in that space. Designers should be especially careful to avoid dead volumes in projects that use gases, and operators should be aware of them in their systems and use the most appropriate purging technique for each case.

**Active and Passive Safety Systems for Gases** It is important to understand the concepts of active and passive safety systems for gas applications because, although both are important, they may contribute in different ways to mitigate the risk under certain circumstances.

**Active Safety Systems** These systems consist of mechanical or electrical devices that must be activated manually or automatically to perform their protection function. Examples: push fan, exhaust fan, sensors.

**Passive Safety Systems** These systems consist of mechanical devices or intelligent designs that perform their protection function continuously without the need to be activated. Examples: ventilation openings, natural convection and ventilation, elevated light roofs, large walls or absence of wall.

## No Ventilation, No Operation Principle

Ventilation by far is the most important defense measure against a potential hydrogen leak accumulation in an enclosed space. Sometimes so-called "natural" or mechanically unforced ventilation works just fine. In certain cases, it may not even require significant influx of fresh air—internal air circulation and exchange due to convection may be sufficient, all subject to engineering calculations and verification by a qualified professional, of course.

In many cases, however, unforced ventilation is not possible or sufficient. Even for containerized hydrogen generation, compression, and storage equipment located outdoors, forced mechanical ventilation is advisable. In most situations with tightly packed enclosures, such ventilation is negative or exhaust. In some cases, such as a large warehouse with high ceiling, an overhead pull fan will be quite appropriate.

One can never go wrong applying a prudent principle of "no ventilation, no operation" to containerized hydrogen generation, compression, and large storage facilities. This essentially implies that some calculated level of mechanical forced (likely exhaust) ventilation must be present inside the containerized equipment at all times. This will ensure that the probability of any hydrogen fugitive emissions accumulation inside the enclosure will be significantly reduced. Such ventilation may also be inter-locked with hydrogen detection that triggers a higher ventilation flow upon detection.

The most critical advantage of continuous ventilation versus the one that is not continuously running and is only engaged upon hydrogen detection is that in the latter case the ventilation engagement depends on the reliability, sensitivity, and accuracy of the hydrogen detector while in the former case ventilation is independent (see defense-in-depth above). Even if the hydrogen detection is working within its calibration range, the timely detection is not guaranteed because a leak can be directed away from the detector. Hence, it may take a significant time for a detector to react and trigger ventilation. By that time a hazardous hydrogen concentration may have already been accumulated and the engagement of the forced ventilation at that moment may even have an adverse effect because of added turbulence if a high concentration hydrogen—air mixture finds an ignition source. To the contrary, under the continuous ventilation condition, any leak will be diluted on a continuous basis and inevitably carried over by the airflow to a hydrogen detector. At that moment, additional ventilation may be triggered or the system may be shut down. The inevitability of hydrogen being carried by the airflow around the enclosure ceiling following the air circulation pattern is due to the fact that hydrogen acts as a "slave" to air—being 14.5 times lighter than air hydrogen cannot resist any forced airflow greater than its buoyancy force. Because, in a regular ISO container, hydrogen buoyancy force is very limited, particularly at the ceiling, even a modest forced air circulation will carry hydrogen with it.

## Gas Purging

Gas purging is a very critical operation to ensure safety of both industrial and research laboratory equipment. Mistakes in gas purging may lead to devastating harm.

In the dictionary, "to purge" means to physically remove something completely. When working with gases, the meaning is quite similar, and purging a tube, cylinder or gas reservoir means most of the time to remove almost all the gas by relieving its pressure to near the atmospheric pressure. However, to completely remove the gas from the cylinder, it must be displaced by a purge gas (vapor or fluid) or by vacuum. This process is used for many purposes: (1) preparative: preparation of a cylinder to receive a different gas or a pure gas; (2) analytical: preparation of a gas sampling reservoir or bag to receive an analytical gas sample; (3) safety: to put a gas mixture out of the flammable range or to decontaminate a room containing toxic gas.

The gas purge process can be divided didactically into three main types: pushing purge, continuous dilution purge, and step-by-step dilution purge.

### Pushing Purge

Typically used for tubes, when the purge gas **B** enters by one side and forces the gas **A** (inside the tube) through the tube to leave by the other side. It is appropriate when the initial pressure inside the tube is near the atmospheric pressure, and the necessary quantity of gas **B** to purge the tube is approximately the internal hydraulic capacity of the tube.

### Continuous Dilution Purge

A system filled with gas A needs to be partially or totally filled with gas B. The purge gas B is continuously fed into the system, and a composition of both gases continuously leaves the system at another point. Depending on the required final concentration of gas A, the flow of the purge gas, and the volume of the system, the process can take a long time and consume considerable amounts of purge gas B. This type of purge is not recommended for complex systems or when there are dead volumes, because the gases do not mix very well unless there is a turbulent flow, and an undesired amount of gas A may remain in the system. Taking the necessary care, this process of purge is very useful, and the same principle applies to the ventilation or decontamination of rooms, reservoirs, and systems.

### Step-by-Step Dilution Purge

This can be the most efficient, safe, and fast type of purge for complex systems, mainly if vacuum is available. The process is accomplished by reducing the reservoir pressure as much as possible to initial pressure ($p_i$) and then raising the pressure with the purge gas to a higher final pressure ($p_f$), repeating this procedure for each step.

For practical results, it is necessary that $p_f \gg p_i$, and the best results are obtained using vacuum. This helps reducing the purge gas consumption, achieving a lower incidence of contamination (important for trace gas applications, for example), and ensuring that every point of a complex system

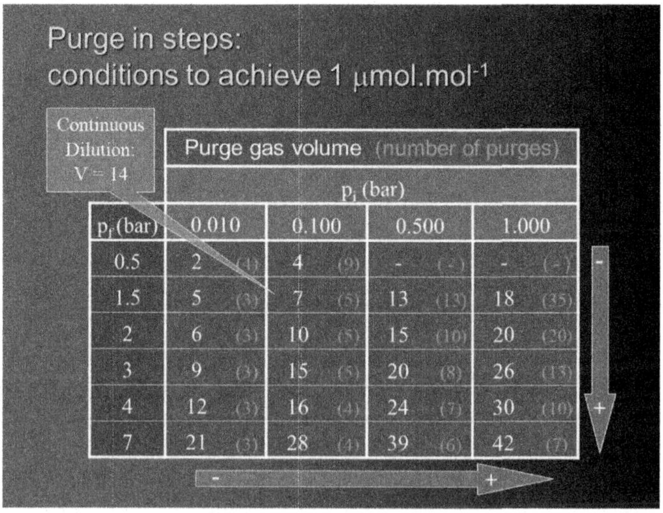

**Purge in steps:**
**conditions to achieve 1 μmol.mol⁻¹**

| Continuous Dilution: V = 14 | Purge gas volume (number of purges) | | | |
| | $p_i$ (bar) | | | |
| $p_f$ (bar) | 0.010 | 0.100 | 0.500 | 1.000 |
| 0.5 | 2 (4) | 4 (9) | - (-) | - (-) |
| 1.5 | 5 (3) | 7 (5) | 13 (13) | 18 (35) |
| 2 | 6 (3) | 10 (5) | 15 (10) | 20 (20) |
| 3 | 9 (3) | 15 (5) | 20 (8) | 26 (13) |
| 4 | 12 (3) | 16 (4) | 24 (7) | 30 (10) |
| 7 | 21 (3) | 28 (4) | 39 (6) | 42 (7) |

**FIGURE 6.25** Results of step-by-step dilution purge of a gas inside the reservoir to reach a concentration of 1 μmol/mol, for different pressures (absolute). In the example shown, the volume of purge gas consumed is equal to 7, expressed in number of times the hydraulic capacity of the reservoir. Accomplishing the same task by continuous dilution doubles the volume of purge gas, 14. *N.P. Neves Jr., International Conference on Hydrogen Safety 2011, 2011. http://conference.ing. unipi.it/ichs2011/*

is reached. However, attention should be paid to the compatibility of the gas with the vacuum pump components, including the oil, where applicable.

Fig. 6.25 shows estimations of the purge gas consumption and the number of steps necessary to purge a reservoir from [A] = 100 %mol/mol to achieve 1 μmol/mol using step-by-step dilution purge [20].

## IN SUMMARY

It is safe to predict that in the next 80 years during a gradual phasing out of hydrocarbon-fueled transportation and power plants, the energy technology landscape will be filled with a wide variety of options. Most of those will be based on significant hybridization of existing and new energy technologies using all conventional and alternative fuels in various combinations dictated by their respective business cases and environmental considerations, where hydrogen and electricity will play more and more significant roles as twinning energy vectors going forward.

Taking passenger vehicles fueling infrastructure as an example, a regular "gas" station will likely include over 10 fueling options depending on regional preferences ranging from conventional gasoline and diesel to CNG and blends with hydrogen, LPG, methanol, ethanol, hydrogen, and fast and regular electric charging outlets. Even today, most of the hydrogen public dispensing stations are being deployed and integrated into the existing gasoline–diesel stations.

From the regulatory and standardization perspective, this will require not only a harmonization of the existing international single fuel station standards to avoid conflicting requirements but also tweaking local regulations to allow multifuel options at a single fueling site.

Tomorrow has arrived and there is no turning back. Exciting and challenging times await!

## ACKNOWLEDGMENTS FROM THE FIRST AUTHOR

I would like to dedicate this space to paying tribute to three outstanding individuals who had a defining and significant influence on my life and professional career. These brilliant visionaries were ahead of their time in their own way and belong to a category that I call *"international founding fathers of the hydrogen energy movement"*. Two of them have timelessly passed away, while the remaining one is still tirelessly leading the international hydrogen energy community.

Academician Prof. Dr. *Valery A. Legasov* (1937–88) was instrumental in leading and directing hydrogen energy research and demonstration programs in the former Soviet Union. He was one of the first to understand the critical importance of hydrogen safety for nuclear energy applications. Valery Legasov is best known to the world as the Head of the Soviet delegation at the historic meeting at the International Atomic Energy Agency (IAEA) in Vienna in August 1986 who presented a 5-hour marathon report on the cause and consequences of the Chernobyl nuclear accident (April 26, 1986). Valery Legasov essentially sacrificed his life to learn the truth about the Chernobyl accident and develop effective postaccident mitigation measures. Unlike other members of the emergency team, who were rotated out of the Chernobyl plant to avoid prolonged radiation exposure, Legasov remained at his post throughout the ordeal. Because of that, he received an estimated four times maximum allowed radiation dose. He died on April 27, 1988. On September 20, 1996, then Russian president Boris Yeltsin posthumously bestowed on Legasov the honorary title of Hero of the Russian Federation for the "courage and heroism" shown in his investigation of the Chernobyl disaster.

I was fortunate and privileged to serve as his assistant on science and technology for the last 5 years of his life. He opened the door for me into the hydrogen world for which I am eternally grateful. Valery Legasov led by example. He set the bar of professionalism, dedication, and personal integrity so high that not many people could even approach it. I am trying living up to his expectations ever since.

Prof. Dr. *T. Nejat Veziroglu* (1924–) or *Dr. V*, how he is lovingly being called these days, does not need an introduction. He has been unwaveringly and enthusiastically leading the global hydrogen community since the first groundbreaking The Hydrogen Economy Miami Energy (THEME) Conference in March 1974. Inventor of the concepts of "Hydrogen Economy" and "Hydrogen Civilization" and founder of the *International Journal for Hydrogen Energy*, Dr. V has been a spiritual and intellectual leader and custodian of the hydrogen energy movement for over 40 years.

I was privileged to first meet Nejat Veziroglu in 1985 when, working under the Legasov's wing, I had been assigned to lead the organization of the seventh WHEC in Moscow in

September 1988. Being appointed as the Executive Secretary of this WHEC, I worked closely with Nejat who served as the WHEC's Co-Chair (along with a Deputy Prime Minister of the USSR). I was enchanted by Nejat's charisma and vision from the first time we met. After Legasov's death, Nejat showed tremendous support in the last months before the WHEC when the burden of the conference preparation and execution fell heavily on my shoulders and members of the Legasov's group at the Moscow State University where the base of the WHEC Secretariat was stationed. We made the conference a great success. Nejat extended his support further and invited me in 1989 to the United States as a visiting scientist to the Universities of Miami, California Riverside, and Hawaii at Manoa. Doing research under his attentive guidance for 4 months at the University of Miami was not only a tremendous learning experience, but it was also an eye opener to the world of endless opportunities. At the end of my visit in May 1990, I knew that not only my life had changed but also I would have an international career in hydrogen, one way or another.

*Alexander (Sandy) Stuart* (1924–2014) was a hydrogen *romantic* in the best and classic sense of this word. Yet, he was also a strong leader in the drive toward the worldwide implementation of clean energy in the form of hydrogen fuel [65]. In 1948, Sandy, with his farther Ted, founded the Electrolyser Corporation with the vision that hydrogen will be produced from water using renewable energy sources such as hydropower, which is plentiful in Canada.

Sandy, a veteran of the Royal Canadian Navy serving as a Lieutenant during World War II, loved the sea. A consummate outdoorsman, he was passionate about wildlife and championed environmental causes long before it was fashionable [65]. His passion for preserving environment led him to become a Director of the World Wildlife Fund Canada and the Chairman of the Nature Conservancy of Canada. He earned the Order of Canada for his work as an environmentalist, hydrogen expert, and exporter [66].

I was privileged to meet Sandy Stuart first briefly at the seventh WHEC in Moscow and then substantially at the end of my US research visit in 1990 (see above) when I stopped in Canada for a week at his invitation. Sandy's vision and quiet passion for green hydrogen future were fascinating. Being previously indoctrinated by Valery Legasov and Nejat Veziroglu, I was ready to absorb and appreciate the hydrogen path that lied ahead. There is no coincidence that 5 years later I joined the Electrolyser Corporation (later renamed Stuart Energy Systems) under Sandy's leadership first as a researcher and eventually an Executive Vice-President. My two professional areas of hydrogen safety research and water electrolysis started under Valery Legasov flourished under Sandy Stuart. The latter, however, would not have been possible without the fateful patronage of Nejat Veziroglu.

I would also like to express sincere gratitude to my contributing coauthors, Sergio Pinheiro de Oliveira and Newton Pimenta Neves Junior, whose valuable input, dedication, and enthusiasm helped making this chapter more complete.

I am also grateful to Paulo Emílio V. de Miranda, the Book Technical Editor, for always friendly and attentive guidance that helped tremendously in writing this chapter.

## REFERENCES

[1] A.V. Tchouvelev, Presentation at the ISO/TC 197 Plenary Meeting in Montreal, February 28, 2013. Stone Pyramid Image, 2013. https://www.123rf.com/profile_snolligoster.

[2] A.V. Tchouvelev, Presentation at the International Energy Agency (IEA) Hydrogen Roadmap North America Workshop in Washington in January 2014, 2014.

[3] The ASME Codes & Standards Writing Guide. http://www.cstools.org/WritingGuide/Cover_to_Documentation_Style_Guide.htm#Writing_Guide/WG_1-3_Codes_and_Standards_-_Definitions_&_Requirments.htm.

[4] https://www.nfpa.org/News-and-Research/News-and-media/Press-Room/Reporters-Guide-to-Fire-and-NFPA/About-codes-and-standards.

[5] Hydrogen Tools Portal: https://www.h2tools.org/hsp.

[6] A.V. Tchouvelev, General Considerations on Harmonization, 2016. Interview by Karen Quackenbush for the Hydrogen and Fuel Cell Safety Report published by the Fuel Cell and Hydrogen Energy Association, September 2016.

[7] Compiled by A.V. Tchouvelev from the material on the ISO website.

[8] A.V. Tchouvelev, in: Presentation at the ISO/TC 197 Strategic Planning Meeting, December 6, 2017, 2017.

[9] www.iso.org/va. FAQs on the Vienna Agreement — August 2016 edition.

[10] IEC website/TC 105.

[11] The International Bureau of Weights, Measures (BIPM), What Is Metrology?, 2004–2017. https://www.bipm.org/en/worldwide-metrology/.

[12] Euramet, Metrology in Short, third ed., 2008.

[13] OIML, International Vocabulary of Terms in Legal Metrology, 2000.

[14] JCGM 200:2012. International vocabulary of metrology — Basic and general concepts and associate terms (VIM).

[15] JCGM 100:2008. Evaluation of measurement data — Guide to the expression of uncertainty in measurement.

[16] P.R.G. Couto, J.C. Damasceno, S.P. Oliveira, Chapter 2 Monte Carlo simulations applied to uncertainty in measurement, in: V.(W.K.) Chan (Ed.), Theory and Applications of Monte Carlo Simulations, INTECH, March 6, 2013, ISBN 978-953-51-1012-5. https://doi.org/10.5772/53014 under CC BY 3.0 license.

[17] S.P. Oliveira, Metrology and quality in measurements of fuel cells variables, in: Presentation at 1st Conference Cycle: Hydrogen and Sustainable Energetic Future of Ceara State. Fortaleza, Brazil, March 15–16, 2011.

[18] BIPM, Member States, 2016.

[19] Signatory to CIPM MRA. BIPM https://www.bipm.org/en/cipm-mra/participation/signatories.html.

[20] N.J. Pimenta, International Conference on Hydrogen Safety 2011, 2011. http://conference.ing.unipi.it/ichs2011/.

[21] ISO/IEC Guide 51: 2014, Safety Aspects — Guidelines for Their Inclusion in Standards, 1999.

[22] J. LaChance, A.V. Tchouvelev, J. Ohi, Risk-informed process and tools for permitting hydrogen fueling stations, Int. J. Hydrogen Energy 34 (2009) (2009) 5855–5861.

[23] RSA's (UK Royal Society for the encouragement of Arts, Manufacturers & Commerce) Risk Commission in its Overview of Risk (RSA, 2004).

[24] ISO/IEC Guide 73: 2009, Risk management — Guidelines for use in standards.

[25] ISO/DIS 19880-1 Gaseous Hydrogen — Fueling Stations — Part 1: General Requirements, 2017.

[26] I.R.C.A. Pinto, I. Cardoso, O. Pinto Jr., N. Geier, 3rd International Lightning Meteorology Conference, April 21–22, 2010.

[27] B. Evarts, Fires at U.S. Service Stations (Rep.), April 2011.

[28] J. Spouge, What Is the Risk of Crossing the Road?, June 22, 2015.

[29] Head of Janus, Vatican museum, Rome. Author: Loudon Dodd, commons wiki, 27 July 2009, under Creative Commons licence.

[30] A.V. Tchouvelev, Presentation at the ICHS2011, September 12, 2011, San Francisco, 2011.

[31] RSA Risk Commission report "Risk & Childhood" (RSA, 2007).

[32] A.V. Tchouvelev, J. LaChance, A. Engebo. IEA HIA Task 19 Hydrogen Safety Effort in Developing Uniform Risk Acceptance Criteria for the Hydrogen Infrastructure. Extended Abstract to WHEC2008, Brisbane, Australia.

[33] A.V. Tchouvelev. Presentation at the IEA HIA Hydrogen Safety Stakeholder Workshop "Sharing Knowledge, Identifying Needs, Celebrating Progress" Bethesda, October 2-3, 2012. Top and bottom images: Shutterstock 8443399 and 509722159 under Standard Licence. Mid image: Stuart Energy (2003).

[34] DNV GL, Risk Acceptance Criteria and Risk Based Damage Stability. Final Report, Part 1: Risk Acceptance Criteria. Report No.: 2015-0165, Rev. 1, February 24, 2015.

[35] R. Skjong, K. Ronold, DNV. So much for safety, in: Presentation at OMAE, Oslo, June 24–28, 2002.

[36] Risk acceptance criteria for hydrogen refueling stations, EIHP 2 (2003).

[37] Determination of Safety Distances, EIGA, IGC Doc 75/07/E.

[38] J. LaChance, A.V. Tchouvelev, Development of risk-informed approach to safety, in: Presentation at US DOE-hysafe Workshop on QRA Tools. Washington DC, June 11–12, 2013.

[39] J. JaChance, Comparison of NFPA and ISO Approaches for Evaluating Separation Distances, 2010.

[40] A.V. Tchouvelev, in: Based on ISO/IEC Guide 73:2002. Topical Lecture at ICHS2007 "Risk Management and Hydrogen Safety". San Sebastian, September 11–13, 2007, 2006.

[41] NRC SECY-98–0144.

[42] A.V. Tchouvelev, K.M. Groth, P. Benard, T. Jordan, A hazard assessment toolkit for hydrogen applications, Proceedings of WHEC2014, Korea (June 2014).

[43] K.M. Groth, A.V. Tchouvelev, A toolkit for integrated deterministic and probabilistic risk assessment for hydrogen infrastructure, Probabilistic Safety Assessment and Management PSAM 12 (June 2014) (Honolulu, Hawaii).

[44] K.M. Groth, A. Harris, Hydrogen Quantitative Risk Assessment Workshop Proceedings. SAND2013–7888, Sandia National Laboratories, 2013.

[45] J.O. Keller, et al., in: HySafe Research Priorities Workshop Report. Summary of the Workshop Organized in Cooperation with US DOE and Supported by EC JRC in Washington DC, November 10–11, 2014. SAND2016-2644, March 2016.

[46] K.M. Groth, in: Presentation at HySafe RPW2014, Washington DC, November 2014.

[47] J. LaChance, A.V. Tchouvelev, A. Engebo, Development of uniform harm criteria for use in quantitative risk analysis of the hydrogen infrastructure, Int. J. Hydrogen Energy 36 (2011) 2381–2388.

[48] ISO/DIS 19880-1, Gaseous Hydrogen — Fueling Stations — Part 1: General Requirements, December 2017.

[49] R. Flynn, P. Bellaby, M. Ricci, The limits of upstream engagement in an emergent technology: lay perceptions of hydrogen energy technologies, in: Empirical Studies of Public Engagement, 2014. Chapter 17 Hydrogen Energy.

[50] T. Ikeda, T. Abe, Research and development for safety improvement of hydrogen refueling stations in Japan, in: Presentation at ICHS2017, Hamburg, September 11–13, 2017.

[51] A. Maruta, Hydrogen information portal site "hydrogen energy navi", in: Presentation at US DOE and NEDO Meeting, June 2017, 2017.

[52] IEC 60079-10-1, Explosive Atmospheres — Part 10: Classification of Hazardous Areas — Explosive Gas Atmospheres, second ed., 2015—09.

[53] J. LaChance, J. Brown, B. Middleton, D. Robinson, Leakage data for the use in quantitative risk analysis of hydrogen refueling stations, in: ISO/TC 197, WG11, the Hague, Netherlands, November 2007.

[54] HyApproval WP4 Safety, Agreement on required modelling tools & techniques for risk assessments and simulations, accident scenarios, credible leak rates, V1.2 Final (May 15, 2007).

[55] Canadian Hydrogen Airport Project, Risk Assessment and FMECA Methodology, 2009—11.

[56] HyQRA Benchmark Hydrogen Fueling Station: Scenario Definition, Workshop, November 28, 2008.

[57] G.W. Howard, A.V. Tchouvelev, Z. Cheng, V.M. Agranat, Defining hazardous zones — electrical classification distances, ICHS2005, Pisa (September 2005).

[58] S. Kikukawa, Consequence analysis and safety verification of hydrogen fueling stations using CFD simulation, IJHE 33 (2008) 1425—1434.

[59] Coward, Jones, Limits of Flammability of Gases and Vapors, U.S. Bureau of Mines, 1952.

[60] C. Moen, J. Keller, Hydrogen testing capability at Sandia national laboratories, in: IEA HIA Task 19 Experts Meeting, Joint Workshop on Hydrogen Safety and Risk Assessment, Long Beach, CA, March 16—17, 2006.

[61] M.R. Swain, P.A. Filoso, M.N. Swain, An experimental investigation into the ignition of leaking hydrogen, Int. J. Hydrogen Energy 32 (2) (2007) 287—295.

[62] R. Corfu, J. DeVaal, G. Scheffler, Development of safety criteria for potentially flammable discharges from hydrogen fuel cell vehicles, in: SAE Technical Paper 2007-01-0437, 2007.

[63] NFPA 55:2013, Figure E.4.1(d).

[64] J.O. Keller, M. Gresho, A. Harris, A.V. Tchouvelev, What is an Explosion? Int. J. Hydrogen Energy 39 (2014) 20426—20433.

[65] Toronto Star, thestar.com, December 2014.

[66] Business Wire, Stuart Energy Announcement, September 24, 2003.

# Chapter 7

# Roadmapping

David Hart[1,2]

[1]*E4tech, London, United Kingdom and Lausanne, Switzerland;* [2]*Imperial College London, London, United Kingdom*

## AN INTRODUCTION TO ROADMAPPING

### What Is a Roadmap and What Is It for?

Roadmapping is a fluid term, often used to describe a plan or strategy that aims to achieve one or more linked objectives. A roadmapping approach can be taken in many diverse situations, such as for general business development or for something as specific as developing a response to an outbreak of a disease [3]. A common form of roadmap is a technology roadmap, which is particularly applicable to the energy sector and to the hydrogen and fuel cells sectors within that. Such a technology roadmap can vary in its coverage from company level up to sector or even country level and provide a framework for long-term planning for the development of a technology in the market and regulatory environment in which it sits.

The International Energy Agency (IEA) defines a technology roadmap as "a dynamic set of technical, policy, legal, financial, market and organisational requirements identified by all stakeholders involved in its development. The effort shall lead to improved and enhanced sharing of and collaboration on all related technology-specific research, design, development and demonstration (RDD&D) information among participants. The goal is to accelerate the overall RDD&D process in order to deliver an earlier update of the specific energy technology into the marketplace" [1]. McDowall and Eames, who have written on hydrogen energy—related roadmaps in particular, point out that roadmaps, and other foresight methods, "can play an important role in the development of shared visions of the future…and mobilising resources necessary for their implementation." [2].

The roadmap states the goal and sets out actions or targets that must be implemented or achieved. The actions and targets in the roadmap, both intermediate and final, should be linked together so that they work in conjunction to achieve the overall desired outcome. They should be clear and implementable, rather than aspirational. A roadmap can cover any period of

Science and Engineering of Hydrogen-Based Energy Technologies. https://doi.org/10.1016/B978-0-12-814251-6.00007-1
**357**

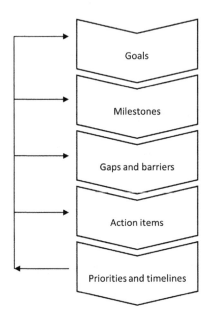

**FIGURE 8.1** The logic of a roadmap.

time, but in the energy sector and in innovation, it is commonly used for mid-to long-term periods, for instance, multiple years to multiple decades. This is a result of the typical slow pace of innovation and change associated with these technologies, the energy system as a whole, and the policy frameworks into which they fit. Fig. 8.1 outlines the logical flow of a roadmap.

The process of roadmapping identifies—and ideally engages—the relevant stakeholders in the industry or sector of focus and takes into consideration how these entities currently interact. This provides an understanding of what change would be possible, over a given period, and how this change could be implemented successfully, from a baseline of the current state of the industry or sector. A baseline for the current development and deployment of the technology at question is also required [4]. These form the basis of how the future is considered in the roadmap. In the original literal sense of a roadmap, it is hard to know what direction to take if you do not know where you currently are.

## Roadmaps and Scenarios Are Not Synonyms

Like roadmaps, no fixed definition exists for scenarios. The two are quite different, however. Scenarios are typically more or less hypothetical, but internally consistent, views of the future, which allow different actions and ideas to be explored. Scenarios can be used to understand the effects of different actions and which actions might be robust in different future sets of conditions. They are often used to help develop roadmaps.

Conversely, roadmaps are not a prediction or a hypothetical evolution of the future from postulated events and drivers of change. Rather, a roadmap lays out the intended or expected future as a goal and then describes the means by which that future may be attained. The goal could be that a technology in question will succeed in penetrating markets (e.g. that 500 fuel cell buses might be introduced into Europe), and actions to support the technology will be needed for this to happen. For example, an increase in efficiency or reduction in cost might be needed, through targeted R&D or the removal of policy barriers. A roadmap is thus a tool that promotes structure in the planning of a future and also stimulates action of this plan—through multiple stakeholder engagement and clear dissemination. Roadmaps can be summarized as a combination of (1) what is likely to happen, (2) what is hoped will happen, and (3) what may be made to happen based on an understanding of (1) and (2) [4].

Identifying what is likely to happen or what might happen is often done through scenario development. If a roadmap is based on one or more scenarios, then the roadmap design must take into consideration the robustness of those scenarios. The further that a scenario is cast into the future, the less accurate it is likely to be, taking into consideration the increasing effects of innovation, changes in external conditions, and other uncertainties over time. Nevertheless, the roadmap itself can be stretched over any period of time, as long as progress toward the goals is monitored, and actions or targets adapted as required when changes occur or more information becomes available [2].

## How Roadmapping Can Consider the Future

Since the purpose of a roadmap is to identify a desired possible future and help guide different actors to achieve it, the clearest possible identification and definition of that future is paramount. To subsequently realize that future, the underlying priorities and timelines must be identified and agreed. Expanding on the earlier example, it might be that the goal is for 500 fuel cell buses to be deployed across 10 different European countries within 5 years. This target can be ambitious but must also be achievable. If no fuel cell buses are currently available, or those in existence fail technically after 1 month, then the plausibility of the goal is undermined. If, however, tens of fuel cell buses are in operation and it is clear, under sensible assumptions, that 500 could be produced and deployed, then the goal is more appropriate. Further analysis would include requirements, in addition to the bus technology, for the buses to be able to be deployed: refueling infrastructure; available hydrogen of the right purity; trained service personnel; appropriate safety regulations; etc. The state of these requirements will then directly influence the design of the roadmap.

The overarching priorities and timelines allow the articulation of specific goals. These goals can then be broken down into milestones—the broader the scope of a roadmap, and the longer the period of time which it covers, the more important it is to have steps along the "road," in the form of milestones and targets. In this way, progress can be tracked and the roadmap can be adjusted accordingly, as milestones may need to be changed in response to events. For example, putting 50 fuel cell buses into a particular city might require a new depot to be built. If the depot takes 2 years to build and requires 1 year to achieve planning permission, a critical path for those 3 years can be sketched out, and specific actors and actions identified. If after 1 year the planning permission is still not achieved, the roadmap will need to be revisited and adjusted—perhaps by stretching the timeline or choosing a different location.

A wide variety of relevant metrics, or Key Performance Indicators (KPIs), can be used to monitor the evolution of a technology or sector. In addition to the examples above, they could vary as widely as technical targets for production cost or efficiency of a technology or its acceptance within a certain segment of the population.

The corollary to this identification of requirements is the identification of barriers that are preventing the goal from being achieved or are slowing down its achievement. The planning permission for the bus depot may be in progress, and suitably pure hydrogen available locally, but the employment regulations in place may not permit people to be trained to operate buses running on gaseous fuel. Such cases may arise where regulations were conceived before such an option was considered possible; they are not designed to exclude but nevertheless require significant effort, and often time, to change. So the different barriers should be methodically considered and determined, allowing the overall pathway to achieve the goal to be built in a structured manner.

The nature of each barrier, ways of overcoming, dismantling or avoiding it, and the actions to take to do so can then be articulated. Ideally, specific actions will be defined for each relevant party. These actions may be in parallel to, overlap with, or in some cases be identical to those identified earlier.

Bringing together the different sets of actions and milestones allows a coherent view of the overall direction of the strategy to be developed.

These roadmaps often include visualization of the strategy. Phaal et al. suggest that graphical output may be an input aid (e.g., a template to complete during a workshop) or an output aid (helping to explain the direction and links) [5]—see, for example, Fig. 8.2. The illustration in Fig. 8.3 shows a schematic representation of the intended progress linked to the UK Hydrogen and Fuel Cells (HFC) Roadmap, which is intended primarily for communication purposes.

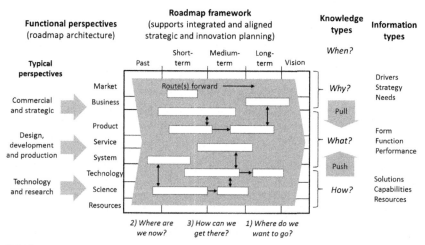

**FIGURE 8.2** A roadmap framework. *Author's work based on Phaal et al. (2009). Redrawn from R. Phaal, C.J.P. Farrukh, D.R. Probert, Visualising strategy: a classification of graphical roadmap forms, Int. J. Technol. Manage. 47, (2009) 286–305.*

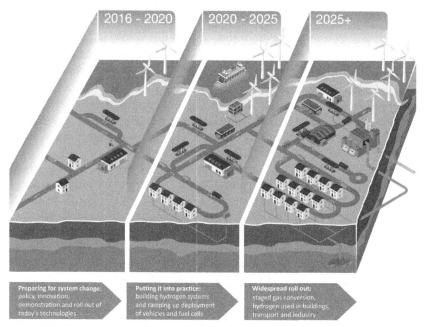

**FIGURE 8.3** Graphical communication of a roadmap vision. *E4tech and Element Energy (2016).*

## Why Is a Roadmap Useful?

A roadmap should not be an end in itself. The process, if conducted well, has value in bringing together different actors and focusing on a particular objective, and the outcome of the exercise may show that the initially hoped-for goal is unachievable. However, the main purpose of a roadmap should be to help enable the goal to be achieved. It is not uncommon either for the roadmap itself to be seen as the achievement, and put on a shelf, or simply for there to be no funds—or no committed organization—to take forward the plan, monitor progress, and coordinate between actors. The roadmap can only be as effective as its implementation allows.

Robert Phaal, at the UK University of Cambridge's Institute for Manufacturing (IfM), cites the following as useful aspects of a roadmap and roadmapping exercise [6]:

- Bringing stakeholders together to develop a shared vision
- Creating a strategic plan based on input from all the relevant stakeholders
- Ensuring investment in R&D and all technology and innovation planning is in line with strategic goals
- Improving communication by providing clear, easy-to-understand outputs

The inclusion of all the relevant stakeholders in the roadmapping process allows the perspectives of each to be considered and creates the potential to form a strategy with a shared outlook. Stakeholders will potentially have competing or at least widely diverging views: They may develop competing technologies; be facing different barriers; or be truly interested in only one aspect of the bigger picture. "Fuel cells and hydrogen," for example, covers a hugely diverse range of technologies and applications. Some fuel cell companies do not use and have no interest in hydrogen, whereas some hydrogen technology companies are indifferent to fuel cells. Likewise, policymakers face different sets of needs and requirements than universities, or early-stage companies. By bringing them together, it should be possible to identify and lay out a vision that is common to all, even if aspects of the detail are more relevant to one set of stakeholders or another.

In addition to the simple participation of the different stakeholders, which ideally allows each to better understand the others' perspectives, their different inputs can help ensure a better strategic plan. Policy experts can guide on policy matters, whereas researchers can give insight into the scientific and technical status of a sector. These different perspectives can be used to build a more complete and robust strategy than would otherwise be the case. A well-run process for creating roadmaps encourages structured, collaborative thinking on how a stated outcome can be achieved.

If the involvement of the different stakeholders can be maintained through the process of developing the roadmap, and not just in gathering initial information and perspectives, they are likely to feel greater ownership of the roadmap. They can help ensure that the steps laid out are appropriate and in line with the strategy already discussed and agreed, and ultimately help to deliver those steps, helping to ensure that the right levels of emphasis and investment are in place.

Finally, the involvement of the different stakeholders can also help achieve a clear and easy-to-follow output. Different organizations and participants will use different language and concepts, and for a complex roadmap that involves scientists, policymakers, business people, and many others, it is important for the purpose, plans, and actions to be clear, simple, and unambiguous.

Developing a mutually agreeable roadmap is not straightforward, and different techniques can be used to achieve it. For example, in an area that has many different but somehow connected sectors, such as hydrogen and fuel cells, it may be useful to create focused roadmaps for relevant technologies and sectors and then bring them together. Such "miniroadmaps" allow a focus on the specific status and needs of a particular application or technology. An overarching roadmap can be built by bringing together all of the common aspects, plus any required specificities to address the most relevant or important of the miniroadmap requirements. Although this approach has the benefit of allowing focus, it requires that the roadmap developers pay close attention to the overlaps between the miniroadmaps, any synergies that can be derived, or mutually exclusive pathways and outcomes, to produce a fully representative final document. In any case, identifying areas of potential conflict or divergence is useful in understanding how future communication or joint working may be encouraged.

The UK HFC Roadmap published in 2016 took this approach, building multiple miniroadmaps, and subsequently knitting them together in a detailed summary of actions by sector and application [7].

## Who Uses Roadmaps, and How?

The production of a roadmap should preferably consider not only the actors directly involved but also others who may either be implicated or interested in the result. In practice, a roadmap can be used in different ways, by different actors. The actors developing the roadmap will vary depending on the type of roadmap, for example, a national-level energy technology roadmap may be developed for a particular reason by a particular actor or set of actors, as suggested in Table 8.1.

## TYPES OF ROADMAP

As discussed above, roadmaps may fall into numerous categories according to their emphasis and purpose. Four relatively distinct types can be identified, within the energy sector at least:

## Public Body Strategy

Typically produced by, or for, a body that represents or has an interest in a specific sector or industry grouping, these roadmaps focus on the implications of a particular sector or subsector in a national or regional context. In the UK, such bodies might include the Committee on Climate Change, an independent nondepartmental public body formed to advise the UK Government, among

**TABLE 8.1** Possible Roadmap Developers and Motivations

| Entity | Example Roadmap Use |
|---|---|
| Industry Association | • Monitoring progress of a technology or sector<br>• Lobbying for particular support<br>• Identifying specific actors or industry sectors who could be supported or targeted |
| Technology company in the sector | • Planning for technology development and commercialization<br>• Identifying barriers to development<br>• Assessing its position against others<br>• Identifying lobbying needs<br>• Changing emphasis or internal investment to maintain direction |
| Large corporation implicated by the sector | • Identifying risks or opportunities from the emerging sector<br>• Assessing timing and impacts of sector development on its own position<br>• Understanding the position of others |
| Government | • Identifying positive or negative impacts of technology uptake<br>• Designing or implementing policy to support the sector<br>• Removing unnecessary barriers to development or implementation |
| Other public body (e.g., city council) | • Identifying local opportunities or benefits of deployment<br>• Identifying local centers of excellence<br>• Guiding planning permissions or regulations |
| International body | • Identifying supranational trends or opportunities<br>• Bringing together or coordinating country-level fora to support a particular technology or sector<br>• Providing a holistic perspective of a multinational sector |

others, on tackling climate change; BEIS—the UK Government Department for Business, Energy and Industrial Strategy; or the Automotive Council—a UK industry-run body that oversees the combined strategy of the UK automotive industry. Each has a different perspective on a given subject, and so is likely to approach a roadmap in a different way.

A wide range of techniques can be used in developing a public body roadmap; however, some techniques are somewhat more likely to be used in particular situations than others. The description below is by no means prescriptive.

Public bodies may develop roadmaps in-house but often lack the resources or capacity to do so. Consultants or others may be engaged to provide support by conducting detailed analysis on relevant technologies, markets, or other aspects of a sector and provide objective advice to the body requiring the roadmap. The consultants' reports may simply feed into the final roadmap or the roadmap may be produced by the consultants directly.

Depending on the type of roadmap and the existing knowledge in the sector, the analysis may take many forms. For a technology or set of technologies that have implications for the energy system, modeling of grid interactions or effects on supply or demand may be required. Cost forecasting and market penetration estimation will also be necessary to understand areas where support may be required in the near or longer term, and interactions or competition between technology options may also figure. In some cases, a technique called "backcasting" may be used. This assumes a plausible future situation, such as the deployment of a technology, and then works backward to identify the steps required to enable that future and the timing of the steps.

The roadmaps developed for public bodies are likely to try to be as objective as possible, providing well-researched and often conservative inputs to avoid the challenge of overoptimism. They will often use a scenario approach based on conventional underlying assumptions of growth and penetration rates, economic discount rates, learning factors, etc. They may include expert input and some level of consultation, but this is not always the case.

## Single Industry Lobbying or Industry-Led Analysis

Industries, both emerging ones such as fuel cells and established ones such as aerospace, may build roadmaps not only to identify the ways in which they may thrive but also more specifically to inform their lobbying positions. They may form part of developing a strategy or be commissioned in response to an opportunity or threat. For example, when a national government is undertaking a review of its energy policy, different sectors will produce a wide variety of evidence, including roadmaps, to try to influence government to consider their sector favorably and even to suggest what measures they believe are appropriate. In other instances a roadmap may be commissioned in response to the publication of another document, which either does not cover the sector of interest for the industry or which is negative about it. The industry may feel that their technology was not covered accurately or that the detail used was insufficient or inappropriate.

These activities may be undertaken by companies, if they are large and well connected, or by industry associations. Although in some instances they are built from the ground up and very detailed, using often confidential internal data from within the industry, in others they may be based on the collation and judicious use of other information sources and contain no new data.

## Coordinated Industrial Strategy

Roadmaps addressing a specific industrial strategy almost inevitably require the industry in question already to be established. An industrial strategy roadmap could be developed for the gas turbine industry or the automotive industry, but it would be much harder for the fuel cell industry, such as it is. This is because such a strategy is usually grounded in a good knowledge of industrial capability and requirements, costs, and barriers. A gas turbine, although it is undergoing continual refinement through R&D, is technologically mature, and the path forward can be mapped with some certainty. A fuel cell technology may still be subject to disruption and invention, with early technologies and manufacturing processes quite different from those that will be needed in a mature market.

If a sector feels the need for an industrial strategy, it is often because it feels under threat. A sector that has an important role within an economy will in almost every case get government support to develop an industrial strategy, or the strategy may even be led by government. In such a case the analysis and roadmap are likely to focus on the relatively near term, are to secure or grow jobs and the industrial base, and will include close participation from all of the potentially affected actors. The roadmaps produced by the UK Automotive Council are a good example—first produced in 2009, they are regularly updated and address very specific technology and product areas within the automotive sector. The breadth of sectors covered includes roadmaps on buses, intelligent connected vehicles, electrical energy storage [8], and lightweight structures, among others.

## International Framework

The final major roadmap archetype discussed here examines the picture from an international perspective. These roadmaps may be regional or even global, considering a technology or sector from one of two primary perspectives. In some cases, they may take the position of a somewhat detached observer, examining the potential for a technology, sector, or industry to develop and the implications if it does—or does not. The IEA's hydrogen and fuel cell roadmap [9] falls roughly into this category. The second perspective is more partisan and could be taken by a multicountry body such as the European Fuel Cell and Hydrogen Joint Undertaking, which is a focal point for the entire Fuel Cells and Hydrogen (FCH) industry across multiple countries, or the Hydrogen Council, which represents global organizations. In both cases, the analysis and roadmapping framework is likely to be focused on the identification of technology potential, industry potential, and the requirements that need to be fulfilled to make them successful. The former may then identify generic issues—for example, that costs are too high to allow rapid penetration—but may not identify specific actions or actors who could address the issues. The latter are directly implicated in trying to help the sector succeed, and so will also do their best to identify where and how to help.

If a very broad geographical frame is considered, the roadmap is likely to be strong in identifying international strengths, weaknesses, and relationships but may be weaker in the detail of how to address them. This weakness stems at least in part from the complexity and time required to analyze all regions in detail, to produce robust and defensible conclusions and recommendations.

## Typical Roadmap Audiences

Roadmap audiences will also vary depending on the type of document being developed and are likely to be in addition to the actors who develop them. For a company developing its own technology roadmap, it is clear that audience is the company itself and possibly some investors or advisors. For national-level energy technology roadmaps, audiences may include

- national government decision-makers in ministries of energy, environment, industry, natural resources, and infrastructure;
- national government decision-makers in ministries of finance or economics;
- state/provincial and local policymakers and national regulators;
- energy sector decision-makers, particularly from industries that produce or consume large amounts of energy (e.g., electricity, natural resources, agriculture, and energy-intensive industry);
- leading scientific, engineering, policy, social science, and business experts involved in researching specific energy technologies and the supporting policies and financing mechanisms needed to accelerate commercialization;
- NGOs engaged in research and advocacy in low-carbon energy.

For certain audiences, roadmaps are "living documents," which means that they are constantly or at least regularly checked, edited, and modified as circumstances change. For example, a breakthrough in fuel cell catalyst technology could advance commercialization in a particular area, requiring a modification to timelines, proposed support areas or mechanisms, and technologies indirectly affected by the development. A series of companies going bankrupt could sufficiently weaken a nascent industry such that its roadmap requires revision.

## THE COMPONENTS OF A ROADMAP

What goes into a roadmap depends on its subject and its scope, as well as the authors and audience described above. If an external, nonexpert, or general audience is likely, then some form of accessible introduction to the roadmap is essential, explaining its purpose, scope, and boundaries. Some roadmaps will then lay out in detail the analytical approach taken, as well as the assumptions and perhaps some of the calculations. Others may not expose the calculation methodology but will move rapidly to the outputs of the roadmap process in terms of conclusions and recommendations. Whether it is laid out in detail or not, roadmaps built on solid, repeatable, and transparent analysis are typically more likely to gain consensus and be easier to promote and defend.

## Data and Analysis Activities

A wide range of data and analysis activities can be undertaken. The list below is not exhaustive but covers many of the options.

### Baseline Research

Projecting a future requires agreement and solid representation of the present situation. This may seem obvious, but establishing a baseline does not have to be a part of roadmap development. When it is, it will consider aspects such as the following *situation analysis*:

- The **technology** or technologies under consideration may be benchmarked and a set of performance criteria established. These will usually include the following:
  - *Development status.* At what technology, manufacturing and commercial readiness level does the technology sit? How much development is still required?
  - *Technology performance.* What is the efficiency of the technology, its lifetime, and durability? What is not yet known about how it behaves in different environments?
  - *Technology costs.* What is the current capital cost for the technology and what are the main influences? Are these subject to significant fluctuation or risk, for example, scarce resources or competing demands? What influences the installation and operating cost, and what range do these cover?
  - *Technology potential.* How much better could the technology perform, in any of the areas above? Is it near to theoretical limits, or is there still room for meaningful advancement? This kind of analysis may be conducted in depth using cost models and based on the latest research and development but may also be carried out primarily by expert interview.
  - *Environmental impacts.* What impact does the manufacturing and use of the technology have on land, air, and water? Does it emit anything of particular concern or which is stringently regulated? Does the supply chain have significant impacts and can these be mitigated?

- *Links to other technology areas.* Is development in the technology dependent on other sectors or areas, or can it benefit from them? Do developments in this technology have the potential to cross over into others?

- The existing or potential **markets** for the technology or application will be examined. For example:
  - *Existing markets.* Is the technology already being sold, and what markets are receptive? How big are they, and how big could they become? Are they growing or shrinking? What influence will changing demographics, competing technologies, or shifts in regulation bring?
  - *New markets.* Are there new geographical regions, customer types, or applications that could be target markets for the technology? Could they be developed, given a particular focus, or do they require a lower cost or better-performing technology?
  - *Actors.* Who are or could be the distributors of the technology, and who are the customers? Is this a B2B or B2C business? How are sales financed, and is there benefit in new financing models?

- The **public policy** environment will have major implications for the technology and its opportunities.
  - *Current policies.* What policies, regulations, laws, and standards currently affect this technology, either positively or negatively? Are they a significant benefit or problem? Are they likely to change, and can they in fact be changed or are they deeply entrenched? Conducting a systematic review of policy can bring light to these different areas and show where improvements may be made.
  - *Policy options.* Are there particular policy options that could be brought to bear to improve the position of the technology or sector? Could these be enacted relatively quickly, and is major consultation required? Are they likely to have knock-on effects in other sectors or for other technologies? If so, who else would need to be involved in a consultation?

- Understanding the different **actors** involved and implicated in the technology or sector can aid in identifying bias.
  - *Current actors.* Who are the actors who benefit directly or indirectly from the development of the sector? Are they all engaged or being consulted as part of the roadmap process? Are they all active (e.g., technology developers) or are some passive recipients of the technology (e.g., local residents in an area where the technology has been deployed)? What specific goals and strategies do these different actors have, and do they coincide or conflict?
  - *Potential actors.* Which other parties might be involved to improve the roadmap development, or help ensure its implementation? Why are they not already engaged, and will it be straightforward to bring them in? In some cases, powerful actors will be uninterested in the new technology, and their behavior may help or hinder its development unwittingly.

## Roadmap Development

Building on the analysis conducted to establish the baseline, either or both modeling and expert judgment can be used to build the roadmap. If the roadmap is an internal document with few stakeholders, such as a company technology strategy, it may be developed relatively simply. The information from existing cost models and supplier quotes, coupled with the identification of key decision points and milestones, may be sufficient to allow a way forward to be identified. It is important to note that no roadmap will be perfect, and indeed reality will soon diverge from most roadmaps, as both foreseen and unexpected factors change. The best designed product roadmap could not have foreseen the terrible events of the Japanese Tohoku tsunami in 2011, and the enormous disruption that would occur as a result, and even much smaller events can have a meaningful impact on the potential for a roadmap to be followed. This means that for a roadmap to be effective, it must not only be implemented by the relevant actors but also be updated and edited as circumstances change or goals are achieved.

## Considering System Effects

For some technologies, there is little or no need to consider system effects—the interaction between different parts of the energy system, for example. For many HFC technologies, it is essential. The 2016 UK Roadmap for Hydrogen and Fuel Cells [7] made note of this: "The biggest benefits of HFC applications are only seen when the whole energy system is considered, and when long-term benefits are taken into account." This statement relates directly to the fact that widespread use of hydrogen is both most likely and most valuable, when it is accepted in many different and interlinking applications. So although hydrogen used in transport can solve problems related to emissions and energy security in particular, restricting it to transport means that its use is likely to be suboptimal and may not in fact be economically viable. The 2016 UK roadmap considered this system interaction from the perspective of clusters of applications, with these grouped into functional blocs depending on whether they were more representative of the supply side or the demand side of the energy system. The arrows in Fig. 8.4 indicate the most direct interactions, such as the opportunity for hydrogen in transport also to be used for providing system services—for example, through using electrolyzers as dispatchable loads.

The development of the roadmap actions and timeline therefore also considered these interactions and illustrated dependencies in a simple way (Fig. 8.5). So, for example, the comment that pipeline conversion to 100% hydrogen should be coordinated with the development of economic bulk hydrogen production may seem self-evident to those involved in the process but may not be appreciated by those working in a specific technology aspect of one or the other application.

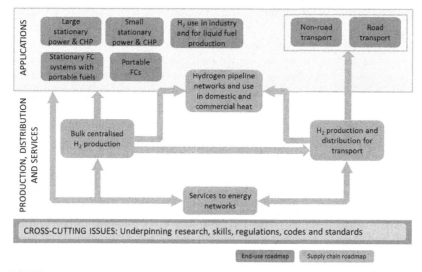

**FIGURE 8.4** Important energy system interactions for hydrogen and fuel cells. *E4tech and Element Energy (2016).*

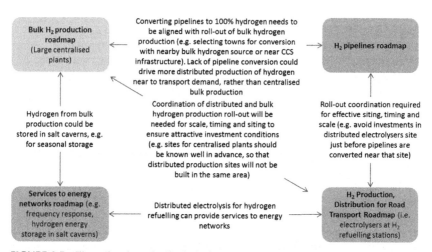

**FIGURE 8.5** Illustrating dependencies in hydrogen and fuel cells roadmap design. *E4tech and Element Energy (2016).*

## Analytical Support

Once the baseline is developed, the forward-looking portion of the roadmap can be built. The specific characteristics of the different technologies or sectors can be used to populate more or less detailed models, to examine the future options. These models may be relatively simple calculations of future capacity or economic value or can include sophisticated system models that attempt to optimize across many different technologies and sectors. However, these models do not predict the future or give the "correct" answer. Best practice should require that the outcomes are sense-checked and tested and that they are used as guidance in developing the final roadmap. This is not least because in emerging sectors—and to some extent in existing sectors—not all of the data will be available, error margins will be wide, and the models may not be able to consider all of the interactions or other conditions that are important. Strong and plausible analysis is likely to help consensus-building, and good analysis can simplify the process of identifying the important milestones and options to be considered.

### Expert Input

With or without a modeling phase, the production of a roadmap will be guided by expert input. This may take the simple form of in-house meetings for a small and specific roadmap goal, or fully facilitated workshops for either a more complex effort or one that involves many different stakeholders. These workshops can be organized by theme: policy, technology, or social impact, for instance; or perhaps by technology or application. Whichever approach is chosen, the purpose of the workshop should be to elicit expert input and, where possible, to obtain consensus on goals and targets from technology providers, business, government, and community leaders.

A workshop may be supported by individual interviews or other information gathering, with different goals: checking and improving assumptions; identifying technical and institutional barriers; defining different pathways and options; and developing and agreeing priorities for action.

## Typical Outputs

The final outputs of a roadmap are often summarized in a highly synthetic manner, as illustrated in Fig. 8.6. Although the underlying details and supporting steps of a recommendation may be described elsewhere, pulling together the main actions, recommendations, and milestones on a single diagram helps to put the roadmap into context. In the example shown here, different sectors of transport are shown on the same map, with actions and targets denoted in different colors. These actions include recommendations for research, development and demonstration, support mechanisms, regulation, and information provision and coordination. All of these recommendations are designed to lead toward specific future outcomes, for example, ensuring a local production capability for certain types of hydrogen vehicles. Some of these actions can be taken in isolation, but others will only work together.

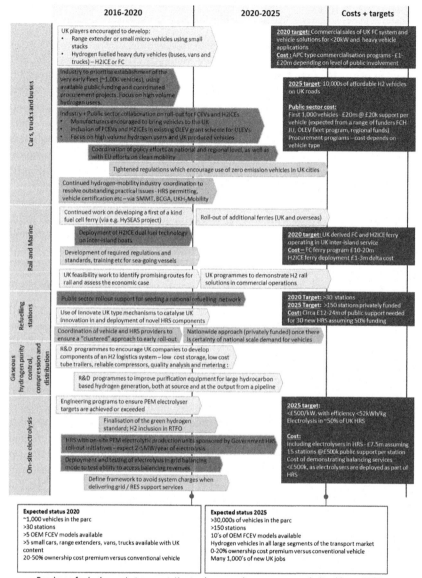

Roadmap for hydrogen in transport. Key to chevron colours: grey – regulation, blue – RD&D, yellow – deployment support, purple – information provision and coordination

**FIGURE 8.6** An individual roadmap depicted graphically. *E4tech and Element Energy (2016).*

## PUBLIC POLICY

### The Importance of Policy

Public policy is an extremely important and powerful component of any roadmap or roadmapping exercise and deserves consideration in its own right.

Public policy sets the context for all technology and applications, which may be explicitly encouraged or discouraged, or may simply be left to develop without active policy engagement. No technology or application can escape policy frameworks, however, as they guide and support not only technologies themselves but also market structures and interactions, profitability, competitiveness, and many other things. Individual behavior and responses will also be influenced by public policy and the social framework that it provides.

The relevance of policy in designing and implementing roadmaps has been discussed earlier. However, even if a roadmap is not under consideration, policy strongly influences where, when, and how technologies become successful, and hydrogen and fuel cell technologies are no exception.

### Where Hydrogen and Fuel Cell Technologies Fit Into Policy

Only very recently have the majority of hydrogen and fuel cell technologies—or at least those used for energy or consumer rather than industrial applications—started to become competitive. This is a result of decades of investment and, at least in part, of supportive policy. It is also the result of developments in adjacent and complementary technologies, such as batteries and photovoltaics. Nevertheless, the majority of fuel cell and hydrogen applications remain more expensive than their competitors.

HFC technologies have been supported for a variety of different reasons through their development. Industrial technologies such as electrolyzers had an early market and little or no competition for hydrogen production. Until it became possible to obtain and manage natural gas, electricity was the best way to produce large volumes of hydrogen for fertilizer and other uses. Natural gas—based steam reforming then rapidly overtook electrolysis as the technology of choice, however. Hydrogen remained prevalent where its chemical characteristics made it essential—in refineries, as a generator cooling fluid, as a clean reducing atmosphere in manufacturing environments—but the majority came from natural gas, and the electrolyzer market remained very small. Fuel cells on the other hand have had almost no true markets, with space missions one of the few areas where they could be afforded. Yet developers have persevered, and much of the largest and longest-standing investment has been a result of supportive policy.

### What Is Special About Hydrogen and Fuel Cells?

Civilization is passing through an era dominated by fossil fuels. Until relatively recently, hydropower, wind, and biomass provided almost all of the energy used worldwide. The discovery and harnessing of fossil resources enabled extraordinary achievements, but fossil resources are finite (in an absolute or at least an economic sense) and their use has consequences. Resource extraction and the transport and use of fossil fuels have greatly benefited our economies but have been responsible for extraordinary environmental problems at the same time.

Some of these problems have been ignored, been considered less important than the benefits, or simply been unknown in the past. Public policy has in general supported resource extraction and use, which has been tied to wealth generation and to improving conditions for some—although not all—of those able to get access to the energy made available. However, slowly these benefits are becoming outweighed by the problems. Some would argue this has been happening for a long time, but the global addiction to cheap energy and to fossil fuel has been too strong to break.

In principle, a return to the renewable energies used in the past, or to some equivalent, would allow mankind to continue development and improve the life of many, while significantly reducing the negative consequences. Solar, wind, hydropower, and biomass, used appropriately, could power everything we do. To make it emissions-free for as much of the chain as possible, zero-emissions energy carriers are required. Or at least, energy carriers whose emissions have no negative impact, with water vapor being an obvious example. Because various carbon compounds are known to have damaging effects on the environment, carbon-containing fuels do not meet the requirements. Electricity and hydrogen have been considered for some time as potentially ideal and complementary "fuels," "energy carriers," "energy vectors," or even "energy currencies." Electricity has already been in common use in the energy sector for well over a hundred years, but hydrogen, hard to handle and relatively expensive, has remained in its industrial niche. And this of course is where strong and coherent policy is essential.

## The Value of Externalities

To realize the benefits of zero-emissions fuels and energy chains requires those benefits to be explicitly valued. Either an actual cash value can be attributed to their use, or policy can erect nonfinancial hurdles for competing and dirtier options. The externalities of fossil fuel use—the costs or value that they provide which are not explicitly accounted for in their market price—need to be calculated and considered as part of the overall cost–benefit equation and different forms of policy brought to bear to account for them.

In addition to environmental benefits, hydrogen and fuel cell technologies can bring a range of other attributes, which are intrinsically or explicitly valued by policymakers. These should also form part of the decision-making process when deciding if and how to provide support for the sector.

Environmental benefits:

- Reduction or elimination of polluting emissions at the point of use.
- Reduction or elimination of carbon dioxide and other greenhouse gas emissions along the length of the fuel chain.

- Elimination of some resource extraction and its negative consequences. Of course, this depends on a great number of variables—where and how the hydrogen is produced and transported, what the primary resource is, and how it is eventually used.

Industrial benefits:

- Support to industrial development
- Generation of employment
- Reviving declining communities and regions

Economic benefits:

- Improving competitiveness
- Reduction of balance of payments deficits (for example, by reducing dependence on imported oil)
- Improvement of energy security (by allowing diversification of sources or allowing indigenous capacity to be used).

Some of these benefits are at a local level, some at a national or regional level, but all have value to policymakers and to society.

## Where Policy Can Act

One of the components of developing an effective roadmap is to assess the potential for the technology, application, or sector being considered. This assessment is often technical and probably economic but may also be environmental. The characteristics displayed should be evaluated in the light of existing policy, and future aims, to understand how support can best be given.

A concrete example of an area in which policy has been inconsistent is in the support, which has been given for several years to biofuels in Europe. These benefit from mandated levels of uptake and may count toward various targets, which a country has to meet, in terms of both emissions reduction and the percentage of renewable fuel that has been required. Fuels produced from biogenic resources, such as biodiesel and biomethane, were included in the legislation. Hydrogen, however, was not eligible if produced by electrolysis from wind or solar power because no biogenic component was included. So despite the zero or negligible emissions related to the fuel chain, hydrogen was not given the same advantage as a competing alternative. More recently, proposals have seen "renewable fuels of nonbiological origin" or RFNBOs, gain credence as alternatives to biofuels. In this case the policy was not intended to exclude such renewables, but when the legislation was written, they had not been considered relevant, and so the wording excluded them almost by accident.

Promotional policy can also be poorly drafted. China's subsidies for battery electric vehicles were originally offered to the manufacturer of the vehicle and gave no incentive for the vehicle to be actually sold or driven, simply for it to be built. Although this inevitably supported manufacturers, it delivered few or no environmental benefits, as the vehicles were not contributing to reducing air pollution.

Careful, evidence-based policymaking is therefore required, to ensure both that policy intended to support and that policy may simply exist already do not have unintended negative consequences. For the HFC sector, this can be complex, as it touches a great variety of other sectors, where policy overlaps in many areas and ways. A further example of this is "sector coupling." This recently coined term refers to any crossover between historically distinct sectors, such as electricity and transport. Because both hydrogen fuel cell and battery *vehicles* can in principle materially affect *electricity* demand, supply, infrastructure, and markets, policy needs to recognize and accommodate that fact as far as possible; however, the requirements and structures of the sectors may simply be hard to align. Electricity markets might require one-way power flows and be uneconomic to manage for only a few kilowatts of generation, whereas electrolyzers in hydrogen markets may also need the ability to provide small but high-value services to grid markets to be economic.

## Effective Policymaking

Making good policy is not straightforward. To do so, all of the primary objectives must be identified and categorized. A primary objective might be to reduce emissions of NOx in built-up areas or to develop a competitive hydrogen bus industry. Secondary objectives must also be considered—should the bus industry be in a particular location to also support jobs in a local community? Some objectives may not be easily compatible with others; exploitation of a primary resource is often likely to conflict with a need to maintain or improve the environment.

Once all of these different aspects have been identified, and their compatibility or overlaps assessed, initial policy design can take place. This can take the form of legislation or may simply provide incentives or disincentives to action. An overarching policy must take into consideration any regulations or other frameworks that already exist, either so that they do not conflict, or in order to modify them appropriately. Relevant actors may also be identified, especially if they are strongly implicated in either enforcing or adapting to the policy.

As far as possible, the design of policy should consider the implications for existing frameworks and regulations, both "above," e.g., supranational, and "below," e.g., regional. This is to try and ensure that it does not conflict. In practice, policymaking usually happens top-down, and so usually regional and local policy has to fit within an international or national policy framework.

## EXAMPLE HFC ROADMAPS

A wide range of hydrogen and fuel cell technology-relevant roadmaps have been produced, over many decades. Some of these are at a global, or at least international level, whereas some are specific only to cities or regions. The selection below is not exhaustive, nor is it intended to represent particular types of approach or attempt to judge the quality of the different outputs. It is, however, intended to provide some indication of where to find relevant literature that expands on the discussion above.

## Global Roadmaps

### IEA HFC Technology Roadmap [9].

The IEA examined a future global energy scenario in which hydrogen played a greater role than in their traditional Energy Technology Perspectives scenarios. This included some estimates of the economic cost and emissions benefit of rolling out hydrogen into the stationary and transport sectors across a selection of global regions. The modeling suggests that a considerable time—15—20 years—would be required to reach a self-sustaining Fuel Cell Electric Vehicle (FCEV) market after the first 10,000 were deployed. This seems to be a long period for manufacturers, policymakers, and others to have to wait. The roadmap suggests some high level actions, particularly in terms of deployment and information gathering.

### The Hydrogen Council—Hydrogen Scaling up [10].

The Hydrogen Council published a vision of hydrogen in the global energy system in 2017. It focuses on the contribution that hydrogen can make in 2050 and suggests roles that hydrogen can play in the energy transition, including as an enabler for large-scale renewables, a buffer to improve system resilience, and as a renewable feedstock.

## Country or Regional Roadmaps

Many roadmaps have been produced for different countries and regions, and the number continues to increase as the concept of hydrogen energy becomes better understood. The late 1990s and early 2000s saw roadmaps built for countries and regions implicated in some way by the technology interest of the time, such as Japan, Canada, and even the UK.

### Japan

Japan's approach has always reflected its reputation for long-term thinking and the early engagement of major corporations in that thinking. A project called WE-NET, World Energy Network, or International Clean Energy System Technology utilizing Hydrogen [11], examined many of the areas now starting to be developed and exploited for HFC technologies globally, including the renewable production of liquid hydrogen or organic hydrogen carriers and their long-distance transport to demand centers. This was primarily a research project, however, and the government subsequently funded development and demonstration work to bring a range of technologies closer to production. The Ministry of Economy, Trade and Industry published a roadmap in 2014 [12], subsequently updated, which laid out at a high level many of the goals and targets for Japan. This roadmap is of particular importance as it has been strongly supported by both national and regional government in Japan, specifically being referenced by the Prime Minster on several occasions. It is extremely rare for such high level buy-in to occur.

## UK

Although several other analyses have been undertaken over a number of years, the UK's most recent Hydrogen and Fuel cell Roadmap was released in 2016 [7]. The focus of the analysis and output was specifically on the most plausible and beneficial opportunities for the UK, and so areas where the UK had no particular competitive advantage were not considered in detail. An initial screening exercise led to the identification of 11 subsectors of interest, for which "miniroadmaps" were produced, each laying out in more detail the technological, economic, and political aspects of particular applications. The different miniroadmaps were compared and contrasted and overarching roadmaps developed for the final summary report. This final report focused on four value propositions for the UK:

1. Hydrogen as a major component of the future energy system
2. Hydrogen in transport, for air quality and decarbonization
3. Fuel cell combined heat and power (CHP) for energy efficiency
4. Stand-alone fuel cell products that have intrinsic benefits

Areas that fell outside of these were not analyzed.

The UK's history of analysis in this area may offer some insight into developments, as it stretches back over several decades. For example, a 2004 analysis for the then UK Department of Trade and Industry considered the strategic framework for the country [13].

A summary "routemap" document from 2004 is also available [14], as is a separate document covering only fuel cells, developed by the industry and its representative association [15].

Regional roadmaps were also developed, including an "Action Plan" from London in 2002 [16]. London has continued to support hydrogen and fuel cells and has deployed both stationary and transport fuel cell applications.

## China

China has only much more recently entered the arena, and in 2016 it produced a Hydrogen Fuel Cell Vehicle Technology Roadmap as part of the Energy

Saving and New Energy Vehicle Technology Roadmap. It focuses on fuel cell stacks, systems, and vehicles and on hydrogen technology, including infrastructure and on-board storage [17].

### The United States

The United States has produced a range of documents addressing the possibilities of hydrogen and fuel cells from a national perspective. These include a Hydrogen Energy Roadmap [18], and a Vision of its transition to hydrogen [19], both developed in 2002. Since that time, similar publications have focused on specific plans for entities such as the US Department of Energy, including a Hydrogen Research, Development, and Demonstration Plan updated in 2013 [20].

Individual states have also produced their own roadmap-type documents, including Hawaii's geothermal Hydrogen Roadmap in 2008 [21] and Renewable Hydrogen Program Plan in 2011 [22].

The US Midwest region released a Hydrogen Roadmap in 2017, centered around Ohio [23].

### Australia

A national Australian Roadmap was written in 2008, but no longer seems to be available online. However, a new version is in development and a much more recent analysis was carried out for South Australia alone and published in 2017 [24].

## ROADMAPS: IMPLICATIONS AND CONCLUSIONS

To be valuable, a roadmap simply needs to help something move forward. To do this, it needs to be appropriate to the sector and actors under consideration, provide clear and actionable recommendations, and be accessible to those who need it. It may be sophisticated or simple, detailed, or sketchy, as long as it is fit for the purpose it was intended for.

The strengths of a roadmap lie in the rigor and inclusivity that *should* go into its development and in the synthetic, approachable, and implementable outputs. It should allow the building of a consensus on how to approach the future, and mutual support for the actions required.

However, roadmaps suffer both from systemic and specific weaknesses, perhaps the most important of which is that they are only documents. For a roadmap to be of any real value, the steps laid out in it must be implemented, conditions and progress monitored, and changes made when necessary to keep the roadmap relevant. All too often this is not the case, particularly with roadmaps that require multiple and diverse stakeholders to take action to move them forward, or where there is no obvious "owner."

Particular issues that can arise include the time span of the roadmap. Something that is rooted in the near future and is fundamental to the survival or growth of a single company is likely to receive the focus and input it deserves. So a company developing fuel cell components will likely follow and update its technology roadmap regularly and frequently. A government looking at the optimization of the energy system in 2050 may struggle to grasp all of the threads necessary to keep a roadmap live and relevant, especially as the external environment may change dramatically over a longer period of time. A relevant example is the much greater drops in prices for renewable electricity than predicted by most analysts, which have affected not only roadmaps directly related to those renewables but also those for other technologies competing with them or using them as resources.

This frames a more general point relevant to roadmaps for immature technologies. Because these will be subject to significant levels of innovation, which is inherently very hard to predict, their roadmaps may be rendered irrelevant in a very short period of time. The need to monitor, update, and modify the roadmap is very high for such technologies, which certainly include HFC technologies.

Roadmapping can sometimes be seen as an alternative to taking action. Governments and other entities are criticized for developing roadmaps instead of "doing something." The reality is more complex. Without a good roadmap, the entity is very likely to put in place inadequate or unhelpful regulations, fund the wrong R&D, or support the chosen technology or sector in ways that are either ineffective or suboptimal. However, the roadmap, as stated above, is only the start of the process. If a roadmap is produced and put on a shelf, but several years later a new roadmap is produced, cynicism is both likely and probably warranted. The roadmap should be a small—but extremely important—catalyst, which enables real action.

## ACKNOWLEDGMENTS

The author would like to thank Chester Lewis, E4tech London for his help in early drafts of this chapter, and the E4tech team in general for their work on roadmapping in different sectors and contexts, which has informed this review.

## REFERENCES

[1] International Energy Agency, Energy Technology Roadmaps — a Guide to Development and Implementation, 2014.

[2] W. McDowall, M. Eames, Forecasts, scenarios, visions, backcasts and roadmaps to the hydrogen economy: a review of the hydrogen futures literature, Energy Pol. 34 (11) (2006) 1236—1250.

[3] World Health Organization, Ebola Response Roadmap, 2014.

[4] W. McDowall, Technology roadmaps for transition management: the case of hydrogen energy, Technol. Forecast. Soc. Change 79 (3) (2012) 530—542.

[5] R. Phaal, C.J.P. Farrukh, D.R. Probert, Visualising strategy: a classification of graphical roadmap forms, Int. J. Technol. Manag. 47 (2009) 286—305.

[6]    R. Phaal, Roadmapping for Strategy and Innovation, 2015.

[7]    E4tech and Element Energy, Hydrogen and Fuel Cells: Opportunity for Growth. A Roadmap for the UK, 2016.

[8]    Automotive Council UK, Advanced Propulsion Centre UK, Electrical Energy Storage Roadmap, 2017.

[9]    International Energy Agency, Technology Roadmap — Hydrogen and Fuel Cells, 2015.

[10]   Hydrogen Council, Hydrogen — Scaling up. A Sustainable Pathway for the Global Energy Transition, 2017.

[11]   M. Chiba, H. Arai, K. Fukuda, WE-net: Japanese hydrogen Program, Int. J. Hydrogen Energy 23 (3) (1998) 159–165.

[12]   Agency for Natural Resources and Energy, Ministry for Economy, Trade and Industry, Japan, Summary of the Strategic Roadmap for Hydrogen and Fuel Cells, 2014.

[13]   E4tech, Element Energy, Eoin Lees Energy, A Strategic Framework for Hydrogen Energy in the UK, 2004.

[14]   UK Department of Trade and Industry, Sustainable Energy Technology Routemaps: Hydrogen, 2004.

[15]   Fuel Cells UK, UK Fuel Cell Development and Deployment Roadmap, 2005.

[16]   London Hydrogen Partnership, London Hydrogen Action Plan, 2002.

[17]   SAE China, Strategy Advisory Committee of the Technology Roadmap for Energy Saving and New Energy Vehicles, Hydrogen Fuel Cell Vehicle Technology Roadmap, 2015.

[18]   United States Department of Energy, National Hydrogen Energy Roadmap, 2002.

[19]   United States Department of Energy, A National Vision of America's Transition to a Hydrogen Economy — to 2030 and beyond, 2002.

[20]   United States Department of Energy, Energy Efficiency and Renewable Energy, Fuel Cell Technologies Office, Multi-year Research, Development and Demonstration Plan, 2013.

[21]   Sentech Inc, Analysis of Geothermally Produced Hydrogen on the Big Island of Hawaii: A Roadmap for the Way Forward, 2008.

[22]   Kolohala Holdings, LLP and Hawaii Natural Energy Institute, Hawaii Renewable Energy Program Plan 2010–2020, 2011.

[23]   CALSTART, Hydrogen Roadmap for the US Midwest Region, 2017.

[24]   Government of South Australia, A Hydrogen Roadmap for South Australia, 2017.

Chapter 8

# Market, Commercialization, and Deployment—Toward Appreciating Total Owner Cost of Hydrogen Energy Technologies

Robert Steinberger-Wilckens, Beatrice Sampson
*University of Birmingham, Birmingham, United Kingdom*

## HYDROGEN IN THE MARKET TODAY

### Applications

Already today, hydrogen is broadly used for various applications in industry. Historically, the production of urea for agricultural fertilizers via ammonia synthesis was the first large-scale application of hydrogen [1], followed by cracking of heavy oil and desulfurization in oil refineries. Further fields of current hydrogen use are [2] as follows:

- ammonia synthesis,
- methanol production,
- hydrogenation of fats in food processing,
- metallurgy,
- cooling in gas turbines,
- production of artificial diamonds,
- chemical industry in general, and, more recently,
- energetic use.

We therefore find the following categories of use:

- nonenergetic in fertilizer production, food industry, chemical industry, metal and glass production, etc.,
- indirect energetic in refineries, and

*Science and Engineering of Hydrogen-Based Energy Technologies.* https://doi.org/10.1016/B978-0-12-814251-6.00008-3
**383**

- direct energetic when used as a replacement for natural gas (e.g., where hydrogen is available free as a by-product).

In the future, the final bullet point will be gaining growing importance in that hydrogen may be used as a substitute for natural gas with the goal of decarbonizing the gas and especially the heating sector [3,4] by either mixing hydrogen into the natural gas and thus continually reducing the contribution of carbon dioxide emissions, or completely converting gas heaters and boilers, cookers, and internal combustion engines (ICEs) to hydrogen fuel. Another future use would be so-called "power to gas" [5] where electricity is used to split water into oxygen and hydrogen, the latter being stored and at a later point in time converted back to electricity via a gas turbine or fuel cell [6]. Hydrogen would therefore serve as an electricity storage medium at large scales, which would be fully compatible with the previously mentioned admixing into natural gas grids. The electricity and gas energy sector and markets would therefore be intimately linked. Any hydrogen produced could also be branched out as a fuel for fuel cell electric vehicles (FCEVs) [7], thus adding the third large energy market: transportation fuels. Hydrogen would therefore act as a "hinge" element between all three main energy markets and add a considerable element of security, resilience, and flexibility to the energy sector [6].

## Refinery Use of Hydrogen

Traditionally, refineries have used hydrogen for desulfurization of oil products using, for example, the process indicated by Eq. (8.1) [8]:

$$R_1 - S - R_2 + 2H_2 \rightarrow R_1 - H + R_2 - H + H_2S \tag{8.1}$$

which is catalyzed by CoMo or NiMo catalysts. $R_x$ indicates (arbitrary) part-chains of the heavy oil. Similarly, when processing crude oil, refineries "crack" the long hydrocarbon chains (C−C bonds) into shorter bits, producing the different commercial fractions (light and heavy oils, naphtha, etc.), for example [9]:

$$RCH_2CH_2CH_2CH_3 + H_2 \rightarrow RCH_3 + CH_3CH_2CH_3 \tag{8.2}$$

Clearly, additional hydrogen is required to saturate the ends of the "chopped" chains. More recently, the thermal energy necessary to crack crude oil is being replaced by adding hydrogen and thus encouraging the long hydrocarbon chains to dissociate into shorter fractions. When producing the hydrogen from natural gas or from partial oxidation of heavy oil residues, the energy balance is more positive than with the pure thermal process. On another note, hydrogenation (Eq. 8.3) is used to purify and stabilize petrochemical products, by removing nitrogen and metals, or specifically by converting alkenes and aromatics to saturated alkanes (paraffins) that will not convert during storage or form gums that will interfere with handling equipment [9]:

$$R-CH{=}CH_2 + H_2 \rightarrow RCH_2CH_3 \tag{8.3}$$

## Ammonia Synthesis

Ammonia is produced predominantly by the Haber—Bosch process from nitrogen (air) and hydrogen with an iron catalyst at high temperatures and pressures ($400-500°C$, $15-20$ MPa) according to Eq. (8.4) [10]:

$$N_2 + 3H_2 \rightarrow 2\,NH_3 \tag{8.4}$$

Ammonia is the basis for products such as fertilizers, explosives, fibers, and plastics, as well as pharmaceuticals. It is also used as a cooling agent in refrigeration with a high efficiency because of a favourable compression and evaporation/liquefaction temperature cycle.

The annual demand for hydrogen from "captive" sources (i.e., produced on-site, see below) in ammonia synthesis is $335 \times 10^9$ Nm$^3$ worldwide (accounting for 55% of total annual captive hydrogen) [11]. These numbers are substantial and we will see a bit further down how this compares with other gas markets. In the 1980s this share was 80%, showing that other applications of hydrogen have since caught up considerably (especially oil refining and methanol production).

## Food Industry

In food processing, hydrogen has predominately fulfilled the role to serve as "hardener" for plant oils when producing margarine. This "hydrogenation" has to be balanced with the requirement of avoiding to create 'trans' partially saturated fats that are today recognised as a source of cardiovascular diseases. In some cases, hydrogen is used as an inert gas in processing delicate foods to avoid any intrusion of oxygen, but mainly in the form of hydrogen peroxide for sterilization [12].

## Metallurgy

Pure hydrogen/oxygen flames are used in metallurgy for cutting and welding because of the high temperatures obtained when using stoichiometric mixtures and the lack of any corrosive reaction products—apart from water, of course. Hydrogen is also used for atmospheres in annealing and brazing, when any oxidation of the treated components has to be avoided (e.g., formation of scales on metal surfaces at high temperatures) [13], for hardening, etc. Metal oxides are converted to pure metals by reducing with hydrogen [14]. This latter application may develop into a major use of hydrogen in industry in the near future [15,16].

## Chemical Industry

In the chemical industry, products such as styrene, ethylene (as a precursor for polyethylene production, PE), polyurethane, olefins, or peroxide are produced with the help of hydrogen [17]. The base product is often synthesis gas (syngas), which is a mixture of hydrogen and carbon monoxide. Besides methanol, plastics, pharmaceuticals, and explosives are also produced from syngas, which again is today mostly derived from methane steam reforming of natural gas. Syngas is equivalent to town gas that fed the predecessor of the natural gas infrastructure we know today and was produced by coal gasification.

## Other Uses: Power Industry, Electronics, Glass, and Others

In the pulp and paper production, hydrogen peroxide is used for chlorine-free bleaching [18] and substantially reduces pollution from the traditional chlorine, hydrosulfite, or sodium hydroxide bleaching. In many industrial applications, hydrogen is used for a variety of reasons and for a large number of processes [19]. Mostly the role is prescribed by the need to retain a reducing environment to avoid any oxidation of components or products. This ranges from generator cooling, where the bearings are kept free of abrasive oxide layers, to inert atmospheres for semiconductor-producing furnaces, artificial gemstone, and diamond production. Hydrogen is added to vehicle fuels to reduce $NO_x$ formation.

Last but not least, hydrogen is also used as a balloon gas. This closes the circle back to the first uses of hydrogen in military reconnaissance balloons at the turn from the 18th to 19th century and today weather and observation balloons [20], children's balloons mostly being filled with helium for safety reasons.

## Energetic Use

The only large-scale energetic use of hydrogen known today is the space industry, which uses hydrogen as a rocket fuel [21]. The main reason for this is the outstanding gravimetric energy density of hydrogen, which is 33 kWh/kg (HHV) (cf. Table 8.1).

# Hydrogen Markets

Hydrogen was first used at large scale with the balloons and airships coming into use at the end of the 19th century—leading to the Zeppelin development [22] and the well-known barrage balloons that protected the United Kingdom in World War II [23]. Many of the now well-known hydrogen-related materials problems were then first encountered (steel embrittlement, requirements for special sealings and washers/gaskets, leakage rates from storage vessels, etc.). Large-scale employment of hydrogen was then spurred by the fertilizer

**TABLE 8.1** Overview of Energetic Properties of Hydrogen and Other Fuels

| Lower Heating Values | Volumetric | Gravimetric (kWh/kg) |
|---|---|---|
| Hydrogen | 3.00 kWh/Nm$^3$ | 33.33 |
| Methane | 9.97 kWh/Nm$^3$ | 13.90 |
| Natural gas | 8.8–10.4 kWh/Nm$^3$ | 10.6–13.1 |
| Propane | 25.89 kWh/Nm$^3$ | 12.88 |
| Butane | 34.39 kWh/Nm$^3$ | 12.70 |
| Town gas[a] | 4.54 kWh/Nm$^3$ | 7.57 |
| Crude oil | 10.44 kWh/L | 11.60 |
| Diesel | 10.00 kWh/L | 11.90 |
| Gasoline | 8.80 kWh/L | 12.00 |
| Methanol | 4.44 kWh/L | 5.47 |

[a]51 vol% $H_2$, 18 vol% CO, 19 vol% $CH_4$, 2 vol% $C_nH_m$, 4 vol% $CO_2$, 6 vol% $N_2$.

industry in the early 20th century when many large-scale hydro energy developments [24] were equipped with electrolyzers to produce the hydrogen gas
necessary for feeding the synthesis according to Eq. (8.4) [25]. Up to the mid-
1990s, many town gas networks operated in Germany (especially East Germany and Berlin) and Sweden, with a town gas supply still in operation in the
city of Stockholm [26]. Town gas is a mixture of hydrogen and carbon
monoxide and is the product of coal gasification, which fed the urban gas
supply networks from the early 20th century to the 1960s [27]. It consists of up
to 50 vol% of hydrogen, which is why many natural gas networks today would
also support hydrogen admixture and even complete conversion to hydrogen
because the steel piping materials, still in place from former days, are just as
suitable for hydrogen as for methane.

## Market Types

The hydrogen market has some peculiarities, which we do not necessarily find
in other markets. For one, the bulk of hydrogen provision occurs in the shape
of "captive hydrogen," indicating hydrogen produced on-site of a customer by
a third party (as a contractor, or via plant leasing, etc.). It is not possible to
identify a market price because of the monopoly structure of delivery once the
plant has been installed [11].

In contrast, "merchant" hydrogen is that hydrogen produced centrally,
transported to the customer by surface transport or pipeline, and used for a

variety of industrial or energy processes. Here, the market price is defined by free trading and competition. This part of the market is currently growing [11].

Two other sources of hydrogen exist, which could be exploited to feed the early stages of the development of a widespread hydrogen infrastructure [28]:

- by-product hydrogen, and
- stranded (surplus) hydrogen.

Many processes in the chemical industry release hydrogen as a by-product, for instance, the production of chlorine from salt (NaCl) by electrolysis, the coating of metals with zinc, and others. Partly, this hydrogen is captured and put to internal use as a raw material (refineries), but most often as a natural gas replacement. This is due to the lack of suitable customers and the cost of upgrading to industrial standards (i.e., 5-point = 99.999 pure hydrogen) for subsequent sale. In some cases, the hydrogen is too impure or too diluted and is simply vented.

Surplus (or "stranded") hydrogen could potentially be captured (e.g., from chlorine production), but can neither be used locally, nor processed, stored, and sold due to a variety of reasons, including distance to marketing hubs, cost of processing, unwillingness to engage in markets outside of the core business etc. It therefore is a resource that could be tapped if an adequate infrastructure can be put in place to collect and process the hydrogen.

The amount of hydrogen theoretically available as a by-product and from stranded hydrogen sources in Europe has been estimated to be between 2 and $10 \times 10^9$ Nm$^3$/year [28] (Table 8.2).

## Market Volumes

Annual worldwide hydrogen production is in excess of $650 \times 10^9$ Nm$^3$ (approximately 2 TWh/a) [29]. The share of the EU is approximately $90 \times 10^9$ Nm$^3$/year (approximately 270 GWh/a) [28]. Consumption had risen by 5%−10% annually since the year 2000 and only slowed following the financial crisis 2008, now settling at around 3.5% annually [11].

**TABLE 8.2** Market Statistics of Hydrogen Deliveries, EU Statistics (c.1997)

| | |
|---|---|
| Captive | $55 \times 10^9$ Nm$^3$ (93%) |
| By-product | $3 \times 10^9$ Nm$^3$ (5%) |
| Merchant | $1.2 \times 10^9$ Nm$^3$ (2%) |
| Direct energetic | $0.5 \times 10^9$ Nm$^3$ |
| Surplus (stranded) | $0.015 \times 10^9$ Nm$^3$ |

For comparison:

- the world natural gas market has a volume of $3.5 \times 10^{12}$ Nm$^3$/year [30],
- the EU natural gas market trades $400 \times 10^9$ Nm$^3$/year (c.5 TWh) [30].

The volume of hydrogen marketed is thus in the order of 10%−15% of the natural gas market; however, hydrogen is not generally observed as a readily available commodity. This is mainly due to the fact that most of the hydrogen deliveries occur from the so-called "captive" sources where contractors produce hydrogen on-site for a single customer, mostly refineries and large chemical plants. Therefore, this hydrogen does not appear as a freely transported and traded resource and is not perceived by the general public. In addition, there is no publicly available market price for these services.

Nevertheless, public statements from refinery companies state that hydrogen could be delivered "off the fence" of refineries for much the natural gas price (partly being a by-product of oil processing and not of methane reforming). Leaving aside the market price fluctuations of natural gas and taking long-term average numbers, this would be equivalent to something around and below €1/kg of hydrogen. With a current selling price ("pump price") of hydrogen at hydrogen refuelling stations (HRSs) of above and around €10/kg H$_2$ [31], this means that much of the selling price of hydrogen is caused by handling. Especially, bottled hydrogen would be much in excess of the stated price. A detailed analysis of current and future cost of hydrogen produced from a variety of sources, including renewable energy, showed that hydrogen can be produced at a very low cost from any type of energy (Table 8.3).

**TABLE 8.3** Current and projected hydrogen production costs (Roads2Hycom 2008 & 2009).

| Process | Feedstock | Feedstock cost [€/kWh] | Hydrogen cost [€/kWh HHV] | Hydrogen cost [€/kg] |
|---|---|---|---|---|
| Steam methane reforming (SMR) | Natural gas | 0.026 | 0.061 | 1.6 |
| Electrolysis | Electricity, base load | 0.02 | 0.049 | 1.617 |
| | Electricity, wind, large scale | 0.091 | 0.156 | 5.148 |
| | Electricity, PV, large scale | 0.13 | 0.203 | 6.70 |
| Biomass gasification | Lignocellulosis | 0.114 | 0.194 | 6.40 |

## CAPEX VERSUS OPEX—TOTAL COST OF OWNERSHIP FOR HYDROGEN TECHNOLOGIES

Renewable energy, fuel cell, and hydrogen projects suffer in economical comparisons from the fact that they all induce a high initial capital investment. This is partly due to the stage of technical and commercial development they are in, with a high amount of research investment needing to be reclaimed from early sales. On the other hand, the market sectors addressed by these technologies are very well-developed markets with long-standing structures and pricing. The energy monopolies keep customer prices high—which is good for comparison with the emerging technologies mentioned—while on the other hand have driven the pricing level down in the past decades to a very competitive level—which is again of disadvantage. All in all the current cost level of fossil fuels corresponds to a mature market based on well-established technologies that have gone through past price squeezes. Investment in energy conversion technology will tend to be low, driving down capital expenditure (CAPEX) and making technologies look very competitive and cost-effective. Little emphasis is laid on the efficiency of the conversion processes because the fuel to be used for conversion appears to be cheap, leading to low operating expenditure (OPEX). Nevertheless, the influence of OPEC on fossil energy pricing has shown in the past (Fig. 8.1) that regulatory and political impacts on the options to purchase and use fossil fuels can be of a decisive and disruptive nature to the national economies of the fossil fuel importing countries.

The "spark spread" for natural gas or "dark spread" for coal-fired power units defines the economic benefit of producing electricity from this fossil

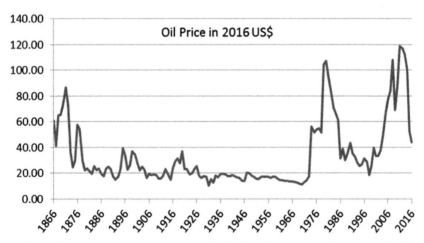

**FIGURE 8.1** Historical oil price development 1866 to 2016 in US$ of 2016 (using US inflation rate). *BP Statistical Review of World Energy June 2017. From the annual energy statistics published by BP on: http://www.bp.com/statisticalreview.*

source of energy [34]. Therefore, technologies such as gas turbines for electricity generation and gas-driven combined heat and power plants (CHP, delivering electricity and heat from decentralized units) have in the past run into severe economic problems [35], when gas prices surged and the ensuing OPEX destroyed the economical case for specific installations. Because energy conversion technologies look at operational lifetimes of one or two decades, price stability with respect to OPEX would be highly attractive but can obviously (cf. Fig. 8.1) not be guaranteed and creates a high risk of operation.

Renewable energies and energy conversion technologies with high efficiency change this picture because they induce lower OPEX owing to lower fuel consumption for the same delivery of energy or even eliminate any fuel costs (e.g., wind, photovoltaics). Apart from maintenance work, the OPEX is thus of a different order of magnitude than with conventional technologies. Therefore, the higher CAPEX will be in part offset by operational savings. Nevertheless, in many conventional economical appraisals, OPEX is not considered and comparisons will be based on CAPEX [36,37].

To have a fair and economically sensible comparison, however, CAPEX and OPEX have to be combined to the "total cost of ownership." This is a method of calculating all costs ensuing from purchasing and operating an energy conversion device during its lifetime (assuming for simplification purposes that it runs at constant power P):

$$\text{TCO} = \text{CAPEX} + t \times 1/\eta \times P \times (c_{\text{fuel}} + c_{CO_2}) + t \times \eta \times P \times c_{O\&M}$$

$$+ n \times L + \text{DECEX} \qquad (8.5)$$

with TCO total cost of ownership, CAPEX capital expenditure for purchasing and installing the unit, t operational lifetime (hours), $\eta$ efficiency of conversion, P net power rating of the unit (kW), $c_{\text{fuel}}$ cost of fuel (currency unit/kWh heating value), $c_{CO_2}$ specific cost of $CO_2$ emissions (in currency unit/kg $CO_2$), $c_{O\&M}$ cost of ordnance and maintenance (O&M) (currency unit/kWh energy delivered), n number of years operated, L any levies or taxes collected on an annual basis and not connected to energy delivered, and DECEX expenditure for decommissioning. In this equation, the operating costs connected to fuel and O&M may be much higher than the original investment. In this case, the cost of fuel becomes a key factor, as does the efficiency of fuel conversion.

## Improved Conversion Efficiencies

This is where hydrogen enters the picture. Used in fuel cells, the conversion efficiency can surpass 60% net efficiency. This is quite far removed from the approximately 20% a road vehicle achieves or even the 40% for large stationary diesel engines (in CHP units). Comparing a diesel or petrol (ICE) with an electric, hydrogen-driven vehicle (FCEV) therefore allows for higher fuel costs to break even in an economic comparison. At the typical €10 per kg of hydrogen at which this fuel is regularly sold at European filling stations, the fuel costs of an ICE vehicle and an FCEV are very similar. Once the pump price falls to the predicted 3 to 2 €/kg, hydrogen would be an immensely competitive fuel. Taking the assumptions generally made for fuel cell maintenance (very low) and decommissioning (very low) for granted for the purpose of this analysis, the post-investment costs for an FCEV would be much lower than for an ICE vehicle.

## Price Stability Effects

As can be seen from Fig. 8.1, the volatility of fossil fuel prices (in this case, crude oil) is high and can go to extremes. Sharp peaks in price are often followed by periods of reduced costs, but inevitably, the next peak will follow. Because the root cause for the price volatility is rarely to be sought in scarcity of energy resources alone, it has to be accepted that other effects such as political influence, regulation, or impacts from the financial markets take their toll. Any energy system relying exclusively or very heavily on crude oil products, such as the world transport system, will be strongly impacted by these price fluctuations and thus be economically highly vulnerable.

Producing bulk amounts of hydrogen from renewable energy sources not only supports a sustainable primary energy supply infrastructure with minimal emissions but also allows investments to remain in the country and contribute to local job growth instead of being exported to the countries selling fossil energy. This is because most renewable energy developments are capital intensive and the investment will occur within the country, thus creating local and national wealth. Even if equipment has to be imported local creation of value will by far surpass that of fossil fuels. [37a] With low cost on the side of operations (quite contrary to fossil energy conversion), and investment in national renewable energies and hydrogen production, hydrogen can contribute to long-term stability of energy prices. Once a renewable hydrogen production facility has been built and financed, there are few mechanisms that influence the cost of production. The only exceptions are schemes that use biomass and waste where shifts in market segments and ensuing volatility of prices have been observed. One example is the recycling business which, as it grows, increases the "demand" for waste. This reduces the income of waste processors who rely on being paid to handle waste. Once waste has a (positive) market value, this business model collapses. Many biomass schemes were rendered unviable when they relied on being paid to "eliminate" waste once this waste could be sold off as a raw material. This happened with the waste paper business or biogas plants processing food waste. A recent move of supermarkets to give away wastes to charities was, for example, not welcomed by waste processors [38].

Any technology that therefore shifts costs from operation to investment at the same TCO will have an advantage in that it reduces any future risks of price volatility. It can be safely assumed that fossil energy costs will have a tendency to increase in the future. As reserves rapidly deplete and the pace at which resources can be turned into reserves is dramatically slowing, the midterm price curve will inevitably point upward. Renewable energies such as solar photovoltaics or wind energy do not have any fuel costs. A large investment is made up-front, upon which the only running costs are the expenses for O&M and any taxes or levies imposed. This means that the price calculated for a unit of service (kWh, km driven, etc.) will be largely predictable and may be considered constant over the lifetime of the installation. Renewable energy therefore brings an element of price stability and security to the energy system.

Hydrogen will need to rely on renewable energy sources to make a positive impact on the environment ("green" hydrogen). Therefore the same argument as developed above will also be applicable to green hydrogen delivered to the energy markets. This will refer to renewable electricity converted to hydrogen in an electrolysis unit or the photocatalytic production of hydrogen. Because of the lack of continuous consumption of fuels in the process, the cost of this hydrogen will be practically constant over the lifetime of the conversion equipment. The only variable in the hydrogen production will be the price of deionized water. For hydrogen produced from biomass by gasification, biogas-reforming or by algae, though, energy crops or nutrients will be necessary to drive the process, thus introducing some potential price volatility.

Nevertheless, as indicated above, hydrogen can be produced from a variety of sources. This creates a high flexibility in the choice of feedstock for hydrogen production and thus increased options for switching from one feedstock to another in case of scarcity or price rises [6]. In this way the dependency on a single source of energy input into the hydrogen production is eliminated and the impact of price volatility of a single energy carrier is reduced.

When building on hydrogen from renewable energies, influences of world market energy price volatility on the economy are significantly reduced. This adds a decisive element of both security of supply and affordability because the risk of an impact of external energy markets and policy developments on the economy is greatly reduced. Transport fuels are an outstanding example of the impact that world politics can take on key aspects of a healthy economic development. With a high dependency of the pricing of processed oil products on international markets, world market price volatility of crude oil and oil products will immediately take a hit at the economic competitiveness of businesses. Successfully introducing hydrogen and synthetic natural gas (SNG, produced from hydrogen and carbon dioxide) fuels in road transportation will reduce the dependency on fuel imports and mitigate the impact of oil price fluctuations, as well as securing long-term price stability in this sector. It also reduces risk and therefore allows to reduce the contingency margins costed into market prices of oil products and the services that depend on them.

## Levies and Taxes

Politics have another way to take strategic decisions on energy use by taking influence on levies and taxes as shown in Eq. (8.5). Many countries tax road vehicles according to their environmental impact (mostly measured in $CO_2$ emissions per 100 km) (factor L in Eq. 8.5). Another measure adopted by the European Union is that of carbon taxation (factor $c_{CO_2}$ in Eq. 8.5). Certificates are released to the energy and industrial sector of a country that put a price tag on carbon dioxide emissions. Initially covering the complete $CO_2$ emissions of the commercial sector, over time, the number of certificates is reduced [39]. The intention is to force the commercial sector to gradually reduce carbon emissions or pay money to buy the necessary certificates off players that do not need them. In this way, carbon emissions receive a "price tag" that allows to add cost to greenhouse gas emissions and in this way create an impact on costings that is aligned with the political will to reduce the environmental impact of energy supply. Unfortunately, the amount of carbon certificates has in the past been far in excess of the required number, resulting in a devaluation of the certificates [40].

The impact of levying a carbon tax on hydrogen production can be seen in Fig. 8.2. It shows the levelized cost of hydrogen for a number of different

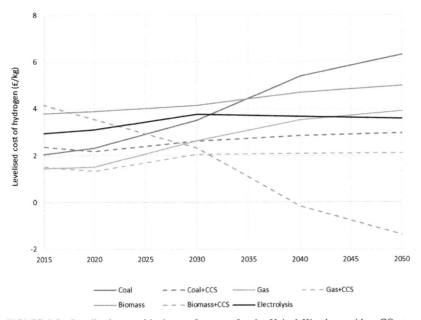

**FIGURE 8.2** Levelized cost of hydrogen forecasts for the United Kingdom, with a $CO_2$ tax increasing from £50/t$CO_2$ in 2020 to £250/t$CO_2$ in 2050 [41]. No tax is levied on electricity in this diagram. Capital cost data are from Ref. [42]; feedstock price forecasts are primarily from Refs. [43] and [44]; and other data are taken from the UK TIMES energy system model [45].

production pathways, fossil and renewable. From 2020 to 2050, it is assumed that carbon taxes increase from £50 t/$CO_2$ to £250/$tCO_2$. Production units with high carbon emissions become substantially more expensive than plants with carbon capture and storage (CCS). CCS costs could be considered as another way of monetizing emission impacts, by the way. With increasing carbon taxes, biomass CCS turns from the most expensive to the cheapest pathway. The effective negative carbon emissions resulting from atmospheric $CO_2$ being sequestered underground via biomass conversion could be considered as a way to pay for actively removing carbon from the atmosphere, a process that was performed millions of years ago by the flora for free, resulting in the habitable planet we (still) live on today. As discussed previously, the cost of hydrogen from electrolysis remains largely constant.

## Pump Price Versus Societal Costs—The Concept of "Externalities"

As mentioned above, renewable energy and hydrogen projects suffer from the fact that they induce a high capital investment, albeit this can be partly offset by operational savings. What makes things worse in the case of these technologies is that they are essentially "very low carbon" but compete with heavily polluting incumbent technologies. The expectation in government policies that low carbon technology should not induce additional costs is misleading in that it ignores the high cost the taxpayer carries for compensation measures caused by the impacts of fossil energy. A large part of the environmental and health costs of energy use results in government expenditure to cover for effects such as the increasing impact of natural disasters, climate change mitigation, emission control programmes, compensation for farmers with reduced crop harvests (because of high concentrations of ozone), damage by acidification of soils and water and to the built environment (e.g., acid rain corroding facades of buildings), etc. In addition to these costs effectively carried by the taxpayer, citizens pay in the way of considerable impacts to health and well-being, for instance, when considering the impact of smog in urban agglomerations on premature deaths [46,47].

These adverse effects have an associated cost, which is mostly not fully borne by those causing the pollution. This refers to effects such as damage to forestry from acid rain, climate change as a result of greenhouse gas emissions, and health impacts from air pollution in cities that are not monetized in the form of levies, such as carbon taxes. Such damage costs are either borne by the government, effectively therefore the tax payer, in the shape of compensations paid, or unacknowledged when encountered in the form of individual adverse effects on health and well-being. These costs are therefore termed "externalities," as they are "external" to market pricing. Monetization of external impacts has been researched for some time, with the first major study by Hohmeyer [48], analyzing externalities of electricity production and showing their impact on the uptake of renewable technologies in the Federal Republic of Germany. This research led to the introduction of feed-in tariffs in Germany in the 1990s as an incentive for renewable energy producers. Effectively they were paid the respective externalities they saved the state by avoiding the externalities incurred by conventional electricity generation.

External costs are defined as a form of economic market failure, where effects and damages brought about by an activity are not included in the costs or prices to undertake the activity, thus leaving them to be paid for by a third party not benefiting from the activity [49]. In the energy markets, there is a strong bias toward polluting technologies, as they are generally cheaper than their clean alternatives since externalities are not charged for today. The costs of these damages are borne by a variety of sources, such as direct costs to the victims, costs to the health system from illnesses, environmental burdens, and costs to agriculture from loss of crops. Internalizing these externalities into prices is crucial to the widespread adoption of clean technologies, as costs would then reflect the real impact of the technology on the macro economy. This would lower the actual cost of clean technologies and increase that of highly polluting ones. Within the framework of Eq. (8.5), this would remedy the lack of adequate pricing of the externalities by integrating them into the equation such as

$$TCO = CAPEX + t \times 1/\eta \times P \times (c_{fuel} + c_{CO_2}) + t \times \eta \times P \times (c_{O\&M} + c_{ext})$$

$$+ n \times L + DECEX \tag{8.6}$$

with $c_{ext}$ being the cost of externalities charged in currency units per kWh of the energy delivered.

Much externality research focuses on transport systems, where external effects, especially on health, have a more direct impact and can be understood by the general population—such as the correlation between exhaust emissions, particulate matter concentrations, and adverse health effects. Some attempts have been made at internalization into prices in this area, such as fuel taxes, vehicle taxation based on carbon dioxide emissions, and the London Congestion Charge; however, often the costs are undervalued. There is little transparency as to where the raised monies are then invested, rendering the internalization meaningless if the funds raised are not used to cover the actual externality costs [50]. Transport is an area in which alternative technologies such as hydrogen fuel cells and batteries have a strong potential to replace current technologies, and externality internalization can have a significant impact on pricing and market uptake.

According to EU rulings, citizens have a statutory right to healthy living conditions, which is largely ignored by councils in the EU urban agglomerations. An increasing number of legal claims are being brought forward to hold councils accountable, causing considerable legal costs. These costs could also be seen as "externalities" and as a corrective impact on energy pricing.

In economic assessments of the viability and competitiveness of technical alternatives to incumbent technologies, a total cost of delivery would have to be used to avoid any bias in the comparisons. Today, this is not the case and zero-carbon technologies are directly compared with highly polluting and damaging technologies. The latter have a history of causing long-term costs to future generations even when they cease operation [51]. The situation result is a biased assessment. A level market field approach is needed where the full costs of service need to be costed into comparisons of different energy technologies.

A fair distribution of costs, where cause and effect are intimately linked, i.e., by energy use being charged with the full societal cost ("polluter pays" principle), would obviously be difficult to implement. Nevertheless, even in the short term, this would trigger the correct incentives to reform the energy market and automatically provide long-term sustainability. Fuel cells and hydrogen fuels are prone to cost more than century-old incumbent, but polluting, technologies, because they are new arrivals to the energy market. Integrating the environmental costs of energy services into market pricing would immediately give these technologies the place in the market that corresponds to their environmental performance.

Much progress has been made worldwide in estimations of environmental and individual costs of energy services, for instance, as estimates of the cost of climate change [52], the external costs of electricity supply [53], and the external costs of transport services [54]. In all cases, a considerable premium is required to level out the cost of conventional and fossil energy provision. In the case of passenger vehicles, this is a surcharge of around 100% on the pump price of petrol and especially diesel (depending on current oil prices). In analyzing the value offer of zero-carbon technologies (at the point of use), including hydrogen, to the economy, the full societal costs of energy services need to be embraced to avoid a completely warped assessment.

## FUTURE COMMERCIALIZATION PROSPECTS OF HYDROGEN: EMERGING BUSINESS CASES

In taking the above on board, we can now define how clean energy markets driven by hydrogen could develop in the future besides the existing markets where hydrogen is essentially traded as a chemical raw material. The release of solely water from electrochemical conversion of hydrogen places it in the realm of zero-carbon technologies at point of use. If this environmental benefit is adequately rewarded, considerable market opportunities can be identified in comparison with polluting, fossil-based energy conversion technologies.

Careful analysis, however, shows [55] that even with supposedly clean fuels such as hydrogen, production from conventional grid electricity causes the environmental impact to increase in comparison with, for instance, a reference diesel or petrol vehicle because of the high primary energy inputs with a considerable carbon footprint. Care has to be taken, therefore, that any implementation of hydrogen fuels actually takes full heed of the environmental and emission issues. Mixing technologies that will deliver zero-carbon services at point of use (e.g., battery electric or fuel cell vehicles) with a supply of fuels from highly polluting sources (e.g., grid electricity and hydrogen from natural gas steam reforming) will cause more damage than the incumbent technologies. This again underpins the need for a full societal cost and environmental impact analysis in going forward to choosing future energy options.

## The Hydrogen Passenger Car

Hydrogen today is sold as a vehicle fuel at €10 per kg of hydrogen and less at HRSs [31,56]. This equates to €0.30/kWh of fuel energy content, which is roughly seven times the price of natural gas. Compared with petrol, this is about double (all taxes and levies included in all prices). In this case, however, the difference is overcompensated for by the higher conversion efficiency of the FCEV. The result is that hydrogen is today already competitive with diesel as a vehicle fuel—as far as the costs of operation (OPEX, excluding the vehicle investment) are concerned (cf. Eqs. (8.5) or (8.6)).

Given that the operation of an FCEV is cheaper than any fossil-fuelled vehicle and the environmental benefits are obvious, a programme of either compensating the FCEV driver for the avoidance of externalities (i.e., allowing for a "feed-in" tariff in the form of a premium paid per kilometer driven) or wrapping up the avoided externalities into a subsidy when purchasing the vehicle can be devised. The complementary measure, however incredibly unpopular, would be to levy the externalities onto the fossil fuel prices following the principle of 'the polluter pays'. In both cases the externalities would be internalized (however, not in the same way and allocated to the correct source) and a level playing field would be created that removes the market bias for polluting technologies.

Adding the positive environmental image of "clean" transport, such an arrangement could in the short- to midterm lead to a considerable market uptake of FCEV, also benefiting from the current negative perception of air quality in urban agglomerations.

## Hydrogen, Synthetic Fuels, and Carbon Certificates

Hydrogen can be reacted with carbon dioxide to form methane (methanation reaction, Sabatier reaction) [57] and oxygen. This could be a substitute for natural gas (synthetic or substitute natural gas, SNG), which could be seamlessly fed into the existing natural gas grid with no changes to be made to the gas market infrastructure. Moreover, if the carbon dioxide were taken from the atmosphere, the balance of carbon emissions would be zero. Unfortunately, the removal or capture of $CO_2$ from the atmosphere is an energy-consuming and not very efficient process [58].

A more probable solution in the near term would be the "reuse" of carbon dioxide issued from, for instance, coal-fired power stations. Currently this is termed "carbon capture and use" (or "utilization," "reuse," or "recycling") (CCU, as opposed to CCS, carbon capture and sequestration) [59]. CCU can be considered a carbon-free process if the original emitter pays for the emission by spending carbon certificates. Then the carbon emission has been accounted for in the energy and emission control system and is free to be used in any other way. The CCU application—for instance, by reacting the carbon dioxide to methane—would then effectively have to be considered carbon-free in the emission bookkeeping.

Alternatively, the primary emitter could pass on the carbon certificate with the carbon release. This would turn the emitter (the coal-fired power station in the initial example) into a carbon-free conversion process and pass on the burden of carbon release to the methane producer. This sounds like a neutral process where the burden is simply passed down the line. Nevertheless, there could be a further benefit in that the carbon now enshrined in methane can be far more efficiently used in a high-temperature fuel cell to produce electricity (with net electrical efficiency above 60%). Therefore the electricity carbon footprint would be considerably reduced over the summary process of coal-fired power station and subsequent reuse of the carbon dioxide. The CCU processor could make a profit out of collecting the carbon certificates and spreading them out over a higher amount of electricity delivered.

In light of the previous discussion in this chapter, the latter process is not ideal because it does not sustainably handle the fossil carbon emissions. An approach that would be more beneficial in the long term is that of capturing carbon dioxide from processes (for example, biogas) to then never release it to the atmosphere again. This would entail that exhaust gases would be always captured and recycled. Although this will not be possible for vehicle applications, it could be an option for stationary power production, CHP units, etc. The fuel gases would need to be captured and sent back to a methanation plant into which hydrogen from renewable sources would be fed to produce fresh methane.

## CONCLUSIONS

Hydrogen markets are developing across the world, with a number of industries backing this development. Although the major part is currently driven by use of hydrogen in refineries and chemical industries (including fertilizer production), growing interest can be identified in several sectors, such as steel making, taking to "greening" their products by avoiding $CO_2$ emissions. The much cited "hydrogen economy," however, might not be developing the way it was predicted to but be pushed forward by developments not envisaged 5 years ago. It also needs to be noted that hydrogen does not always appear in its pure form but may also be wrapped up in zero-carbon methane (from CCU) that is synthesized from hydrogen and carbon dioxide.

The "traditional" way of seeing the hydrogen economy emerge builds on replacement of conventional energy services by those delivered with less or no emissions, subsequently based on hydrogen fuels. This approach is thwarted by the inherently CAPEX-leaning economic analysis that will not recognize the benefits of reducing emissions as long as these are not costed. In promoting hydrogen-based technologies, alternative approaches to determining total cost of ownership, namely total societal cost of services, have to be adopted. These will change not only the way hydrogen technologies are perceived but also the way governments and taxpayers account for damage done in the energy system. As far as the tax payer is concerned, the allocation of cost to the polluter is only fair and relieves the public from costs incurred by individual sources. The government, on the other hand, retains the same amount of income and expenses while allocating levies and damage compensations based on a much more transparent and democratic basis.

## ACKNOWLEDGMENTS

We gratefully acknowledge the PhD stipend support by EPSRC for the Centre for Doctoral Training in Fuel Cells and their Fuels under contract no. EP/L015749/1.

## REFERENCES

[1] L.B. Nelson, History of the U.S. Fertilizer Industry, Tennessee Valley Authority, Muscle Shoals, 1990.

[2] C.-J. Winter, J. Nitsch, Hydrogen as an Energy Carrier, Springer, Berlin/Heidelberg, 1988, https://doi.org/10.1007/978-3-642-61561-0.

[3] D. Sadler, A. Cargill, M. Crowther, A. Rennie, J. Watt, S. Burton, M. Haines, H21 Leeds City Gate, Northern Gas Networks, Leeds, 2016.

[4] W. McDowall, F. Li, I. Staffell, P. Grünewald, T. Kansara, P. Ekins, P. Dodds, A. Hawkes, P. Agnolucci, in: P. Dodds, A. Hawkes (Eds.), The Role of Hydrogen and Fuel Cells in Providing Affordable, Secure Low-carbon Heat, H2FC SUPERGEN, London, 2014.

[5] N. Brandon, Z. Kurban, Clean Energy and the Hydrogen Economy, Philosophical Transactions of the Royal Society, London, 2017.

[6] R. Steinberger-Wilckens, P.E. Dodds, Z. Kurban, A. Velazquez Abad, J. Radcliffe (Eds.), The Role of Hydrogen and Fuel Cells in Delivering Energy Security for the UK, H2FC Supergen, London, UK, 2017.

[7] R. Steinberger-Wilckens, K. Stolzenburg, Hydrogen based traffic — an option for introducing wind energy to the transport market, in: Proceedings of the European Wind Energy Association Conference, Kassel, September 2000.

[8] R. Javadli, A. de Klerk, Desulfurization of heavy oil, App. Petrochem. Res. 1 (2012) 3—19.

[9] S. Parkash, Refining Processes Handbook, Elsevier, Amsterdam, 2003.

[10] Ullmann's Encyclopedia of Industrial Chemistry, Wiley-VCH, Weinheim, 2002.

[11] D. Brown, Hydrogen supply and demand — past, present and future, Gasworld (April 2016) 36—40.

[12] J.G. Brennan (Ed.), Food Processing Handbook, Wiley-VCH, Weinheim, 2006.

[13] D. Garg, P.T. Kilhefner, Moisture-free atmosphere system for brazing ferrous metals, Ind. Heat. 66 (1999) 49—52.

[14] 1970—79, The Great Soviet Encyclopedia, third ed., The Gale Group, Farmington Hills, 2010.

[15] S. Lechtenböhmer, L.J. Nilsson, M. Ahman, C. Schneider, Decarbonising the energy intensive basic materials industry through electrification — implications for future EU electricity demand, Energy 115 (2016) 1623—1631.

[16] M. Arens, E. Worrell, W. Eichhammer, A. Hasanbeigi, Q. Zhang, Pathways to a low-carbon iron and steel industry in the medium-term — the case of Germany, J. Clean. Prod. 163 (2017) 84—98.

[17] Z.J. Schiffer, K. Manthiram, Electrification and decarbonisation of the chemical industry, Joule 1 (2017) 10—14.

[18] M. Spiro, W.P. Griffith, The mechanism of hydrogen peroxide bleaching, Text. Chem. Colorist 29 (1997) 12.

[19] E. Almqvist, History of Industrial Gases, Kluwer, New York, 2003, pp. 47—56.

[20] S. Dunn, History of hydrogen, in: Encyclopedia of Energy, Elsevier, 2004, pp. 241—252, https://doi.org/10.1016/B0-12-176480-X/00039-5.

[21] C. Cleveland, C. Morris, Hydrogen, in: Handbook of Energy, Elsevier, 2014, pp. 309—322, https://doi.org/10.1016/B978-0-12-417013-1.00017-0.

[22] U. Schmidtchen, Fuels — safety | hydrogen: transportation, in: Encyclopedia of Electrochemical Power Sources, Elsevier, 2009, pp. 528—532.

[23] G. Petrangeli, Barrage balloons against aircraft threat: a well proven concept revisited, Nucl. Eng. Des. 240 (2010) 886—890.

[24] A.P. Fickett, F.R. Kaihammer, Water electrolysis, in: K.E. Cox, K.D. Williamson (Eds.), Hydrogen - its Technology and Implications, Hydrogen Production Technology, vol. 1, Chemical Rubber Co., Cleveland, 1977, pp. 3—41.

[25] Anonymous, Power from the Aswan dam to be used to increase still further the cotton crop in Egypt, N. Y. Times (July 27, 1913).

[26] J. Lindblom, A. Elfwén, B. Klasman, E. Grahn, J. Karlsson, J. Roupe, L. Goding, K. Abrahamsson, The Swedish Electricity and Natural Gas Market 2016, Report Ei R2017:06, Swedish Energy Markets Inspectorate, Eskilstuna, 2017.

[27] M.W.H. Peebles, Evolution of the Gas Industry, Macmillan, London/Basingstoke, 1980.

[28] G. Maisonnier, J. Perrin, R. Steinberger-Wilckens, S.C. Truemper, Industrial surplus hydrogen and markets and production, in: Pt. 2 of the European Hydrogen Infrastructure Atlas. Deliverable 2.1 of the Project Roads2HyCom, Document No. R2H2006PU, Oldenburg, 2007.

[29] M. Ball, M. Wietschel, The future of hydrogen — opportunities and challenges, IJHE 34 (2009) 615—627.

[30] BP Statistical Review of World Energy June 2017. Annual energy statistics published by BP on: http://www.bp.com/statisticalreview.

[31] Multi-annual Working Plan 2014—2020, FCH JU, Brussels, 2014.

[32] D. Raine, R. Williams, H. Stroem, G. Maisonnier, S. Vinot, J. Linnemann, S.C. Truemper, Analysis of the current hydrogen cost structure, in: Deliverable 2.4—1 of the Project Roads2HyCom, Document No. R2H2019PU.1, Oldenburg, 2009.

[33] R. Steinberger-Wilckens, J. Linnemann, S.C. Truemper, Cost models for current and future hydrogen production, in: Deliverable 2.4—2 of the Project Roads2HyCom, Document No. R2H2020PU.1, Oldenburg, 2008.

[34] CDC Climate Research: Tendances Carbone — the Monthly Bulletin on the European Carbon Market, I4CE, Paris, January 2013.

[35] R.K. Skinner, Heat rates, spark spreads, and the economics of tolling agreements, Nat. Gas Electr. 28–32 (December 2010).

[36] S. Samsatli, I. Staffell, N.J. Samsatli, Optimal design and operation of integrated wind-hydrogen-electricity networks for decarbonising the domestic transport sector in Great Britain, IJHE 41 (2016) 447–475.

[37] M. Melaina, M. Penev, Hydrogen Station Cost Estimates. Technical Report NREL/TP-5400–56412, NREL, Golden, 2013.

[37a] M.J. Smith, K. Turner, J.T.S. Irvine (Eds.), The Economic Impact of Hydrogen and Fuel Cells in the UK - A Preliminary Assessment based on Analysis of the replacement of Refined Transport Fuels and Vehicles, H2FC SUPERGEN, London, UK, 2017.

[38] M. Judkis, How one company eliminated food waste: 'The landfill can no longer be an option', Wash. Post (January 09, 2017). Washington, D.C., USA.

[39] J. Jaffe, M. Ranson, R.N. Stavins, Linking tradable permit systems: a key element of emerging international climate policy architecture, Ecol. Law Q. 36 (2009) 789–808.

[40] A. Nelson, European carbon market reform set for 2019, Guardian (February 24, 2015).

[41] P.E. Dodds, Energy security impacts of introducing hydrogen to the UK energy system, in: R. Steinberger-Wilckens, P.E. Dodds, Z. Kurban, A. Velazquez Abad, J. Radcliffe (Eds.), The Role of Hydrogen and Fuel Cells in Delivering Energy Security for the UK, H2FC Supergen, London, 2017.

[42] D. Hart, J. Howes, F. Lehner, P.E. Dodds, N. Hughes, B. Fais, N. Sabio, M. Crowther, Scenarios for deployment of hydrogen in contributing to meeting carbon budgets and the 2050 target, Committee on Climate Change, London, 2015. Available at: https://documents.theccc.org.uk/wp-content/uploads/2015/11/E4tech-for-CCC-Scenarios-for-deploymentof-hydrogen-in-contributing-to-meeting-carbon-budgets.pdf.

[43] BEIS Fossil Fuel Price Projections, Department of Business, Energy and Industrial Strategy, London, 2016.

[44] A.L. Stephenson, D.J.C. MacKay, Life Cycle Impacts of Biomass Electricity in 2020, Department of Energy and Climate Change, London, UK, 2014. Available at: www.gov.uk/government/uploads/system/uploads/attachment_data/file/349024/BEAC_Report_290814.pdf.

[45] P.E. Dodds, W. McDowall, Methodologies for representing the road transport sector in energy system models, IJHE 39 (2014) 2345–2358.

[46] COMEAP, The Mortality Effects of Long-term Exposure to Particulate Air Pollution in the United Kingdom, Public Health England, Didcot, 2010. Available at: www.gov.uk/government/uploads/system/uploads/attachment_data/file/304641/COMEAP_mortality_effects_of_long_term_exposure.pdf.

[47] N. Massey, Air in 44 UK cities and towns too dangerous to breathe, UN pollution report finds, Independent (October 31, 2017).

[48] O. Hohmeyer, Social Costs of Energy Consumption - External Effects of Electricity Generation in the Federal Republic of Germany, Springer, Berlin/Heidelberg, 1988, https://doi.org/10.1007/978-3-642-83499-8.

[49] T.L.T. Nguyen, B. Laratte, B. Guillaume, A. Hua, Quantifying environmental externalities with a view to internalising them in the price of products, using different monetisation methods, Resour. Conserv. Recycl. 109 (2016) 13–23.

[50] J.A.J.Aaronson: What does road tax pay for? Tax Guide Online. Available from: http://www.thetaxguide.co.uk.

[51] D. Drollette Jr., The rising cost of decommissioning a nuclear power plant, Bull. Atom. Sci. 70 Years Speaking Knowledge Power (2014).

[52] N. Stern, Stern Review on the Economics of Climate Change. London, 2006.

[53] P. Bickel, R. Friedrich, Externalities of energy - methodology 2005 update, EXTERNE Project Deliver. (2005). Available at: www.externe.info/externe_d7/sites/default/files/methup05a.pdf.

[54] C. Schreyer, C. Schneider, M. Maibach, W. Rothengatter, C. Doll, D. Schmedding, External Costs of Transport, INFRAS, 2004.

[55] O.M. Odgen, R.H. Williams, E.D. Larson, Societal lifecycle costs of cars with alternative fuels/engines, Energy Pol. 32 (2004) 7–27.

[56] The Coalition, A Portfolio of Power-trains for Europe: A Fact-based Analysis − the Role of Battery Electric Vehicles, Plug-in Hybrids and Fuel Cell Electric Vehicles, FCH JU, Brussels, 2010.

[57] P. Sabatier, J.-B. Sendersen, Comptes rendus des séances de l'académie des sciences, section VI − chimie, Imprimerie Gauthier-Villars, Paris, 1902.

[58] E. Kintisch, Can sucking $CO_2$ out of the atmosphere really work? MIT Technology Review (October 7, 2014).

[59] Pembina Institute, Carbon Capture and Utilization. ICON Fact Sheet, Pembina Institute, Calgary, 2015.

# Index

Printed in the United States
By Bookmasters